EDIBLE INSECTS OF
THE WORLD

EDIBLE INSECTS OF THE WORLD

Jun Mitsuhashi

Tokyo University of Agriculture
Tokyo
Japan

CRC Press
Taylor & Francis Group
Boca Raton London New York

CRC Press is an imprint of the
Taylor & Francis Group, an **informa** business
A SCIENCE PUBLISHERS BOOK

CRC Press
Taylor & Francis Group
6000 Broken Sound Parkway NW, Suite 300
Boca Raton, FL 33487-2742

First issued in paperback 2020

© 2017 by Taylor & Francis Group, LLC
CRC Press is an imprint of Taylor & Francis Group, an Informa business

No claim to original U.S. Government works

ISBN-13: 978-1-4987-5657-0 (hbk)
ISBN-13: 978-0-367-78290-0 (pbk)

Visit the Taylor & Francis Web site at
http://www.taylorandfrancis.com

and the CRC Press Web site at
http://www.crcpress.com

Preface

In 2013, FAO issued a report "Edible insects. Future prospects for food and feed security". This report has been seemed to give considerable impact to mass media in many countries. Then, topics on entomophagy were taken up often by TV, newspapers or journals in various countries. It is said that in some countries, eating insects became a boom. Based on the enhancement of general people's interest to entomophagy, 140 companies dealing with insect foods have been reportedly established in France. Events such as exhibition, symposium, workshop, meeting for trial to eat insects, were held in many places. In Japan demand for insect foods is said to increase twice. The author has interested in human consumption of insects for long time, and published a book "Edible insects of the world" in 1984, in Japanese. Since then, he could collect information on entomophagy in fairly large amount. The compilation of the collected information was published in two volumes of books, "Complete World Entomophagy" and "Encyclopedia on the Culture of Entomophagy", both written in Japanese. Up to present, Bodenheimer's book "Insects as human food" is only one book generalized entomophagy in English. Information after this book has not been published as a monograph. DeFoliart collected enormous amount of information on entomophagy, and in 2002 opened his notes to the public on internet home page as "The human use of insects as a food resource: A bibliographic account in progress". Unfortunately, he could not complete this work and the information was not published. Then the author thought that it will be worthwhile to publish a book on generalized entomophagy of the world in English. This is the reason of the publication of this book.

The number of edible insect species is one of frequent questions about entomophagy. It is very difficult to answer for such questions, if not impossible. Because, there are many unidentified species including new species in inland of developing countries, where entomophagy is common. Also, most vernacular names of insects in such places do not correspond with scientific names. In many cases, a vernacular name covers several species which have different scientific names. A vernacular name, therefore, cannot be used for counting number of insect species. These factors make counting of edible insect species extremely difficult.

This book contains about 2,140 edible insect species which were identified with scientific names. However, actually eaten species is further more. In this book, species, whose genus name only was identified, is also included. For scientific names, there seems some confusion. There were some synonym. Also, there were some misprinting or misspelling, both of which had been used by several authors. For such cases, in this book both names are cited as they printed in original paper, because the present author could not determine which name or spelling was authentic. For counting species, the one only whose genus name was known were excluded. In other words, genus name plus sp. or spp. was not counted. When a single species was described with more than two different scientific names, description was made on one of them, and the name selected as representative was indicated with an arrow. The selected name is, however, not necessarily authentic one. Brief description was made about name of country or district in which insects are eaten, what kind of people eat insects, how to collect edible insects, how to eat the insects, and so on. However, information on these items was not sufficiently supplied. Readers, therefore, will find frequently an expression saying just "This species is eaten in this country". In this book, some medicinal insects, which are administered orally, are also taken in. For such insects, target disease or usage was described.

Many readers may be interested what kind of insects are eaten in some selected countries or districts. For such demand, lists of scientific names of edible insects are compiled by country or district in Part II.

For readers, who want to see original papers, cited literatures are listed in Part III. There were many literatures, which present author could not see directly. In these case, the literatures, which the author actually read and cited from, were indicated in [ ].

Contents

Preface   v

Part I: The List of Edible Insects in Order of Taxonomic Group

1. Thysanura   1
2. Ephemeroptera   1
3. Odonata   3
4. Plecoptera   8
5. Orthoptera   9
6. Phasmida   39
7. Isoptera   40
8. Blattaria   47
9. Mantodea   50
10. Anoplura   52
11. Hemiptera   53
12. Coleoptera   77
13. Neuroptera   133
14. Siphonaptera   133
15. Diptera   133
16. Trichoptera   139
17. Lepidoptera   141
18. Hymenoptera   178
19. Numbers of Identified Species in Each Order and Family   220

Part II: List of Edible Insects by Countries or Districts

A. Countries   223
  1. Europe   223
    Austria, Czech Republic, France, Greece, Italy,
    Romania, Rome Antiqua, Russia, Sweden, Ukraine,
    United Kingdom

2. North America   224
   Canada, United States of America (USA)

3. Central area of America   226
   Guatemala, Honduras, Mexico, Nicaragua, Panama

4. West Indies   229
   Cuba, Dominica and Haiti (Hispaniola Island), Jamaica,
   Martinique, Trinidad and Tobago

5. South America   230
   Argentina, Barbados, Bolivia, Brazil, Chile, Colombia, Ecuador,
   Guyana, Paraguay, Peru, Suriname, Venezuela

6. Africa   234
   Angola, Benin, Botswana, Bourbon, Burkina Faso, Cameroon,
   Cape Coast, Central African Republic (CAR), Democratic Republic
   of Congo (DRC), People's Republic of the Congo (PRC), Côte
   d'Ivoire, Gabon, Ghana, Guinea, Kenya, Madagascar, Malawi,
   Mali, Mauritius, Mozambique, Namibia, Niger, Nigeria,
   Rwanda, Sahel (Sub-Saharan Africa), Sao Tome and Principe,
   Senegal, Sierra Leone, Republic of South Africa (RSA), Sudan,
   Tanzania, Uganda, Zambia, Zimbabwe

7. The Middle and Near East   242
   Algeria, Egypt, Arab, Iran (Persia), Iraq, Israel, Kuwait, Levant,
   Morocco, Saudi Arabia, Syria, Tunisia, Turkey

8. Oceania   244
   Australia, Papua New Guinea (PNG), New Caledonia,
   New Zealand, Solomon Islands

9. Asia   245
   Cambodia, China, India, Indonesia, Japan, Laos, Malaysia,
   Mongolia, Myanmar, Nepal, North Korea, Philippines, Singapore,
   South Korea, Sri Lanka, Taiwan, Thailand, Tibet, Timor, Viet Nam

B. Districts   **259**

1. Europe   259
   Europe general, South Europe, Lapland

2. North America   260
   North America general

3. West Indies   260
   West Indies general

4. South America   260
   South America general, Amazonia, Bolivia-Brazil,
   Bolivia-Peru-Brazil-Colombia, Brazil-Colombia, Brazil-Guyana,
   Brazil-Venezuela, Colombia-Peru, Colombia-Venezuela

5. Africa   261
   Africa General, Central area of Africa, Congo Basin,
   Eastern area of Africa, Equatorial area of Africa,
   Northern area of Africa, Northern central area of Africa, Southern
   area of Africa, Southern central area of Africa, Southern Sahara,
   Tropical area of Africa, Western area of Africa

6. The Middle and Near East   263
   Barbary, Crimea, Kalahari Desert, Near East

Part III

References   265

PART I

THE LIST OF EDIBLE INSECTS IN ORDER OF TAXONOMIC GROUP*

1. THYSANURA (Silverfish) [2]

Lepismatidae (Silverfish) **(2)**

Ctenolepisma villosa (oriental silverfish) **(PL I-1) China:** This species is used for curing child paralysis, inducing abortion (Li 1596). **South Korea:** The roasted powder of the adults are used for treating lumbago, difficulty in urination, etc. in North Chungcheong and North Gyeongsang Provinces (Okamoto and Muramatsu 1922).

Lepisma saccharina (silverfish, bookworms) **China:** This species is used as a remedy for rheumatism (Zimian et al. 1997). **Japan:** The Japanese name is *seiyō-shimi*. This species is used for diuretic purposes, treatment of paralysis, etc. (Umemura 1943, Shiraki 1958).

2. EPHEMEROPTERA (Mayflies) [7]

Baetidae (Small mayflies) **(2)**

Baetis **sp. Mexico:** The Mexican name is *moscas de mayo*. The larvae are eaten in Estado de Mexico (Ramos-Elorduy et al. 1998).

Cloeon dipterum (pond olive) **Japan:** The Japanese name is *futaba-kagerō*. The larvae are roasted with other aqueous insects. They are rich in fat, and most people like to eat them (Kuwana 1930).

---

*([ ] : Number of identified edible insects).

Plate I. 1. *Ctenolepisma villosa* adult. Body length: 8–9 mm (cf. p.1) (Photo: K. Umeya, Tsukuba, Japan); 2. *Anax parthenope julius.* a: Adult ♂. Length of abdomen: 50 mm. b: Larva. Body length: 43 mm (cf. p.4); 3. *Locusta migratoria* adult. Body length: 45 mm (cf. p.17); 4. *Crocothemis servilia* adult. Length of abdomen: 28 mm (cf. p.6); 5. A girl selling cooked grasshoppers at Oaxaca, Mexico. Insert: A close-up of the grasshopper, *Sphenarium* sp. Body length: 30 mm (cf. p.26); 6. *Acrida cinerea* adult ♀. Body length: 80 mm (cf. p. 10).

Cloeon kimminsi The larvae and the adults are eaten roasted by the Nishi people in Arunachal Pradesh State, **India** (Singh et al. 2007).

Caenidae (Small square-gill mayflies) **(2)**

Caenis kungu In **Malawi,** people living in the northern part of Lake Malawi catch the adults, and make blocks called *"kungu"* by pressing them. These blocks are said to taste like caviar or salted grasshoppers (Ealand 1915).

Povilia adusta In **Malawi,** people collect the adults and press them to make insect blocks, which also contain several other species such as chironomids (van Huis 2008).

Ephemerellidae (Spiny crawler mayflies) **(1)**

Ephemerella jianghongensis The larvae are frizzled or fried with meat of frogs or fish in the Yunnan Province, **China**. The adults are frizzled with eggs (Chen and Feng 1999).

Ephemeridae (Burrowing mayflies) **(2)**

Ephemera danica The larvae and the adults are roasted by the Nishi people in Arunachal Pradesh State, **India** (Singh et al. 2007). The roasted or boiled larvae are used as a remedy for stomach disorders (Chakravorty et al. 2011).

Ephemera strigata The Japanese name is *mon-kagerō*. In **Japan,** the larvae were roasted with other aqueous insects. They were rich in fat, and most people liked to eat them (Kuwana 1930).

Ephemera sp. **Mexico:** The Mexican name is *mosca de mayo*. The larvae are eaten in Mexico (Ramos-Elorduy et al. 1998).

3. ODONATA (Dragonflies) [40]

Aeschnidae (Hawker dragonflies) **(8)**

Aeschna brevifrons The larvae are consumed by the Quichua people in **Ecuador** (Onore 1997).

Aeschna marchali The larvae are consumed by the Quichua people in **Ecuador** (Onore 1997).

Aeschna mixta The larvae and adults are eaten raw by the Karbi and Rengma Naga people in Assam State, **India** (Ronghang and Ahmed 2010).

Aeschna multicolor (common blue darner) **Mexico:** The Mexican name is *padrecitos*. The larvae are eaten (Ramos-Elorduy et al. 1998). **USA:** The larvae were eaten by the native Americans of the arid regions of the West (Ebeling 1986).

Aeschna peralta The larvae are consumed by the Quichua people in **Ecuador** (Onore 1997).

Aeschna **spp. India:** The larvae whose vernacular name is *anga-mechep* are consumed by the Ao-Naga people (Meyer-Rochow 2005). **Mexico:** Mexican name is *padrecitos*. The larvae are eaten (Ramos-Elorduy et al. 1998). **Thailand:** People living in Northeast Thailand eat the larvae called *darner* (Hanboonsong et al. 2000).

Anax guttatus The larvae and adults are boiled or roasted in **Thailand** (Bristowe 1932).

Anax parthenope julius (lesser emperor dragonfly) **(PL I-2a,b)** Japanese name is *gin-yanma*. In **Japan**, the larvae were roasted and the adults were spit-roasted in the Nagano Prefecture (Takagi 1929a). **China:** The larvae are fried without coating (Mao 1997, Umeya 2004). This species is also used as an antidote (Zimian et al. 1997).

Anax **sp. Indonesia:** The larvae and adults of some species belonging to this genus are fried, boiled or roasted. It is also used as an ingredient of *botok* cuisine (Clausen 1954, Pemberton 1995). **Mexico:** Mexican name is *padrecitos*. The larvae and the adults are consumed in Toluca (Ramos-Elorduy and Pino 1990).

Coryphaeschna adnexa The larvae are consumed by the Oyavalo and Quichua people in **Ecuador** (Onore 1997).

Rhinoaeschna → *Aeschna*

Agrionidae (Narrow-winged damselflies) **(1)**

Ceriagrion **sp.** The larvae of a species belonging to this genus are eaten in **Laos, Myanmar** and **Vietnam** (Yhoung-Aree and Viwatpanich 2005) and **Thailand** (Hanboonsong et al. 2000).

Calopterygidae (Broad-winged damselflies) **(0)**

Hetaerina **sp.** The Yekuana people in Alto Orinoco area, **Venezuela** eat the larvae of a species belonging to this genus (Araujo and Beserra 2007).

Mnesarete **sp.** The Yekuana people in Alto Orinoco area, **Venezuela** eat the larvae of a species belonging to this genus (Araujo and Beserra 2007).

Coenagrionidae (Damselflies) **(0)**

Argia **sp.** The Yekuana people in Alto Orinoco area, **Venezuela** eat the larvae of a species belonging to this genus (Araujo and Beserra 2007).

Enallagma **sp.** In Arunachal Pradesh State, **India**, the larvae and adults of a species, which is called *esh tat tani* by the Galo tribes or *soko yoyo* by the Nishi tribes, are eaten. People prefer the larvae to the adults. They eat the dragonflies raw with bamboo shoots. When the adults are eaten, their wings are removed before eating (Chakravorty et al. 2011).

Cordulegasteridae (Biddies) **(1)**

Anotogaster sieboldii The Japanese name is *oni-yanma*. The larvae are eaten roasted and the adults are eaten spit-roasted in the Nagano Prefecture, **Japan** (Takagi 1929a).

Cordulegaster **sp.** The larvae are eaten fried in oil by the Meeteis people in Nagaland State, **India** (Meyer-Rochow 2005).

Corduliidae (Green-eyed skimmers) **(2)**

Epophtalmia vittigera bellicose The larvae of this species are boiled or fried in **Thailand** (Hanboonsong et al. 2000).

Lauromacromia dubitalis The Yekuana people in Alto Orinoco area, **Venezuela** eat the larvae (Araujo and Beserra 2007).

Macromia **sp.** A species belonging to this genus is boiled or fried in **Thailand** (Vara-asavapati et al. 1975).

Gomphidae (Club-tails) **(3)**

Agriogomphus **sp.** The Yekuana people in Alto Orinoco area, **Venezuela** eat the larvae of a species belonging to this genus (Araujo and Beserra 2007).

Dabidius nanus The Japanese name is *dabido-sanae*. In **Japan,** the larvae are a member of *zazamushi* (see Tricoptera-Leptoceridae-*Parastenopsyche sauteri*). The larvae are boiled with soy sauce in the Nagano Prefecture (Nakai 1988).

Gomphus cuneatus The larvae are fried in **China** (Chen and Feng 1999).

Ictinogomphus rapax In Arunachal Pradesh State, **India,** the larvae and adults are eaten by the Galo people. People prefer the larvae to the adults. They eat the dragonflies raw with bamboo shoots. When the adults are eaten, their wings are removed before eating (Chakravorty et al. 2011).

Progomphus **sp.** The Yekuana people in Alto Orinoco area, **Venezuela** consume the larvae of a species belonging to this genus (Araujo and Beserra 2007).

Stylurus **sp.** The larvae are eaten boiled, raw and as roasted paste in Arunachal Pradesh State, **India** (Chakravorty et al. 2011).

Zonophora **sp.** In **Venezuela,** the Yekuana people eat the larvae of a species belonging to this genus raw (Paoletti and Dufour 2005, Araujo and Beserra 2007).

Lestidae (Emerald damselflies) **(1)**

Lestes praemorsa The larvae are fried in **China** (Chen and Feng 1999).

Libellulidae (Common skimmers) **(24)**

Acisoma parnorpaides The larvae were eaten by the Ao-Naga people in Nagaland State, **India** (Meyer-Rochow 2005).

Brachythemis contaminata In Arunachal Pradesh State, **India**, the larvae and adults are eaten by the Galo people. People prefer the larvae to the adults. They eat the dragonflies raw with bamboo shoots. When the adults are eaten, their wings are removed before eating (Chakravorty et al. 2011).

Brechmorhoga **sp.** The Yekuana people in Alto Orinoco area, **Venezuela** consume the larvae of a species belonging to this genus (Araujo and Beserra 2007).

Cratilla lineate assidua This species is regarded as food in Bali Island, **Indonesia** (Césard 2006).

Crocothemis servilia **(PL I-4) China:** The larvae and adults are fried. (Mao 1997, Chen and Feng 1999). **Indonesia:** This species is eaten in Bali Island (Sésard 2006). **Japan:** The Japanese name is *shōjō-tonbo*. The roasted adults were used for treating diphtheria and other diseases (Miyake 1919). **South Korea:** The dried powder of the adults are used as a fortifier in North Jeolla Province (Okamoto and Muramatsu 1922).

Crocothemis **sp. Indonesia:** The larvae and adults of a species belonging to this genus are fried, boiled or roasted. They are used as an ingredient of *botok* (a kind of cuisine) (Clausen 1954, Pemberton 1995). **Laos, Myanmar, Thailand** and **Vietnam:** People eat a species belonging to this genus (Yhoung-Aree and Viwatpanich 2005).

Dasythemis **spp.** In **Venezuela**, people eat the larvae of some species belonging to this genus raw (Paoletti and Dufour 2005). The Yekuana people in Alto Orinoco area esteemed the larvae of some species belonging to this genus (Araujo and Beserra 2007).

Diplacodes trivialis The larvae and adults are eaten roasted by the Nishi people in Arunachal Pradesh State, **India** (Singh et al. 2007).

Diplacodes **sp.** The larvae are considered edible by the Nishi people in Arunachal Pradesh State, **India** (Chakravorty et al. 2011).

Libellula carolona The adults are appreciated by the Karbi and Rengma Naga people in Assam State, **India** as a chutney or roasted (Ronghang and Ahmed 2010).

Libellula pulchella The larvae are fried, boiled, stir-fried and as part of salads in **Thailand** (DeFoliart 2002).

Neurothemis fluctuans The larvae and adults are eaten by the Karbi and Rengma Naga people in Assam State, **India** (Ronghang and Ahmed 2010).

Neurothemis ramburii This species is used as food item in Bali Island, **Indonesia** (Césard 2006).

Neurothemis sp. In **Indonesia**, people consume a species belonging to this genus in the same way as *Crocothemis* sp. (DeForiart 2002).

Orthecum glaucum This species is eaten in Bali Island, **Indonesia** (Césard 2006).

Orthetrum albistylum The larvae are used as food article in **China** (Hu and Zha 2009).

Orthetrum japonicum The Japanese name is *shiokara tonbo*. The charred adults were used for treating asthma in the Tokushima Prefecture, **Japan** (Miyake 1919).

Orthetrum sabina This species is eaten in Bali Island, **Indonesia** (Césard 2006).

Orthetrum triangulare melania **China:** The larvae are considered edible (Hu and Zha 2009). **Japan:** The Japanese name is *ōshiokara-tonbo*. The larvae were roasted and the adults were spit-roasted in the Nagano Prefecture (Takagi 1929a).

Orthetrum **sp.** Some species belonging to this genus are consumed in Sabah State, **Malaysia** (Chung et al. 2002).

Pachydiplax **sp.** In Arunachal Pradesh State, **India,** the larvae and adults are eaten by the Galo people. People prefer the larvae to the adults. They eat the dragonflies raw with bamboo shoots. When the adults are eaten, their wings are removed before eating (Chakravorty et al. 2011).

Pantala flavescens (globe trotter) **China:** The larvae are eaten (Mao 1997). **Indonesia:** The Javanese name is *capung ciwet*. This species is considered as foodstuff in Bali Island (Lukiwati 2010).

Potamarcha obscura This species is eaten in Bali Island, **Indonesia** (Césard 2006).

Rhyothemis **sp.** A dragonfly species belonging to this genus is boiled or roasted in **Thailand** (Vara-asavapati et al. 1975).

Sympetrum darwinianum The Japanese name is *natsu-akane*. The larvae were roasted and the adults were spit-roasted in the Nagano Prefecture, **Japan** (Takagi 1929a). The infusion of adults with some herbs was used as a medicine for a cough, throat swelling, fever, etc. (Miyake 1919).

Sympetrum eroticum eroticum The Japanese name is *mayutate-akane*. The adults are roasted after removing the heads in the Nagano Prefecture, **Japan** (Mukaiyama 1987). The adults are used for treating diphtheria, tonsillitis, etc. (Umemura 1943).

Sympetrum frequens The Japanese name is *aki-akane*. The adults are roasted after removing the heads in the Nagano Prefecture, **Japan** (Mukaiyama 1987). The adults are used for treating diphtheria, tonsillitis, etc. (Umemura 1943).

Sympetrum infuscatum The Japanese name is *noshime-tonbo*. The adults are roasted after removing the heads in the Nagano Prefecture, **Japan** (Mukaiyama 1987). The adults are used for treating diphtheria, tonsillitis, etc. (Umemura 1943).

Sympetrum pedemontanum elatum (banded darter) The Japanese name is *miyama-akane*. The larvae were roasted and the adults were spit-roasted in the Nagano Prefecture, **Japan** (Takagi 1929a). The infusion of the dried and pulverized adults was used as medicine for a cough (Miyake 1919).

Sympetrum sinense → *Sympetrum darwinianum*

Sympetrum uniforme The larvae are consumed in **China** (Hu and Zha 2009).

Sympetrum **sp.** In Arunachal Pradesh State, **India**, the larvae and adults are eaten by the Galo people. People prefer the larvae to the adults. They eat the dragonflies raw with bamboo shoots. When the adults are eaten, their wings are removed before eating (Chakravorty et al. 2011).

Trithemis arteriosa This species is eaten in **DRC** (Malaisse and Parent 1997a).

Trithemis aurora This species is regarded as foodstuff in Bali Island, **Indonesia** (Césard 2006).

Urothemis **sp.** In Arunachal Pradesh State, **India**, the larvae and adults of a species belonging to this genus are eaten by the Galo people. People prefer the larvae to the adults. They eat the dragonflies raw with bamboo shoots. When the adults are eaten, their wings are removed before eating (Chakravorty et al. 2011).

Megapodagrionidae (Damselflies) **(0)**

Oxystigma **sp.** The Yekuana people in Alto Orinoco area, **Venezuela** eat the larvae of a species belonging to this genus (Araujo and Beserra 2007).

4. PLECOPTERA (Stone-flies) [8]

Nemouridae (Spring stoneflies) **(0)**

Nemoura **sp.** The larvae are eaten roasted or boiled by the Nishi people in Arunachal Pradesh State, **India** (Singh et al. 2007).

Perlidae (Common stoneflies) **(4)**

Kamimuria tibialis The Japanese name is *kawagera*. The larvae are a member of *zazamushi* (see Tricoptera-Leptoceridae-*Parastenopsyche sauteri*). The larvae are simmered or boiled with soy sauce in the Nagano Prefecture, **Japan** (Torii 1957, Katagiri and Awatsuhara 1996). The larvae were used for treating stomach ailments (Miyake 1919).

Oyamia gibba The Japanese name is *ōyama-kawagera*. The larvae are consumed boiled with soy sauce in the Nagano Prefecture, **Japan**. The larvae are a member of *zazamushi* (see Tricoptera-Leptoceridae-*Parastenopsyche sauteri*) (Nakai 1988).

Paragnetia tinctipennis The Japanese name is *ōkawagera* or *ōkurakake-kawagera*. The larvae are a member of *zazamushi* (see Tricoptera-Leptoceridae-*Parastenopsyche sauteri*). The larvae were treated as food in the Nagano Prefecture, **Japan** (Takagi 1929a).

Perlatibialis → *Kamimuria tibialis*

Perla tinctipennis → *Paragnetia tinctipennis*

Perlodes frisonana The Japanese name is *amime-kawagera*. The larvae were roasted in the Nagano and Fukushima Prefectures, **Japan** (Takagi 1929a).

Perlodidae (Predatory stoneflies) **(0)**

Isoperla **sp.** The larvae and adults of a species belonging to this genus are consumed in the **USA** (Ebeling 1986).

Pteronarcidae (Giant stoneflies) **(4)**

Pteronarcys californica (California salmon fly) The adults are eaten by steaming to make a kind of cake in the **USA** (Essig 1947 and 1965, Sutton 1985).

Pteronarcys dorsata **India:** The larvae are roasted or boiled by the Nishi people in Arunachal Pradesh State (Singh et al. 2007). The **USA:** The adults are steamed to make a kind of cake (Sutton 1988).

Pteronarcys princes The adults are steamed to make a kind of cake in the **USA** (Sutton 1988).

Pteronarcys reticulata The larvae are consumed mostly in the Fukushima and Nagano Prefectures, **Japan** (Takagi 1929a).

5. ORTHOPTERA (Grasshoppers) [305]

Acrididae (Grasshoppers) **(210)**

Abracris → *Osmilia*

Acanthacris ruficornis This species is a food item in **Malawi** (DeFoliart 2002), **PRC** (Bani 1995), **Sahel** (Bani 1995), **Zambia** (Mbata 1995) and **Zimbabwe** (Chavanduka 1975).

Acorypha clara This species is eaten in **Africa** (Barreteau 1999).

Acorypha glaucopsis This species is treated as food in **Cameroon** (Barreteau 1999).

Acorypha nigrovariegata The adults are roasted, fried or boiled-dried in **Zambia** (Mbata 1995).

Acorypha picta The adults are used as food in **Cameroon** (Barreteau 1999).

Acrida bicolor This species is treated as food item in **Cameroon** (Barreteau 1999) and **Zimbabwe** (Chavanduka 1975).

Acrida chinensis The adults are relished **China** (Mao 1997).

Acrida cinerea **(PL-I-6)** (long-headed grasshopper) **Cameroon:** This species is a food resource (Barreteau 1999), and in some other countries in **Africa** (van der Waal 1999). **China:** The larvae and adults are consumed (Hu and Zha 2009). **Japan:** The Japanese name is *shōryō-batta*. The larvae and the adults are cooked or roasted with soy sauce, fried or pickled in the Nagano and Yamanashi Prefectures (Miyake 1919, Takagi 1929b). **Laos:** The Laotian name is *takkataen*. This grasshopper is roasted or fried (Nonaka 1999a). **North Korea:** This species is consumed as a food item (Okamoto and Muramatsu 1922).**South Korea:** The adults are used as a food article (Pemberton 1994). **Thailand:** This species is a significant dietary factor (Hanboonsong et al. 2000).

Acrida exaltata This species is treated as a food item in Arunachal Pradesh State, **India** (Singh and Chakravorty 2008).

Acrida gigantea This species is eaten by the Ao-Naga people in Nagaland State, **India** (Meyer-Rochow 2005).

Acrida lata → Acrida cinerea

Acrida oxycephala The larvae and the adults are fit for food in **China** (Hu and Zha 2009).

Acrida sulphuripennis The adults are roasted, fried, or boiled-dried in **Zambia** (Mbata 1995).

Acrida turrita → Acrida cinerea

Acrida willemsei This species is eaten in **Laos, Myanmar, Vietnam** and **Thailand** (Yhoung-Aree and Viwatpanich 2005).

Acrida **sp. Malaysia:** A species belonging to this genus is eaten in Sabah State (Chung et al. 2002). **Sri Lanka:** A species belonging to this genus is used as a foodstuff (Nandasena et al. 2010). **Thailand:** A species belonging to this genus is consumed (Hanboonsong et al. 2000).

Acridites lineola → Schistocerca gregaria

Acridium aerigonosum This grasshopper is a food item in **Indonesia** (van der Burg 1904).

Acridium manilense This grasshopper is is a food article in the **Philippines** (Simmonds 1885).

Acridium melanocorne (brown locust) This grasshopper is of possible food value to Ao-Naga people in Nagaland State, **India** (Meyer-Rochow 2005).

Acridium peregrinum **Algeria:** The locusts are boiled in salt water and then dried for preservation (Künckel d'Herculais 1891). **Morocco:** The locusts collected at the time of their outbreak, are eaten boiled in salt water and then fried (Simmonds 1885).

Acridium perigrinum (migratory locust) **India:** This grasshopper is consumed by the Ao-Naga people in Nagaland State (Meyer-Rochow 2005). The **USA:** Digger Indians eat the adults boiled (Simmonds 1885).

Acridium ranunculum The larvae and adults are fried in the **Philippines** (Gibbs et al. 1912).

Acridium rubescens The larvae and adults are eaten fried in the **Philippines** (Gibbs et al. 1912).

Acridoderes strenua → *Acridoderes strenuus*

Acridoderes strenuus This species is used as a food widely in **Niger** (van Huis 1996).

Acrotylus blondelli This species is treated as a food item in **Niger** (Lévy-Luxereau 1980).

Acrotylus longipes This species is an article of foods in **Niger** (Lévy-Luxereau 1980).

Aeolopus tamulusus The adults are roasted in **Thailand** (Bristowe 1932).

Afroxyrrhepes procera The adults are an eatables in **PRC** (Roulon-Doko 1998, Moussa 2004).

Afroxyrrhepes sp. The adults of a species belonging to this genus are eaten in **Zambia** (DeFoliart 2002).

Agridium melanocorne → *Acridium melanocorne*

Aidemona azteca In **Colombia** and **Venezuela**, Yukpa people wrapped the adults in leaves and roasted (Ruddle 1973).

Aiolopus thalassinus This species is found to be consumed as food in **Africa** (van der Waal 1999).

Aiolopus sp. A species belonging to this genus is good to eat in Sabah State, **Malaysia** (Chung et al. 2002).

Amblyphymus **sp.** A species belonging to this genus is eaten in **Africa** (van der Waal 1999).

Anacridium burri This species is eaten in southern **Africa** (Malaisse 1997).

Anacridium melanorhodon (Sahelian tree locust) This species is a food item in **Cameroon** (Barreteau 1999) and **Niger** (Lévy-Luxereau 1980).

Anacridium moestum This species is consumed as a food item in **Africa** (van der Waal 1999).

Anacridium wernerellum (Sudanese tree locust) This species is used as a food article in **Niger** (Lévy-Luxereau 1980).

Arcyptera fusca The larvae and adults are an article of foods in **China** (Hu and Zha 2009).

Arphia fallax The larvae and adults are preferred edible insects in Oaxaca State, **Mexico** (Ramos-Elorduy et al. 1997).

Arphia pseudonietana The adults are roasted in the **USA** (Sutton 1988).

Atractomorpha bedeli The Japanese name is *onbu-batta*. The adults are cooked with soy sauce in the Nagano and Yamanashi Prefectures, **Japan** (Miyake 1919, Takagi 1929b, Katagiri and Awatsuhara 1996).

Atractomorpha lata → *Acrida lata*

Atractomorpha psittacina This species is eaten in Sabah State, **Malaysia** (Chung et al. 2002).

Atractomorpha sinensis The larvae and adults are a food of people in **China** (Hu and Zha 2009).

Atractomorpha **sp.** A species belonging to this genus is used as food in **Thailand** (Hanboonsong 2010).

Aularches miliaris This species is a food of people in **Thailand** (Leksawasdee 2008).

Boopedon flaviventris The Mexican name is *chapulín*. The larvae and adults are a preferred edible insect in Oaxaca State, **Mexico** (Ramos-Elorduy et al. 1997).

Boopedon **sp. affin.** *flaviventris* The Mexican name is *chapulín*. The larvae and adults are a preferred edible insect in Oaxaca State, **Mexico** (Ramos-Elorduy et al. 1997).

Borborothis brunneri → *Phymateus viridipes*

Brachycrotaphus tryxalicerus This species is a food of people in **Cameroon** (Barreteau 1999).

Bryodema gebleri The larvae and adults are used as a food item in **China** (Hu and Zha 2009).

Caelifera **sp.** The Laotian name is *takkaten*. This species is used as food in **Laos** (Boulidam 2008).

Calliptamus abbreviates The adults are fried after removing the wings and legs in **China** (Chen and Feng 1999).

Calliptamus barbarous cephalotes The larvae and the adults are a food stuff in **China** (Hu and Zha 2009).

Calliptamus italicus The larvae and the adults are a food stuff in **China** (Hu and Zha 2009).

Caloptenopsis nigrovariegata This species is eaten in **Africa** (Mbata 1995).

Calopterus italicus Digger Indians eat the adults boiled in the **USA** (Simmonds 1885).

Camnula pellucida (clearwinged grasshopper) The adults are roasted in the **USA** (Essig 1934, 1965).

Cantatops spissus The adults are a food item in **PRC** (Bani 1995).

Cardeniopsis guttatus The adults are a food item in **Zambia** (DeFoliart 2002).

Caredeniopsis nigropunctatus The adults are used as food in **Zambia** (DeFoliart 2002).

Cardenius guttatus → *Cardeniopsis guttatus*

Cataloipus congnatus This species is a food source in **Africa** (van der Waal 1999).

Cataloipus cymbiferus This species is used as a food in **Cameroon** (Barreteau 1999).

Cataloipus fuscocoeruleipes This species is an eatable in **Sahel** (van Huis 1996).

Catantops annexus This species is an article of food in Arunachal Pradesh State, **India**, by frying in oil after removing wings, and with a bit of salt (Singh and Chakravorty 2008).

Catantops axillaris This species is treated as a food item in **Niger** (Lévy-Luxereau 1980).

Catantops haemorrhoidalis This species is eaten in **Niger** (Lévy-Luxereau 1980).

Catantops ornatus The adults are an article of food in **Zambia** (DeFoliart 2002).

Catantops stylifer This species is used for meal in the Maradi region, **Niger** (Lévy-Luxereau 1980).

Ceracris nigricornis nigricornis This species is eaten in Arunachal Pradesh State, **India**, by frying in oil after removing wings, and with a bit of salt (Singh and Chakravorty 2008).

Ceraeri kiangsu In **China,** this species is consumed as food item (Mao 1997), and used as an anodyne or a medicine for a cough (Zimian et al. 1997).

Chirista compta The adults are a food item in **PRC** (van Huis 1996, Moussa 2004).

Chondacris rosea **China** The larvae and adults are used as food stuff (Zhū 2003, Hu and Zha 2009). **India:** The adults are eaten boiled, fried or as a paste (chutney). Some insects are smoked for further use. Wings, appendages and lower portion of abdomen are removed before eating (Chakravorty et al. 2011). **Thailand:** This species is used as a food item (Hanboonsong et al. 2000).

Chondacris rosea brunneri People living in the southern part of **Thailand** eat this species (Lumsa-ad 2001).

Chondracris rosea pbrunner The Thai name is *tukkatan-E-moh.* People eat this species in **Laos, Myanmar, Thailand** and **Vietnam** (Yhoung-Aree and Viwatpanich 2005).

Chondacris **sp.** A species belonging to this genus is used as food in **Thailand** (Leksawasdi 2008).

Choroedocus robustus This species is used as food in **India** (Sachan et al. 1987).

Chorthippus **sp.** People eat a species belonging to this genus in **Thailand** (Hanboonsong 2010).

Chortoicetes terminifera (Australian plague locust) In Queensland State, **Australia**, the women gather large quantities of the grasshoppers when swarms of this species occur, and remove the wings and legs, and then roaste them (Lumholtz 1890).

Chrotogonus senegalensis This species is a food item in **Cameroon** (Barreteau 1999).

Crytacanthacris → *Cyrtacanthacris*

Curtilla africana In the **Philippines**, people eat grasshoppers of this species fried (Gibbs et al. 1912).

Cyathosternum **spp.** Some species belonging to this genus are consumed as a food stuff in **Zambia** and **Zimbabwe** (Gelfand 1971).

Cyrtacanthacris aeruginosa The adults are roasted, fried or boil-desiccated in **Malawi** and **Zambia** (Mbata 1995).

Cyrtacanthacris aeruginosus unicolor: People of the Ibira and Nupe tribes in **Nigeria** eat the adults raw, or roasted (Fasoranti and Ajiboye 1993).

Cyrtacanthacris septemfasciata (red locust) **Kuwait:** The adults are treated as a food item by Bedouin people. This species of locusts is eaten mostly by the inhabitants of Arabia. People eat only the females, which are quite good when they are fried in butter with salt, or boiled (Dickson 1949). **Saudi Arabia:** The women gather great heaps of the locusts, and singe them in shallow pits, with a weak fire of herbs. The locust meat can be stored for long time (Doughty 1923). **Tanzania:** The adults are boil-fried after removing the wings and legs (Harris 1940). **Uganda:** The grasshoppers of this species are favored as food by many people. They are usually fried, but may be pounded and added to sauces (DeFoliart 2002). In **Zambia** and **Zimbabwe:** The larvae and the adults are boiled in salt water (Chavanduka 1975). This species is also eaten in **DRC, Madagascar, PRC** (Moussa 2004), **RSA** (Quin 1959), and the southern regions of **Sahara** (Chavanduka 1975).

Cyrtacanthacris tatarica (black spotted grasshopper) **Thailand:** The adults are used as food while the head is removed. They can be stir-fried, used in curry as a meat-substitute, or made into a spicy sauce (Vara-asavapati et al. 1975). **Zambia:** The adults are roasted, fried or boil-dried (Mbata 1995). This species is also eaten in **Botswana** (Nonaka 1998) and **India** (Sachan et al. 1987).

Cyrtacanthacris **sp.** A species belonging to this genus is used as a food item in **Thailand** (Hanboonsong et al. 2001).

Diabolocatantops axillaris This species is an article of foods in **Cameroon** (Barreteau 1999) and **Niger** (Lévy-Luxereau 1980).

Diabolocanthops innotabilis The adults are consumed as a food item in Arunachal Pradesh State, **India**. They are fried and boiled or smoked after removing wings and the antennae (Chakravorty et al. 2011).

Dociostaurus moroccanus In the **Near East**, the adults are eaten (Heimpel 1996).

Dociostaurus kraussi nigrogeniculatus The larvae and adults are used as foods in **China** (Hu and Zha 2009).

Ducetia japonica This species is treated as food items in **Laos, Myanmar** and **Vietnam** (Yhoung-Aree and Viwatpanich 2005) and **Thailand** (Hanboonsong 2010).

Encoptolophus herbaceous The larvae and the adults are preferred edible insects in Oaxaca State, **Mexico** (Ramos-Elorduy et al. 1997).

Encoptolophus otmitus → *Encoptolophus herbaceous*

Euprepocnemis shirakii **China:** People eat the adults fried after removing the legs and wings (Chen and Feng 1999). **Thailand:** This species is eaten (Hanboonsong et al. 2001).

Exopropacris modica This species is used as a food item in **Cameroon** (Barreteau 1999).

Eyprepocnemis plorans This species is widely used as food stuffs in **Africa** (Roulon-Doko 1998).

Gastrimargus africanus This species is used as food in **Cameroon** (Barreteau 1999), **Niger** (Lévy-Luxereau 1980) and **PRC** (van Huis 1996, Mousa 2004).

Gastrimargus determinatus This species is eaten in **Africa** (Barreteau 1999).

Gastrimargus proceus This species is a food resource in **Cameroon** (Barreteau 1999) and **Niger** (Lévy-Luxereau 1980).

Gomphocerus sibiricus The larvae and adults are consumed as a food stuff in **China** (Hu and Zha 2009).

Gryllus aegyptius **South Europe, Barbary,** and **Egypt:** People ate this grasshopper by baking after removing its wings (Cuvier 1827/35).

Gryllus lineola **South Europe, Barbary,** and **Egypt:** People ate this grasshopper by baking after removing its wings (Cuvier 1827/35).

Gryllus locust **South Europe, Barbary,** and **Egypt:** People ate this grasshopper by baking after removing its wings (Cuvier 1827/35).

Gryllus tataricus **South Europe, Barbary,** and **Egypt:** People ate this grasshopper by baking after removing its wings (Cuvier 1827/35).

Harpezocatantops stylifer This species is used as food items in **Cameroon** (Barreteau 1999).

Heteracris coerulescens This species is used as a food sources in **Africa** (Bahuchet 1985).

Heteracris guineensis The adults are treated as a food item in **PRC** (van Huis 1996, Mousa 2004).

Heteracris **sp.** This species is good to eat in **Africa** (van der Waal 1999).

Hieroglyphodes assamensis This species is good to eat in Arunachal Pradesh State, **India** (Singh and Chakravorty 2008).

Hieroglyphus banian This species is used as a food source in **Thailand** (Jamjanya et al. 2001).

Hieroglyphus concolour This species is an eatable in Arunachal Pradesh State, **India** (Singh and Chakravorty 2008).

Hieroglyphus oryzivorus This species is fit for food in Arunachal Pradesh State, **India** (Singh and Chakravorty 2008).

Hieroglyphus **sp.** The fried adults of a species belonging to this genus are used with boiled vegetables as a paste (chutney) to take with alcoholic beverages in **India**. Antennae and appendages are removed (Chakravorty et al. 2011).

Homocoryphus prasimus The larvae and the adults are dietary factors in **Mexico** (Ramos-Elorduy and Pino 2002).

Homoxyrrhepes punctipennis This species is an eatable in **Cameroon** (Barreteau 1999) and **DRC** (Heymans and Evrard 1970).

Humbe tenuicornis This species is considered to be edible in **Niger** (Lévy-Luxereau 1980) and **Africa** (van der Waal 1999).

Krausella amabile This species is fit for food in **Cameroon** (Barreteau 1999).

Kraussaria angulifera This species is used for meal in **Cameroon** (Barreteau 1999).

Lamarckiana bolivariana This species is edible in **Africa** (van der Waal 1999).

Lamarckiana cucullata The adults are roasted in **Botswana** (Nonaka 1996).

Lamarckiana punctosa This species is eaten in **Africa** (van der Waal 1999).

Leptysma marginicollis This species is an eatables in Arunachal Pradesh State, **India,** by frying in oil after removing the wings, and with a bit of salt (Singh and Chakravorty 2008).

Leptisma **sp.** The adults are favored boiled, roasted or as a paste in Arunachal Pradesh State, **India**. Anal cirri and antennae are removed (Chakravorty et al. 2011).

Locusta cernensis **Madagascar:** The adults are fried in oil after removing the wings and the legs. Native people consider this species as one of their finest dish (Hope 1842).

Locusta danica → *Locusta migratoria*

Locusta devastator **Southern Africa (Kaffraria):** In the 19th century, Bosjesmans or Wood Hottentots collected the grasshoppers by gathering them into long and deep trenches to eat (Hope 1842).

Locusta gregaria (red skipper) **Saudi Arabia:** Natives ate this locust fried (Hope 1842).

Locusta japonica → *Tettigonia orientalis orientalis* (Tettigonidae)

Locusta mahrattarum **India:** This species is used for meal by the Mussulmauns (Hope 1842).

Locusta migratoria (migratory locust) **(PL I-3) China:** This species is used as an anodyne or a medicine for a cough (Zimian et al. 1997). **Crimea:** This species was commonly eaten in 19th century (Hope 1842). **Japan:** The Japanese name is *daimyo-batta*. The adults were cooked with soy sauce in the Nagano and Yamanashi Prefectures (Miyake 1919, Katagiri and Awatsuhara 1996). The charred adults were used for treating convulsive fits (Miyake 1919). **Mexico:** This species is raised as food (Ramos-Elorduy 1997). The **Philippines:** The adults are boiled, cooked or fried (Gibbs et al. 1912, Adalla and Cervancia 2010). **PNG:** The Kiriwinians eat the locusts raw or cooked (Meyer-Rochow 1973). This species is also eaten in following countries:

RSA (DeFoliart 2002), **Thailand** (Hanboonsong et al. 2000), **Ukraine** (Hope 1842), **Zambia** (Mbata 1995) and **Zimbabwe** (Dube et al. 2013).

Locusta migratoria capito The adults are fried, or sun-dried in **Madagascar** (Decary 1937).

Locusta migratoria manilensis **China:** The adults are used as a food items fried after removing the wings and legs (Chen and Feng 1999). **Malaysia:** This species is eaten in Sabah State (Chung et al. 2002).

Locusta migratoria migratorioides (African migratory locust) The adults are roasted, fried, or dried after boiling in **Cameroon** (Barreteau 1999), **PRC** (Moussa 2004), **Tanzania** (Owen 1973), **Uganda** (Harris 1940, Bouvier 1945), **Zambia** (Mbata 1995) and **Zimbabwe** (DeFoliart 2002).

Locusta onos **Mongolia** and **India:** This species was a food source by the Mongols and Indians (Hope 1842).

Locusta pardalina This species is used as good diet in **RSA** (Quin 1959) and **Zambia** (Mbata 1995).

Locusta persarum **Persia (Iran):** The adults were dried and salted (Hope 1842).

Locusta tartarica **Crimea:** This locust was commonly eaten in Crimea (Hope 1842). **Kalahari Desert:** People ate this species by frying them after removing the wings and legs (Dornan 1925).

Locusta viridissima **Europe:** This species was occasionally eaten, and was reported not to possess the flavor belonging to the migratory locusts (Hope 1842).

Locusta **sp.** A species belonging to this genus is used as a food items in Arunachal Pradesh State, **India,** by frying in oil after removing the wings, and with a bit of salt (Singh and Chakravorty 2008).

Locustana pardalina This species is edible in **RSA** (Quin 1959) and **Zambia** (Mbata 1995).

Lophacris **sp.** This species is foods of people in **Brazil** (Setz 1991).

Melanoplus atlanis The roasted adults are relished in the **USA** (Essig 1934 and 1965).

Melanoplus bivittatus (two-striped grasshopper) The adults are roasted in the **USA** (Essig 1934 and 1965).

Melanoplus devastator (devastating grasshopper) The adults are roasted in the **USA** (Essig 1934 and 1965).

Melanoplus differentialis (differential grasshopper) The adults are roasted in the **USA** (Essig 1934 and 1965).

Melanoplus femurrubrum (redlegged grasshopper) **Mexico:** The Mexican name is *chapulín*. The larvae and adults are an article of foods (Essig 1934 and 1965). The **USA:** The roasted adults are eaten (Essig 1934 and 1965).

Melanoplus mexicanus The Mexican name is *chapulín*. The larvae and adults are used as a food items in Oaxaca State, **Mexico** (Ramos-Elorduy et al. 1997). The hind legs are pulverized with water, and used as a strong diuretic (Ramos-Elorduy de Conconi and Pino 1988).

Melanoplus mexicanus mexicanus → Melanoplus sanguinipes

Melanoplus sanguinipes (migratory grasshopper) The **USA:** The adults are roasted (Essig 1934 and 1965). This species seemed to be eaten by aboriginal people since long time ago, as large amounts of this species were found in Lakeside Cave, Utah state (BC 2600) (Madsen and Kirkman 1988).

Melanoplus sumichastri The larvae and adults are consumed as a foodstuffs in **Mexico** (Ramos-Elorduy and Pino 2002).

Melanoplus **spp.:** The larvae and adults of some species belonging to this genus are used as a food items in Oaxaca State, **Mexico**. These species is raised as food for humans (Ramos-Elorduy 1997, Ramos-Elorduy et al. 1997).

Mesopsis abbreviates This species is used for meals in **Africa** (Roulon-Doko 1998).

Metaxymecus gracilipes → Tylotropidius gracilipes

Nomadacris septemfasciata → Cyrtacanthacris septemfasciata

Ochrotettix cer salinus The larvae and adults are foodstuffs in Oaxaca State, **Mexico** (Ramos-Elorduy et al. 1997).

Oedaleus carvalhoi This species is used as a food items in **Africa** (van der Waal 1999).

Oedaleus decorus The larvae and adults are food items in **China** (Hu and Zha 2009).

Oedaleus flavus This species is an article of foods in **Africa** (van der Waal 1999).

Oedaleus infernalis The Japanese name is *kuruma-battamodoki*. In **Japan**, the adults are roasted in the Nagano and Yamanashi Prefectures (Mukaiyama 1987).

Oedaleus nigeriensis This species is treated as a foodstuffs in **Cameroon** (Barreteau 1999).

Oedaleus nigrofasciatus The adults are roasted, fried, or boiled-desiccated in **Zambia** (Mbata 1995).

Oedaleus senegalensis This species is used as a food items in **Africa** (Mignot 2003).

Oedaleonotus enigma (valley grasshopper) The boiled adults are a food sources in the **USA** (Essig 1934 and 1965).

Oedipoda corallipes In **West Indies,** the adults are used for meals (Bodenheimer 1951).

Oedipoda migratoria **USA:** Digger Indians eat the adults roasted (Simmonds 1885).

Oedipoda subfasciata This species is foods of people in the **Philippines** (Simmonds 1885).

Opeia **sp.** The Mexican name is *chapulí.* The larvae and adults are a dietary factor in **Mexico** (Ramos-Elorduy et al. 1998).

Ornithacris cyanea (magnifica) The adults are food items in **Zimbabwe** (Gelfand 1971).

Ornithacris turbida The adults are a food article in **PRC** (Bani 1995).

Ornithacris **sp.** A species belonging to this genus is treated as a foodstuffs in **Zimbabwe** (Weaving 1973).

Orphula azteca The larvae and adults are used as food in **Mexico** (Ramos-Elorduy and Pino 2002).

Orphulella **spp.** The adults of some species belonging to this genus are smothered or roasted in **Colombia** (Ruddle 1973). **Venezuela:** The adults of some species belonging to this genus are also eaten by the Yukpa people (Ruddle 1973).

Orthacanthacris humilicrus This species is treated as a food items in **Niger** (Lévy-Luxereau 1980).

Orthochtha magnifica This species is an article of foods in **Africa** (Mbata 1995).

Orthochtha turbid This species is used as a food items in **Africa** (Bani 1995).

Orthochtha venosa This species is an eatables in **Cameroon** (Barreteau 1999).

Osmilia flavolineata **Colombia:** The adults are smothered or roasted (Ruddle 1973). **Mexico:** The larvae and adults are eaten in Oaxaca State (Ramos-Elorduy et al. 1997). **Venezuela:** The adults are used as a food items by the Yukpa people (Ruddle 1973).

Osmilia **spp. Columbia:** The adults of some species belonging to this genus are smothered or roasted (Ruddle 1973). **Venezuela:** The Yukpa people eat the adults of some species belonging to this genus (Ruddle 1973).

Oxya chinensis **China:** The adults are fried after removing the wings and legs (Chen and Feng 1999). The desiccated bodies are used for curing whooping cough, acute diarrhea, etc. (Nanba 1980). This species is raised as food (Zimian et al. 1997). **Japan:** The Japanese name is *sina-inago.* The adults are imported from China, and are cooked with soy sauce and sugar (Mitsuhashi 2008).

Oxya intricata This species is edible in **China** (Mao 1997), and also used as an anodyne or a medicine for a cough (Zimian et al. 1997).

Oxya japonica japonica **China:** The larvae and adults are used as an anodyne or as medicine for a cough (Zimian et al. 1997). **Japan:** The Japanese name is *hanenaga-inago*. The adults are cooked with soy sauce, or roasted in the Fukushima, Niigata, Gunma, Yamanashi, Nagano, Wakayama, Tottori, Kōchi, and Kumamoto Prefectures (Miyake 1919). This grasshopper was medicinally used as an antifebrile, etc. (Kuwana 1930). **Malaysia:** This species is eaten in the Sabah State (Chung et al. 2002). **North Korea:** The adults are used for treating gonorrhea, pneumonia, swelling, etc. in the Hwanghae, North Hamgyong and South Hamgyong Provinces (Okamoto and Muramatsu 1922). **South Korea:** This species is also consumed as a food items (Pemberton 1994). The adults are used for treating gonorrhea, pneumonia, swelling, etc. in the Gyeongii, North Gyeongsang, Kangwon, North Jeolla, South Jeolla, North Chungcheong and South Chungcheong Provinces (Okamoto and Muramatsu 1922). **Thailand:** This species is a preferred edible insect, and is pounded into a paste, pan-roasted, deep-fried (Sungpuag and Puwastien 1983). **Vietnam:** This grasshopper is the only species eaten in large quantities in this country. The adults are cooked with salt water or sauteed in pork fat after removing the wings, and sometimes the head, intestines and first two pairs of legs (Tiêu 1928).

Oxya ninpoensis The Japanese name is *ninpo-inago*. In **Japan** the adults are occasionally found in the cooked *inago* sold in markets (Fukuhara 1986).

Oxya sinuosa **North** and **South Korea:** The adults are roasted or dried (Pemberton 1994). **Japan:** The Japanese name is *chosen-hanenaga-inago*. The adults, mainly imported from Korea, and are cooked with soy sauce and sugar (Mitsuhashi 2008).

Oxya vicina → *Oxya yezoensis*

Oxya velox → *Oxya japonica japonica*

Oxya yezoensis **(PL II-1)** The Japanese name is *kobane-inago*. In **Japan** this species is the most popular food insects. Miyake (1919) reported that the adults were cooked with soy sauce in the Kochi, and Nagasaki Prefectures. There are, however, many other *inago* dishes, that are consumed in most prefectures. The consumption of *inago* by Japanese exceeds the amount of *inago* collected in Japan. Therefore Japan imports *inago* from China and Korea although the species is different. At present (2015), cooked *inagos* are sold in many food stores. Canned *inago* cooked with soy sauce, sugar and Japanese sweet wine is also available in the Nagano and other prefectures (Mitsuhashi 1997, 2003 and 2005) **(PL II-2)**. The adults of this species have

Plate II. 1. *Oxya yezoensis* adult. Body length: 35 mm (cf. p.21); 2. Cooked *O. yezoensis* (cf. p.21); 3. Mass rearing of *Acheta domestica* in Thailand. Insert: The last instar crickets (body length: 25 mm) (cf. p.29); 4. *Extatosoma tiaratum* adult. Body length: max. 200 mm (cf. p.40); 5. *Brachytrupes portentosus*. An adult fried. Body length: 35 mm (cf. p.29); 6. *Macrotermes falciger* soldier fried. Body length: 25 mm (cf. p.42).

been used for treating various diseases, such as coughing, fever, anemia, peritonitis, etc. (Koizumi 1935, Umemura 1943, Shiraki 1958).

Oxya **sp. India:** A species belonging to this genus is eaten (Singh and Chakravorty 2008). **Thailand:** The adults of a species belonging to this genus are roasted (Hanboonsong et al. 2000).

Oxycantatops congoensis The adults are considered good for eating in **PRC** (Bani 1995).

Oxycantatops spissus The adults are treated as a foodstuff in **Cameroon** (Barreteau 1999) and **PRC** (Bani 1995).

Pachytylus danica → *Locusta migratoria*

Paracinema tricolor This species is used as a food item in **Cameroon** (Barreteau 1999).

Parapleurus **sp.** People eat a species belonging to this genus in **Thailand** (Rattanapan 2000).

Pararcyptera microptera The larvae and adults are eaten in **China** (Hu and Zha 2009).

Patanga avis People eat this species in **Thailand** (Rattanapan 2000).

Patanga japonica **China:** The larvae and adults are considered to be edible (Mao 1997), and is used as an anodyne or a medicine for a cough (Zimian et al. 1997). **Thailand:** People eat this species roasted (Hanboonsong et al. 2000).

Patanga succincta (Bombay locust) **India:** This species is a food item (Sachan et al. 1987). **Thailand:** The Thai name is *tak-ka-taen-mo*. The fried adults and matured larvae are common foods. People also eat this species roasted with salt (Mungkorndin 1981). **Laos, Myanmar** and **Vietnam:** This species is eaten (Yhoung-Aree and Viwatpanich 2005).

Phlaeoba antannata This species is preferred edible insects in Arunachal Pradesh State, **India** (Singh and Chakratorty 2008).

Phymateus viridipes This species is a food item in **PRC** (Bergier 1941, Chavanduka 1975).

Plectrottetra nobilis → *Rhammatocerus viatorius*

Poecilocerastis tricolor This species is used as food in **Zambia** (Malaisse 2005, DeFoliart 2002).

Pseudocephalus litan This species is a food source in **Thailand** (Yhoung-Aree and Viwatpanich 2005).

Pycnodictya flavipes This species is an article of food in **Africa** (van der Waal 1999).

Pyrgomorpha cognata This species is fit for eating in **Cameroon** (Barreteau 1999).

Pyrgomorpha cognate → *Pyrgomorpha cognata*

Pyrgomorpha vignaudi → *Pyrgomorpha vignaudii*

Pyrgomorpha vignaudii This species is an effective food source in **Africa** (Bahuchet 1985).

Rammeacris kiangsu The adults are fried after removing wings and legs in **China** (Chen and Feng 1999).

Ratanga avis → *Patanga avis*

Rhammatocerus maturius The Mexican name is *chapulín*. The larvae and adults are used as provisions in **Mexico** (Ramos-Elorduy et al. 1998).

Rhammatocerus schistocercoides The larvae and adults are used as food in **Brazil** (Embrapa 2000).

Rhammatocerus viatorius The larvae and adults are considered good for eating in **Mexico** (Ramos-Elorduy and Pino 1989 and 2002).

Rhammatocerus **sp. Brazil:** The larvae of a species belonging to this genus are eaten by the Nhambiquara people in Mato Grosso State (Costa-Neto and Ramos-Elorduy 2006). **Venezuela:** A species belonging to this genus is roasted by the Guajibo people (Paoletti and Dufour 2005).

Roduniella insipid In **Africa**, this species is a food item (Bahuchet 1985).

Schistocerca cancellata cancellata **Brazil:** This species is used as a food in Brazil (DeFoliart 2002). **Guyana-Suriname-(Colombia):** Native people of the Arawak (Lokono) tribe eat the adults (Gilmore 1963, Goldman 1963, Posey 1978, Paoletti and Dufour 2005).

Schistocerca cancellata paranensis The larvae and adults are consumed as a foodstuff in **Brazil** and **Colombia** (Gilmore 1963).

Schistocerca gregaria (desert locust) **India:** The larvae and adults are important food article by the Karbi and Rengma Naga people in Assam State, as a chutney or roasted (Ronghang and Ahmed 2010). The dried locusts form an ingredient of curries in Kolkata (Maxwell-Lefroy 1906). **Iraque:** Forbes described this species as *Acridites lineola*. This species is consumed with relish by the Iraqis (Forbes 1813). **Mexico:** This species is raised as food (Ramos-Elorduy 1997). This species is also eaten in **Cameroon** (Barreteau 1999), **Morocco, PRC** (Moussa 2004), **Tanzania** (Bodenheimer 1951) and **Uganda** (Hope 1842).

Schistocerca gregaria flaviventris In **Zambia** and southwestern **Africa**, the adults are used as food. People make foods by roasting, frying, boiling-desiccation of them (DeFoliart 2005, Mbata 1995).

Schistocerca gregaria gregaria This species is widely consumed as a food source in northern **Africa** (DeFoliart 2005).

Schistocerca nitens → *Schistocerca vaga vaga*

Schistocerca paranensis **Guyana-Suriname-(Colombia):** Native people of the Arawak (Lokono), and Tucano tribes eat the adults (Gilmore 1963, Goldman 1963, Posey 1978, Paoletti and Dufour 2005). **Mexico:** The larvae and adults are considered good for eating in Veracruz, Tabasco, Campeche and Yucatan States (Ramos-Elorduy de Conconi et al. 1984). The powdered grasshoppers are used as a nutritional supplement (Ramos-Elorduy de Conconi and Pino 1988).

Schistocerca peregrinatoria In **Angola**, the adults are roasted, dried after boiling (Wellman 1908).

Schistocerca shoshone → *Schistocerca venusta*

Schistocerca vaga vaga In **Mexico**, the Lacandone and Tojolabale people eat the larvae and adults (Ramos-Elorduy and Pino 2002).

Schistocerca venusta (green valley grasshopper) The adults roasted are preferably consumed in the **USA** (Essig 1934 and 1965).

Schistocerca **spp. Brazil:** Some grasshoppers belonging to this genus are used as food by the Nhambiquara people in Mato Grosso State (Setz 1991). **Colombia:** The adults are smothered or roasted (Ruddle 1973). **Ecuador:** The adults of a species belonging to this genus are eaten by the Quichua people (Onore 1997). **India:** People consume the adults of some species belonging to this genus. They are fried and boiled with some leafy vegetables. The wings, antennae, appendages and lower portion of abdomen are removed (Chakravorty et al. 2011). **Mexico:** Native people of the Chontale, Lacandone, Mestizo, Tarasco, Tlapaneca, Totonaca, Tlapaneca and Totonaca tribes eat the larvae and adults of some species belonging to this genus (Ramos-Elorduy et al. 1997, Ramos-Elorduy and Pino 1989). **SAR:** The adults of some species belonging to this genus are treated as a foodstuff (DeFoliart 2002). **Venezuela:** The adults of a species belonging to this genus are eaten by the Yukpa people (Ruddle 1973).

Sherifura hanoingtoni This species is used as a food in **Africa** (Barreteau 1999).

Shirakiacris shirakii → *Euprepocnemis shirakii*

Spharagemon aequale The larvae and adults are an effective food source in Michoacan State, **Mexico** (Ramos-Elorduy de Conconi et al. 1984).

Sphenarium bolivari → *Sphenarium histrio*

Sphenarium histrio The Mexican name is *chapulín*. The larvae and adults are considered good for eating in Oaxaca and Guerrero States, **Mexico** (Ramos-

Elorduy et al. 1997). The hind legs are pulverized with water, and used as a strong diuretic (Ramos-Elorduy de Conconi and Pino 1988).

Sphenarium magnum In **Mexico**, native people of the Zapoteco tribe eat this species (Ramos-Elorduy and Pino 1989). The larvae and adults are eaten in Oaxaca State (Ramos-Elorduy et al. 1997). The hind legs are pulverized with water, and used as a strong diuretic (Ramos-Elorduy de Conconi and Pino 1988).

Sphenarium mexicanum The larvae and adults are treated as a foodstuff by the Tzeltzales Tzoltziles people in **Mexico** (Ramos-Elorduy and Pino 2002).

Sphenarium purpurascens The Mexican name is *chapulín*. In **Mexico**, the larvae and adults are considered edible in Oaxaca State. This species is raised as food for humans (Ramos-Elorduy 1997, Ramos-Elorduy et al. 1997). The hind legs are pulverized with water, and used as a strong diuretic (Ramos-Elorduy de Conconi and Pino 1988).

Sphenarium **spp.** The larvae and adults of several species of this genus are considered good for eating in Morelos, Puebra, and Oaxaca States, **Mexico** (Ramos-Elorduy de Conconi et al. 1984) **(PL I-5)**.

Sphingonotus **spp.** The larvae and adults of some species belonging to this genus are used as food in **China** (Hu and Zha 2009).

Stauroderus scalaris The larvae and the adults are consumed as a foodstuff in **China** (Hu and Zha 2009).

Stenobothrus epacromioides → *Stenohippus mundus*

Stenohippus mundus This species is fit for eating in **Africa** (Mignot 2003).

Taeniopoda auricornis The larvae and adults are considered edible in **Mexico** (Ramos-Elorduy and Pino 2002).

Taeniopoda **sp.** In **Mexico,** native people of the Chontale, Lacandone, Mestizo, Tarasco, Tlapaneca and Totonaca tribes eat grasshoppers of this genus (Ramos-Elorduy de Conconi et al. 1984, Ramos-Elorduy and Pino 1989).

Titanacris albipes The adults are edible in **Brazil** (Setz 1991).

Trilophidia annulata This species is used as food in **Thailand** (Hanboonsong et al. 2000).

Trimerotropis pallidipennis The Mexican name is *chapulín*. **Mexico:** This species is an article of food in D.F. (Ramos-Elorduy de Conconi 1991). The Hñähñu people use this species to prevent the delay of adult tooth development (Aldasoro Maya 2003).

Trimerotropis **sp.** The larvae and adults of a species called *chapulín* in Mexican, are edible in Hidalgo State, **Mexico** (Ramos-Elorduy de Conconi et al. 1984).

Tristria conops This species is an article of food in **Africa** (Roulon-Doko 1998).

Tristria discoidalis This species is consumed as food in **Africa** (Roulon-Doko 1998).

Tropidacris c. cristata **Colombia:** The adult grasshoppers are preferred edible insects, and are smothered or grilled by the Yukpa people (Ruddle 1973). **Venezuela:** The adult grasshoppers are considered good for eating by the Guajibo people (Paoletti and Dufour 2005).

Tropidacris collaris The adults are used as food in **Brazil** (Setz 1991).

Tropidacris latreillei → *Tropidacris c. cristata*

Tropinotus mexicanus The larvae and adults are used as food in Oaxaca State, **Mexico** (Ramos-Elorduy et al. 1997).

Truxalis burtti This species is eatable in **Africa** (van der Waal 1999).

Truxalis johnstoni This species is consumed as a foodstuff in **Cameroon** (Barreteau 1999).

Truxaloides constrictus The adult grasshoppers are fit for eating in **Zimbabwe** (Gelfan 1971).

Tryxalis nasuta → *Acrida cinerea*

Tylotropidius didymus This species is an article of food in **Africa** (Roulon-Doko 1998).

Tylotropidius gracilipes This species is an edible insect in **Cameroon** (Barreteau 1999).

Valanga irregularis (large coast locust) In **PNG,** the Onabasulu people eat this species roasted (Meyer-Rochow 1973).

Valanga nigricornis (Short-horned grasshopper) **Indonesia:** The Javanese name is *belalang kayu*. This species is eaten generally by roasting after removing the wings and legs. Often, seasonings such as onion, garlic, chili or soy sauce is added (Lukiwati 2010). **Malaysia:** This species is a food item in Sabah State (Chung et al. 2002).

Valanga sp. A species belonging to this genus is considered edible by the Onabasulu and Kiriwina people in **PNG** (Meyer-Rochow 2005).

Xanthippus corallipes → *Oedipoda corallipes*

Xanthipus corallipes zapotecus **Mexico:** The Mexican name is *chapulín*. The larvae and adults are food source (Ramos-Elorduy et al. 1998). **West Indies:** This species is consumed in great quantities (DeFoliart 2002).

Zonocerus elegans The adults are considered edible in **Mozambique** and **RSA** (Quin 1959).

Zonocerus variegates **Nigeria:** The adults are roasted (Roulon-Doko 1998). The Ibira, Nupe, and Yoruba people eat the adults raw, or roasted (Fasoranti and Ajiboye 1993). This species is also eaten in **Cameroon** (de Colombel 2003) and **CAR** (Barreteau 1999).

Catantopidae (10)

Catantops melanostichus This species is edible in **Africa** (van der Waal 1999).

Catantops quadrates This species is used as a food item in **Africa** (Bahuchet 1985).

Catantops stramineus This species is a foodstuff in **Africa** (Barreteau 1999).

Diabolocantatops axillaris This species is used as food in **Africa** (van Huis 1996).

Exoporpacris modica This species is a food source in **Africa** (Barreteau 1999).

Harpezocantatops stylifer This species is considered to be edible in **Africa** (Barreteau 1999).

Oxycantatops congoensis This species is used as a good meal in **Africa** (Bani 1995).

Oxycantatops spissus This species is a food item in **Africa** (Bani 1995).

Parapropacris notata This species is treated as a foodstuff in **Africa** (Roulon-Doko 1998).

Phaeocantatops decorates This species is used as a food item in **Africa** (van der Waal 1999).

Gryllacrididae (wingless long-horned grasshoppers) (2)

Gryllacris africana This species is consumed in **Africa** (Bahuchet 1985).

Stenopelmatus fuscus (Jerusalem cricket) In the **USA**, this species is consumed as food by the native Americans of the arid area of the West (Ebeling 1986).

Stenopelmatus **sp.** The adults are eaten by the Lacandones people in **Mexico** (Ramos-Elorduy and Pino 2002).

Gryllidae (True crickets) (32)

Acanthoplus **sp.** A species of crickets of this genus is used as food in **Botswana** (Nonaka 1996).

Acheta assimilis → *Gryllus assimilis*

Acheta bimaculatus → *Gryllus bimaculatus*

Acheta confirmata People eat this species in **Thailand** (Vara-asavapati et al. 1975).

Acheta domesticus (house crickets) **(PL II-3)** In **Canada** and the **USA,** this species is eaten (Taylor and Carter 1992). **Laos:** The larvae and adults of this species are fit for eating (Boulidam 2008). **Mexico:** This species is used as a food item in the Guanajuato, Morelos and Veracruz States (Ramos-Elorduy 2009), and raised as food (Ramos-Elorduy 1997). The hind legs are pulverized with water, and used as a strong diuretic (Ramos-Elorduy de Conconi and Pino 1988). **Thailand:** The Thai name is *maeng-sa-ding,* or *jing-reed-khao.* The larvae and adults of this species are consumed (Mitsuhashi 2008).

Acheta mitrata → *Teleogryllus emma*

Acheta smeathmanni **Africa:** People like to catch this cricket from the nest underground, and to eat it (Hope 1842).

Acheta testacea The Thai name is *jing-rheed.* This species is eaten in **Thailand** (Hanboonsong et al. 2000).

Acheta **spp.** People eat crickets of this genus by roasting, frying or sun-drying in **Zambia** (Mbata 1995) and in **Zimbabwe** (Chavanduka 1975).

Ampe **sp.** In **PRC,** a cricket of this genus is eaten (Moussa 2004).

Brachytrupes achatinus → *Brachytrupes portentosus*

Brachytrupes membranaceus **Benin:** The Fon people mainly in the southern area eat this species (Tchibozo et al. 2005). **Nigeria:** The Yoruba people eat this species roasted (Fasoranti and Ajiboye 1993, Dalziel 1937, Ene 1963). **Zimbabwe:** People catch the cricket in the following manner. When a stinging ant at the end of a piece of grass is pushed down into a hole in the ground, the cricket jumps out from its tunnel, and people catch it. Crickets are grilled with a pinch of salt added, and are eaten as relish with stiff porridge (Jackson 1954, Gelfand 1971). This species also eaten in following countries: **Angora** (Wellman 1908), **Cameroon** (Seignobos et al. 1996), **DRC** (Adriaens 1951), Malawi, **PRC** (Bani 1995), **Tanzania** (Bodenheimer 1951), **Uganda** (Owen 1973) and **Zambia** (Mbata 1995).

Brachytrupes portentosus **(PL II-5) China:** The adults are consumed by cooking with sugar and soy sauce (Chen and Feng 1999). This species is used as an antidote, an antifebrile, or as a suppressor of swelling (Zimian et al. 1997). People of the Thai tribe living in the western areas make a kind of paste from the larvae and adults by crushing them after removing legs, wings, and intestines, and then mixing with various food stuffs (Shu 1989). **India:** This species is consumed by the Ao-Naga people in Nagaland State (Sachan et al. 1987, Meyer-Rochow 2005). **Indonesia** (Bodenheimer 1951), **Laos:** The Laotian name is *ching-reep-ong.* The crickets are eaten in the area of Dong Makkhai village near Vientiane (Boulidam 2008). **Thailand:** The Thai name is *ji-pom* or *ji-konke*. People eat this big cricket by frying without coating, tempura or boiling (Leksawasdi 2010, Kuwabara 1997a and b). **Vietnam:** These crickets are highly esteemed by gourmets of the northern parts of the country (Tiêu 1928).

Brachytrupes **sp. India:** A species belonging to this genus is eaten in Arunachal Pradesh State by frying simply in a hot pan or in oil after removing the wings, stomach and head (Singh and Chakravorty 2008). **Nigeria:** A species belonging to this genus is used as a foodstuff in western Nigeria (Banjo et al. 2006).

Brachytrypes → *Brachytrupes*

Chrotogonus senegalensis This species is consumed in **Cameroon** (Seignobos et al. 1996).

Curtilla africana In **Uganda**, the adults are used as a food item (Ealand 1915).

Gryllodes berthelius → *Velarifictorus aspersus*

Gryllodes melanocephalus This species is an article of food by the Ao-Naga people in Nagaland State, **India** (Meyer-Rochow 2005).

Gryllodes mitratus → *Teleogryllus emma*

Gryllus assimilis (black field cricket) **Brazil:** The boiling water of the cricket's entire body is said to be good for difficulty in urination (Costa-Neto and de Melo 1998). **India:** This species is a food source in Arunachal Pradesh State, by frying simply in a hot pan or in oil after removing the wings, stomach and head (Singh and Chakravorty 2008). **Mexico:** This species is eaten by the people of Mazahua, Mixteca, Nahua and Otomí tribes (Ramos-Elorduy 2009). **USA:** The Nishinam name is *hallli*. The Nishinam people living along the Sacramento River or San Joaquin River eat the adults roasted (Essig 1965).

Gryllus bimaculatus (two spotted cricket) **China:** The adults are cooked with sugar and soy sauce (Chen and Feng 1999). **India:** This species is consumed as food by the Ao-Nag people in Arunachal Pradesh State as simply frying in a hot pan with or without oil after removing the wings, stomach and head (Meyer-Rochow 2005, Singh and Chakravorty 2008). **Laos, Myanmar, Vietnam** (Yhoung-Aree and Viwatpanich 2005). **Thailand:** The adults are eaten by skewer-roasting (Hanboonsong et al. 2000). **Uganda:** The adults of this species are eaten (Ealand 1915). **Zambia:** The adults are roasted, fried, boiled or dried (Mbata 1995).

Gryllus campestris **India:** The larvae and adults are used as food by the Karbi and Rengma Naga people in the Assam State as a chutney or roasted

(Ronghang and Ahmed 2010). **Jamaica:** The adults are consumed as food source (Sloane 1725).

Gryllus chinensis The adults are considered edible in **China** (Hu and Zha 2009).

Gryllus domesticus → *Acheta domesticus*

Gryllus lineola In **Old Barbaries**, this cricket is roasted (Simmonds 1885).

Gryllus mitratus → *Teleogryllus mitratus*

Gryllus ritzemae **Japan:** The Japanese name is *kurotsuya-kōrogi*. This cricket was used as an antifebrile in the Tokushima Prefecture (Yamada 1952).

Gryllus testaceus **China:** This species is used as an antidote, or as a suppressor of swelling (Zimian et al. 1997). **Thailand:** The Laotian name is *ching-reep-sigh*. This species is eaten (Vara-asavapati et al. 1975).

Gryllus **sp. India:** A species belonging to this genus is used as a foodstuff in Arunachal Pradesh State, by frying simply in a hot pan or in oil after removing the wings, stomach and head (Singh and Chakravorty 2008). **Thailand:** A species belonging to this genus is consumed as food (Vara-asavapati et al. 1975).

Gymnogryllus elegans → *Gymnogryllus leucostictus*

Gymnogryllus leucostictus This species is eaten in **Africa** (Malaisse 2005).

Gymnogryllus **spp. Thailand:** At least two species belonging to this genus are used as food (Hanboonsong 2010). **Zambia:** The Bemba people eat a cricket belonging to this genus roasted (Sugiyama 1997).

Homeogryllus japonicas **Japan:** The Japanese name is *suzumushi*. The adults were used as a tonic (Shiraki 1958). **South Korea:** The roasted adults are used as a tonic in South Chungcheong Province (Okamoto and Muramatsu 1922).

Homeoxipha **sp.** A species belonging to this genus is eaten in **Thailand** (Hanboonsong 2010).

Liogryllus → *Gryllus*

Loxoblemmus arietulus The Japanese name is *haraokame-kōrogi*. The adults are used for treating dysentery, convulsive fits, or reducing fever in **Japan** (Umemura 1943).

Loxoblemmus doenitzi **China:** This species is used as a diuretic (Nanba 1980). **Japan:** The Japanese name is *mitsukado-kōrogi*. The larvae and adults were roasted, or cooked with soy sauce and sugar in the Nagano, Niigata, Fukushima and Yamagata Prefectures (Takagi 1929c). The adults were used for treating dysentery, convulsive fits, or reducing fever (Umemura 1943).

Madasumma **sp.** A species belonging to this genus is eaten in **Thailand** (Leksawasdi 2008).

Metioche **sp.** The adults of a species belonging to this genus are used as an article of food by the Kiriwina people in Trobriand Island, **PNG** (Meyer-Rochow 1973).

Modicogryllus confirmatus → *Acheta confirmata*

Modicogryllus nipponensis The Japanese name is *hime-kōrogi*. The larvae and adults are roasted, cooked with soy sauce and sugar in the Nagano, Niigata, Fukushima and Yamagata Prefectures in **Japan** (Takagi 1929c).

Nisitrus vittatus This species is treated as a provision in Sabah State, **Malaysia** (Chung et al. 2002).

Pteronemobius **sp. Thailand:** A species belonging to this genus is eaten (Hanboonsong 2010).

Scapsipedus asperses In **China,** this species has been used medicinally for more than 2000 years (Shennung Ben Ts'ao King: Read 1941). Even at present, people use this species as an antidote, an antifebrile, or as a suppressor of swelling (Zimian et al. 1997).

Scapsipedus parvus The Japanese name is *tanbo-kōrogi*. This cricket was used as an antifebrile in the Tokushima Prefecture, **Japan** (Yamada 1952).

Tarbinskiellus orientalis The adults are fried or toasted in India (Chakravorty et al. 2011).

Tarbinskiellus portentosus → *Brachytrupes portentosus*

Teleogryllus comma In **PNG,** the Kiriwina people eat this cricket (Meyer-Rochow 1973).

Teleogryllus commodus (black field cricket) This species is used as food by the Walbiri people in **Australia** (Meyer-Rochow 2005).

Teleogryllus emma (field cricket) **Japan:** The Japanese name is *enma-kōrogi*. The adults were roasted with soy sauce and sugar in the Niigata and Nagano Prefectures (Miyake 1919, Katagiri and Awatsuhara 1996). This cricket has been used as an antifebrile in the Tokushima Prefecture (Yamada 1952). **North Korea:** The dried adults are used for treating dysentery, etc. in North P'yongan Province (Okamoto and Muramatsu 1922). **South Korea:** The dried adults are used for treating dysentery, etc. in North Gyeongsang and Kangwon Provinces (Okamoto and Muramatsu 1922).

Teleogryllus mitratus **Laos:** The Laotian name is *chi lor*. This species is considered good for eating (Boulidam 2008). **Thailand:** People eat this species (Hanboonsong 2010), **Myanmar** and **Vietnam:** This species is also consumed (Yhoung-Aree and Viwatpanich 2005).

Teleogryllus taiwanemma **China:** This species is used as a diuretic (Nanba 1980).

Teleogryllus testaceus → *Teleogryllus mitratus*

Teleogryllus **spp. PNG:** The adults of a species belonging to this genus are eaten raw or grilled on hot ashes by the Chuave and Kiriwina people (Meyer-Rochow 1973). **Thailand:** People eat a species belonging to this genus (Hanboonsong 2010).

Teleoqzyllus (*Teleogryllus* **?**) *derelictus* The larvae and adults are eaten in **China** (Hu and Zha 2009).

Velarifictorus aspersus This species is used as a diuretic in **China** (Inagaki 1984).

Velarifictorus micado The Japanese name is *tsuzuresase-kōrogi*. The adults are used for treating dysentery, convulsive fits, or reducing fever in **Japan** (Umemura 1943).

Velarifictorus **sp.** People eat a species belonging to this genus in **Thailand** (Hanboonsong 2010).

Gryllotalpidae (Mole crickets) **(7)**

Gryllotalpa africana (African mole crickets) **China:** This species is used as an antidote, an antifebrile, or as a suppressor of swelling (Zimian et al. 1997). **India:** The larvae and adults are eaten by the Karbi and Rengma Naga people in Assam State as a chutney or roasted (Ronghang and Ahmed 2010). **Japan:** The Japanese name is *kera*. The adults are roasted after removing their wings (Kuwana 1930). The larvae and adults, which are roasted, infused or raw, are used as remedies for various diseases. The adults are said to be good for cerebral hemorrhage, constipation, urination, etc. (Miyake 1919). **Laos:** The Laotian name is *maeng xone*. This species is consumed by roasting or frying (Nonaka 1999a). **North Korea:** The adults raw or cooked are used for treating various diseases such as gonorrhea, lumbago, etc. in North P'yongan, South P'yongan, Hwanghae, and North Hamgyong Provinces (Okamoto and Muramatsu 1922). **South Korea:** The adults are fit for eating in North Jeolla, South Jeolla, North Chungchong, South Chungchong, North Gyeongsang and Gyeonggi Provinces (Okamoto and Muramatsu 1922). This species is also consumed in the following countries: **Indonesia,** the **Philippines,** (DeFoliart 2002), **PNG** (Meyer-Rochow 1973), **Sri Lanka** (Nandasena et al. 2010), **Thailand** (Lumsa-ad 2001), **Uganda** (Bodenheimer 1951) and **Zimbabwe** (Chavanduka 1975).

Gryllotalpa africana microphtalma (mole crickets) Thai name is *maeng-gi-son*. People eat this subspecies in **Thailand** (Chen et al. 1998).

Gryllotalpa formosana **China:** This species is used as an antidote, an antifebrile, or as a suppressor of swelling (Zimian et al. 1997).

Gryllotalpa fossor → *Gryllotalpa africana*

Gryllotalpa gryllotalpa (Europian mole cricket) **China:** This species is said to be effective against hepatic cancer (Zimian et al. 1997). **India:** The larvae and adults are eaten roasted or fried in Arunachal Pradesh State (Singh and Chakravorty 2008).

Gryllotalpa hirsute → *Gryllotalpa longipennis*

Gryllotalpa longipennis This species is eaten in Sabah State, **Malaysia** (Chung et al. 2002).

Gryllotalpa orientalis In **China,** the adults are roasted with salt, then dried (Chen and Feng 1999).

Gryllotalpa unispina (giant mole cricket) In **China,** the adults are roasted with salt, then dried (Chen and Feng 1999). This species is said to be effective against hepatic cancer (Zimian et al. 1997).

Gryllotalpa spp. **China:** Some species were used to reduce difficulty in delivery in the olden times (Li 1596). **India:** The adults of a species belonging to this genus are consumed boiled, roasted and as a paste (Chakravorty et al. 2011). **Indonesia:** A species belonging to this genus is used in West Java (van der Burg 1904). The **Philippines:** A species, whose vernacular name is *aro-aro* is a very popular food among villagers in Central Luzon. People fry or sautée the adults with vegetables (Adalla and Cervancia 2010). **PNG:** Some species belonging to this genus are eaten by the Chuave people (Meyer-Rochow 2005). **Mexico:** A type of mole crickets is also eaten (Ramos-Elorduy and Pino 2002).

Hemiacrididae (3)

Acanthoxia gladiator This species is considered good for eating in **Africa** (Roulon-Doko 1998).

Hieroglyphus daganensis This species is considered edible in **Africa** (Mignot 2003).

Mazaea granulos This species is considered as a foodstuff in **Africa** (Roulon-Doko 1998).

Raphidophoridae (Camel crickets) (0)

Diestrammena sp. The adults of a species belonging to this genus are roasted in Arunachal Pradesh State, **India** (Singh and Chakravorty 2008).

Schizodactylidae (Sand crickets) (2)

Schizodactylus monstrosus In **India,** the adults are consumed fried or roasted in Arunachal Pradesh State (Chakravorty et al. 2011). The Karbi and Rengma Naga people in Assam State eat the larvae and adults usually making a chutney or roasted (Ronghang and Ahmed 2010).

Schizodactylus tuberculatus This species is eaten in Arunachal Pradesh State, **India** (Singh and Chakravorty 2008).

Stenopelmatidae (Jerusalem crickets) **(1)**

Stenopelmatus fuscus (Jerusalem crickets) In the **USA,** this species is consumed (Ebeling 1986).

Stenopelmatus **sp.** In **Mexico,** the adults are consumed (Ramos-Elorduy and Pino 2002).

Tettigoniidae (Bush crickets; Long-horned grasshoppers) **(38)**

Anabrus simplex (Mormon crickets) **Africa:** People eat this species (Egan 1917). The **USA:** Native Americans belonging to the following tribes eat this species; Achumawi (Olmsted and Stewart 1978), Gosiute (Malouf 1951, Chamberlain 1911), Surprise Valley Paiute (Kelly 1932), Bannock, Washoe, northern Paiute, western Shoshone, northern Shoshone, Utah southern Paiute, western Ute, Southern Ute and northern Ute (Fowler 1986). The adults are roasted and pulverized (Sutton 1988). This species seems to have been eaten by native Americans for long time ago, as a lot of the crickets were found in plant food collections excavated from Leigh Cave, on the east side of the Great Basin (BC 2170 ± 150) (Jones 1948).

Anabrus coloradus → *Anabrus simplex*

Anabrus nigra → *Anabrus simplex*

Anabrus purpurascens → *Anabrus simplex*

Anabrus similis → *Anabrus simplex*

Anoedopoda erosa This species is eaten in **Africa** (Bahuchet 1985).

Arachnacris **sp.** In India, the adults of a species belonging to this genus are consumed fried, boiled or roasted. Appendages and antennae are removed (Chakravorty et al. 2011).

Caedicina **sp.** A species belonging to this genus is considered edible by the Kiriwina people in Trobriand Island, **PNG** (Meyer-Rochow 1973).

Chloracris brullei In **India**, the adults are consumed boiled or as a paste. Wings and antennae are removed (Chakravorty et al. 2011).

Chromacris colorata → *Romalea colorata*

Conocephalus angustifrons (meadow grasshopper) **Colombia:** The larvae and adults are eaten by the Yukpa people (Ruddle 1973). **Venezuela:** The adults are eaten roasted by the Yukpa people (Ruddle 1973).

Conocephalus gladiatus (meadow grasshopper) The Japanese name is *onaga-sasagiri*. In **Japan**, the larvae and adults are skewered and roasted. Or they are boiled with soy sauce in the Nagano Prefecture (Takagi 1929c).

Conocephalus maculatus (meadow grasshopper) **Japan:** The Japanese name is *hoshi-sasakiri*. The larvae and adults are skewered and roasted. Or they are boiled with soy sauce in the Nagano Prefecture (Takagi 1929c). **Thailand:** People eat this species (Hanboonsong et al. 2000).

Conocephalus thumbergii The Japanese name is *kubikirigirisu* or *kubikiri-batta*. In **Japan**, the adults and larvae are cooked with soy sauce and sugar (Takagi 1929c).

Conocephalus triops (meadow grasshopper) The larvae and adults are consumed in **Mexico** (Ramos-Elorduy and Pino 2002).

Conocephalus spp. (meadow grasshopper) **Africa:** A species belonging to this genus is widely eaten (Bahuchet 1985). **India:** The adults are consumed fried, as a paste (chutney) or boiled. The antennae are removed (Chakravorty et al. 2011). **Thailand:** A species belonging to this genus is eaten (Hanboonsong et al. 2000).

Damalacantha vacca sinica The larvae and adults are eaten in **China** (Hu and Zha 2009).

Ducetia japonica (bamd-wing) People eat this species in **Thailand** (Yhoung-Aree and Viwatpanich 2005).

Eucocephalus pallidus **India:** The adults are fried or added to curry (Pathak and Rao 2000).

Euconocephalus incertus People eat this species in **Thailand** (Hanboonsong et al. 2000).

Euconocephalus thunbergii (cone-head) The Japanese name is *kubikiri-batta*. In **Japan**, the larvae and adults were skewer-roasted or boiled with soy sauce in the Nagano Prefecture (Takagi 1929c).

Euconocephalus spp. **Laos, Myanmar** and **Thailand:** Some species belonging to this genus are used as a food item (Yhoung-Aree and Viwatpanich 2005). **Vietnam:** The adults of some species belonging to this genus are used to make a type of soup (Tiêu 1928).

Gampsocleis buergeri (Japanese katydids) The Japanese name is *kirigirisu*. The adults are roasted, or cooked with soy sauce and sugar in the Nagano Prefecture, **Japan** (Mukaiyama 1987). The adults were used as a tonic, an antifebrile, etc. (Shiraki 1958).

Gompsocleis mikado **North Korea:** The larvae and adults are used as an antifebrile or a tonic in the North Hamgyong and Hwanghae Provinces (Okamoto and Muramatsu 1922). **South Korea:** The larvae and adults are used as an antifebrile or a tonic in the North Chungcheong, South Chungcheong and South Jeolla Provinces (Okamoto and Muramatsu 1922).

Hexacentrus japonicus japonicus The Japanese name is *hatakeno-umaoi*. The adults are roasted, or cooked with soy sauce and sugar in the Nagano Prefecture, **Japan** (Mukaiyama 1987).

Hexacentrus unicolor This species is eaten in Sabah State, **Malaysia** (Chung et al. 2002).

Holochlora albida **India:** This species is eaten by the Ao-Naga people in Nagaland State (Meyer-Rochow 2005). **Sri Lanka:** This species is edible (Nandasena et al. 2010).

Holochlora indica This species is an article of food in **India** (DeFoliart 2002).

Homocoryphus → *Homorocoryphus*

Homorocoryphus ineosus The Japanese name is *kusakiri*. The adults are cooked with soy sauce in **Japan** (Tanaka 1980).

Homorocoryphus nitidulus → *Ruspolia differens*

Homorocoryphus nitidulus vicinus → *Ruspolia differens*

Lanista **sp.** A species belonging to this genus is eaten in **Africa** (Roulon-Doko 1998).

Leprocristus **sp.** A species belonging to this genus is consumed in **Africa** (Roulon-Doko 1998).

Lima cordid This species is eaten by the Ao-Naga people in Nagaland, **India** (Meyer-Rochow 2005).

Locusta japonica → *Tettigonia orientalis orientalis*

Lophacris **sp.** The adults of a species belonging to this genus are eaten by the Nhambiquara people in Mato Grosso State, **Brazil** (Costa Neto and Ramos-Elorduy 2006).

Macrolyristes imperator This species is a food article in Sabah State, **Malaysia** (Chung et al. 2002).

Mecopoda elongatae **China:** This species is used as an antidote, an antifebrile, or as a suppressor of swelling (Zimian et al. 1997). **India:** This species is consumed by the Ao-Naga people in Nagaland State (Meyer-Rochow 2005). **Laos:** This species is eaten (Nonaka et al. 2008). **Malaysia:** This species is a food source in Sabah State (Chung et al. 2002). **Sri Lanka:** This species is eaten (Nandasena et al. 2010). **Thailand:** People eat this species (Rattanapan 2000).

Mecopoda **sp.** This species is consumed in **Thailand** (Chen and Wongsiri 1995).

Microcentrum rhombifolium This species is eaten in Arunachal Pradesh State, **India,** by roasting or frying in oil after removing wings (Singh and Chakravorty 2008).

Microcentrum **sp. India:** The adults are consumed boiled with vegetables. He wings are removed (Chacravorty et al. 2011). **Mexico:** The larvae and adults are consumed (Ramos-Elorduy et al. 1997).

Neoconocephalus incertus : This species is consumed as a foodstuff in **Thailand** (Hanboonsong et al. 2000).

Neoconocephalus **sp. India:** A species belonging to this genus is treated as a foodstuff in Arunachal Pradesh State, by roasting or frying (Singh and Chakravorty 2008). **Colombia:** The Tukanoan people eat the adults of a species belonging to this genus (Hugh-Jones 1979).

Onomarchus **sp**. People eat this species in **Thailand** (Hanboonsong et al. 2000) and in Sabah State, **Malaysia** (Chung et al. 2002).

Petaloptera zandala The Mexican name is *esperanzas*. The larvae and adults are eaten in **Mexico** (Ramos-Elorduy et al. 1998).

Phaneroptera falcate (sickle-bearing bush cricket) The Japanese name is *tsuyumushi*. In **Japan**, the adults are roasted or cooked with soy sauce and sugar in the Nagano Prefecture (Mukaiyama 1987).

Pseudophyllus titan People eat this species in **Thailand** (Hanboonsong et al. 2000). This species is also eaten in **Laos, Myanmar** and **Vietnam** (Yhoung-Aree and Viwatpanich 2005).

Pseudophyllus **sp**. A species belonging to this genus is considered edible in **Thailand** (Leksawasdi 2008).

Pseudorhynchus antennalis → *Pseudorhynchus japonicus*

Pseudorhynchus japonicus The Japanese name is *kayakiri*. In **Japan**, the larvae and adults were skewered and roasted, or boiled with soy sauce in the Nagano Prefecture (Takagi 1929c).

Pseudorhynchus lanceolatus This species are consumed in **Africa** (Roulon-Doko 1998).

Pyrgocorypha **sp**. In **Mexico**, the larvae and adults of a species belonging to this genus are eaten (Ramos-Elorduy et al. 1981).

Romalea colorata The larvae and adults are used as food item in **Mexico** (Ramos-Elorduy et al. 1997).

Romalea **sp**. The larvae and adults of a species belonging to this genus are eaten in **Mexico** (Ramos-Elorduy and Pino 2002).

Ruspolia differens **DRC:** The adults are used as food in **Tanzania** (Bodenheimer 1951), **Uganda, Zambia** (Mbata 1995) and **Zimbabwe** (Chavanduka 1975, Heymans and Evrard 1970, Owen 1973).

Ruspolia ineosus → *Homorocoryphus ineosus*

Ruspolia nidula This species is eaten widely in **Uganda**. It is also a good income generating insect (Agea et al. 2008).

Ruspolia viridulus In **Uganda**, this species is a food source (van Huis 2008).

Ruspolia **sp.** In **Cameroon**, a grasshopper of this genus is used as food (Seignobos et al. 1996).

Scudderia **sp.** A species belonging to this genus is considered edible in **Thailand** (Vara-asavapati et al. 1975).

Stilpnochlora azteca The Mexican name is *esperanzas*. The larvae and adults are eaten in **Mexico** (Ramos-Elorduy et al. 1998).

Stilpnochlora toracica The Mexican name is *esperanzas.* The larvae and the adults are eaten in **Mexico** (Ramos-Elorduy et al. 1998).

Tettigonia orientalis orientalis The Japanese name is *yabukiri.* In **Japan**, the larvae and the adults were skewer-roasted, or broiled with soy sauce in the Nagano Prefecture (Takagi 1929c).

Tettigonia **sp.** A species belonging to this genus is eaten in Arunachal Pradesh State, **India,** by roasting or frying in oil after removing the wings (Singh and Chakravorty 2008).

Thylotropides ditymus This species is consumed by the Ao-Naga people in Nagaland State, **India** (Meyer-Rochow 2005).

Xiphidium gladiatum → *Conocephalus gladiatus*

Xiphidium maculatum → *Conocephalus maculatus*

Tetrigidae (Ground hoppers) **(0)**

Euparatettix **sp.** People eat a species belonging to this genus in **Thailand** (Hanboonsong et al. 2000).

6. PHASMIDA (Stick insects) [7]

Phasmatidae (Walking sticks) **(6)**

Ectatosoma **spp.** → *Extatosoma* **spp.**

Eurycantha horrida In **PNG,** the Kiriwina people eat this species raw or cooked (Meyer-Rochow 1973).

Eurycnema versifasciata In **Malaysia,** the feces of this species are used as a medicine for asthma, stomach disorders, etc. The feces are also consumed as a type of tea (Nadchatram 1963).

Eurycnema versirubra The excrement of this species is used to make a very flavorful tea in **Thailand**. The excrement is sold in Chinese stores in the central parts of Thailand (Jolive 1971).

Eurycnema **sp.** In **Thailand,** people collect the feces of a species belonging to this genus, and make drink by drying the feces and then adding boiling water to them. This species can be fed with leaves of guava. The infusion is used as a medicine for stomach disorders (Yasumatsu 1965).

Extatosoma tiaratum (giant prickly stick insect) **(PL II-4)** The adults are eaten roasted in **PNG** (Stone 1992).

Extatosoma **spp. Indonesia:** A species belonging to this genus is a food source in West Papua State (Tommaseo-Ponzetta and Paoletti 1997). **Mexico:** This species is raised as food (Ramos-Elorduy 1997).

Haaniella echinata This species is roasted in Sabah State, **Malaysia** (Chung 2008, Chung et al. 2002).

Haaniella grayi grayi People of the Bideyuh tribe in Sarawak State, **Malaysia**, eat the eggs of this species. They take the eggs out of the female abdomens, and boil them in water for 30 seconds. After boiling, the egg shells are easily removed (Bragg 1990).

Necrosciidae [1]

Platycrana viridana The adults are eaten after removing the legs and wings in **Malaysia** (Kevan 1991).

7. ISOPTERA (Termites) [67]

Kalotermitidae (Dampwood termites, Drywood termites) **[1]**

Kalotermes flavicollis **Brazil:** The soldiers and adults are a food article (Wallace 1852/53, Posey 1987). **Thailand:** This species is consumed as a foodstuff (Sangpradub 1982).

Rhinotermitidae (Subterranean termites) **[4]**

Coptotermes formosanus The larvae and alates are considered good for eating in **China**, (Mao 1997, Chen and Feng 1999).

Coptotermes havilandi This species is eaten after roasting in **Thailand** (Hanboonsong et al. 2001).

Reticulitermes flavipes **India:** The female and male alates are used as food fried or as a chutney by the Karbi and Rengma Naga people in Assam State

(Ronghang and Ahmed 2010). **Thailand:** The winged adults and queens are eaten. The termites are eaten by roasting with a bit of salt (Bristowe 1932).

Reticulitermes tibialis (subterranean termite) This termite seems to have been eaten by aboriginal people living in western area of the **USA,** for a long time (1100BP), as this species was found in the coprolites excavated from Dirty Shame Rockshelter in Oregon State (Hall 1977).

Termitidae (Higher termites) **[59]**

Acanthotermes acanthothorax In **Tanzania** and **Uganda,** people eat this species raw, or after desiccation (Harris 1940).

Acanthotermes militalis → *Pseudacanthotermes militalis*

Acanthotermes spiniger → *Pseudacanthotermes spiniger*

Acanthotermes **spp.** Some species belonging to this genus are eaten in **Tanzania** (Bodenheimer 1951).

Apicotermes **sp.** A species belonging to this genus is used as a food source by the Aka people in **CAR** (Bahuchet 1985).

Bellicositermes natalensis This species is a food item in CAR (Hoare 2007).

Bellicositermes **sp.** In **Côte d'Ivoire,** the alates of a species belonging to this genus is an important food for the Yafobas people, and the people eat the termites raw, fried or grilled (Villiers 1947).

Cornitermes **sp.** In **Brazil,** native people of the Desâna tribe eat the alates, soldiers and queens raw, or after roasting (Ribeiro and Kenhiri 1987).

Cubitermes **spp.** The alates, soldiers, and the queens of several species of termites belonging to this genus are consumed as food in **Africa** (Moussa 2004).

Labiotermes labralis **Colombia:** The pupae and alates are eaten raw by the Tukanoan people (Paoletti and Dufour 2005). **Brazil:** The larvae are an effective food source (Posey 1979). **Venezuela:** The larvae, pupae and the alates are consumed by the Yanomamo people (Paoletti and Dufour 2005).

Macrotermes acrocephalus The larvae, alates and the nests are fried in **China** (Cheng and Feng 1999, Hu and Zha 2009).

Macrotermes annandalei The larvae, the alates and the nests are eaten fried in **China** (Cheng and Feng 1999, Hu and Zha 2009).

Macrotermes barneyi The larvae, alates and the nests are consumed fried in **China** (Cheng and Feng 1999, Hu and Zha 2009).

Macrotermes bellicosus **Nigeria:** The alates are eaten roasted in several parts of western Nigeria (Ukhun and Osasona 1985, Moussa 2004). **PRC:** The alates, soldiers, and the queens are consumed as food (Ukhun and Osasona

1985, Moussa 2004). This species is also used as food in **CAR** (Roulon-Doko 1998) and **DRC** (Bequaert 1921).

Macrotermes denticulatus The larvae, alates and the nests are fried in **China** (Cheng and Feng 1999, Hu and Zha 2009).

Macrotermes falciger (**PL II-6**) **Benin:** Mainly in southern area, people of the Fon and Nagot tribes eat this species (Tchibozo et al. 2005). **Zimbabwe:** The alates and soldiers are used as a foodstuff by the Shona people (Chavanduka 1975, Bodenheimer 1951). This species is also eaten in **Uganda** and **Zambia** (Chavanduka 1975, Bodenheimer 1951).

Macrotermes gilvus **Malaysia:** This species is consumed as a food item raw, stir-fried without oil, or cooked in porridge or rice, in Sabah State (Chung et al. 2002). The **Philippines:** This species is considered edible (Starr 1991). **Thailand:** The alates are eaten fried (Hanboonsong et al. 2001).

Macrotermes goliath → *Macrotermes falciger*

Macrotermes jinhongensis The larvae, alates and the nests are fried in **China** (Cheng and Feng 1999, Hu and Zha 2009).

Macrotermes menglongensis The larvae, alates and the nests are consumed fried in **China** (Cheng and Feng 1999, Hu and Zha 2009).

Macrotermes mossambicus This species is eaten in **Zambia** and **Zimbabwe** (Silow 1983, Logan 1992).

Macrotermes muelleri This species is a food item in **DRC** (DeFoliart 2002). Pregnant women eat the mud, which forms the nest (Hunter 1984).

Macrotermes natalensis **Angola:** Pregnant women eat the mud, which forms the nest (Hunter 1984). **Cameroon:** This species is also eaten by the Masa people (Mignot 2003). **DRC:** People eat the alates, queens, soldiers, and the workers boiled. They also extract oil from the termites (Bequaert 1921). **Nigeria:** The Yoruba people eat the alates, and the queens roasted with a bit of salt (Fasoranti and Ajiboye 1993). **RSA:** Pregnant women eat the mud, which forms the nest (Hunter 1984). **Zimbabwe:** This species is also eaten (Dube et al. 2013).

Macrotermes nobilis **Congo Basin:** Pregnant women eat the mud, which forms the nest (Hunter 1984).

Macrotermes subhyalinus **Angola:** The alates are eaten fried by the Gingas people (Oliveira et al. 1976). **Cameroon:** The Mofu people living at the foot of Les Monts du Mandara eat this species (Seignobos et al. 1996). **Eastern Africa:** Pregnant women eat the mud, which forms the nest (Hunter 1984). This species is also consumed in **Uganda, Zambia** (Silow 1983) and **Zimbabwe** (Dube et al. 2013).

Macrotermes swaziae This species is eaten in **RSA** (DeFoliart 2002).

Macrotermes vatriolatus → *Macrotermes vitrialatus*

Macrotermes vitrialatus **Angola:** This species is used as a food item by the Mbunda people (Silow 1983). **Zambia:** The alates of this species are consumed as a foodstuff (Silow 1983).

Macrotermes yunnanensis The larvae, alates and the nests are fried in **China** (Cheng and Feng 1999, Hu and Zha 2009).

Macrotermes **spp. Colombia:** The adults of several species belonging to this genus are food source (Dufour 1987). **DRC:** This species is fit for eating (Takeda 1990). **India:** The adults and the eggs of some species belonging to this genus are roasted by the Nishi people in Arunachal Pradesh State (Singh et al. 2007). **Indonesia:** In Java, the winged adults are eaten. The termites burn their wings on the light to which they are attracted, and are then relished (Bodenheimer 1951). **Tanzania:** The alates, queens and soldiers of several species belonging to this genus are eaten raw or dried (Harris 1940). **Zimbabwe:** This species is eaten (Gelfand 1971).

Megagnathotermes katangensis This species is considered good for eating by the Sanga people in **DRC** (Malaisse 1997).

Microcerotermes dubius This species is consumed raw, stir-fried without oil, or cooked them in porridge or rice, in Sabah State, **Malaysia** (Chung et al. 2002).

Microtermes obesus **India:** This species is considered edible in some districts of Meghalaya State. People frizzle the adults in oil, and make a chutney (sauce) after removing the wings (Pathak and Rao 2000, Meyer-Rochow 2005).

Microtermes **spp.** In **Kenya,** the Nandi people eat the alates raw or dried (Yagi 1997).

Nasutitermes corniger **Bolivia:** This species is a food item (Paoletti and Dufour 2005). **Venezuela:** The larvae are eaten raw or roasted by the Yanomamo people around Alto Orinoco (Araujo and Beserra 2007).

Nasutitermes ephratae In **Venezuela**, the larvae are consumed as raw or roasted by the Yanomamo people around Alto Orinoco (Araujo and Beserra 2007).

Nasutitermes ephrateae In **Venezuela**, the swarming alates and pupae are roasted by the Yanomamo people (Cerda and Torres 1999).

Nasutitermes macrocephalus In **Venezuela,** the larvae are consumed as raw or roasted by the Yanomamo people around Alto Orinoco (Araujo and Beserra 2007).

Nasutitermes surinamensis The larvae are relished raw or roasted by the Yanomamo people around Alto Orinoco, **Venezuela** (Araujo and Beserra 2007).

Nasutitermes sp. **Brazil:** The adults of a species belonging to this genus are considered edible (Mendes dos Santos 1995). **Venezuela:** The pupae and alates of a species belonging to this genus are used as a foodstuff by the Yanomamo people (Paoletti and Dufour 2005).

Odontotermes angustignathus The larvae, alates and the nests are eaten fried in **China** (Cheng and Feng 1999, Hu and Zha 2009).

Odontotermes annulicornis The larvae, alates and the nests are fried in **China** (Cheng and Feng 1999, Hu and Zha 2009).

Odontotermes badius **RSA:** The Pedi people eat the alates (Quin 1959). **Zambia:** This species is a food source (Silow 1983).

Odontotermes conignathus The larvae, alates and the nests are fried in **China** (Cheng and Feng 1999, Hu and Zha 2009).

Odontotermes faveafrons The larvae, alates and the nests are fried in **China** (Cheng and Feng 1999, Hu and Zha 2009).

Odontotermes feae **India:** This species forms one of the important insect foods (Gope and Prasad 1983). **Sri Lanka:** This species is considered to be edible (Nandacena et al. 2010).

Odontotermes formosanus **China:** The larvae, alates and the nests are eaten fried (Cheng and Feng 1999, Hu and Zha 2009). **India:** Native people of South India, such as the Kannikaran, Paniyan, Palliyan, Sholaga, Irular and Kota tribes, use this species as a medicine for asthma, and as food to enhance lactation in women (Wilsanand 2005).

Odontotermes gravelyi The larvae, alates and the nests are fried in **China** (Cheng and Feng 1999, Hu and Zha 2009).

Odontotermes hainanensis The larvae, alates and the nests are fried in **China** (Cheng and Feng 1999, Hu and Zha 2009).

Odontotermes magdalense This species is eaten by the Mofu people in **Cameroon** (Seignobos et al. 1996).

Odontotermes obesus This species is fried by the Ao-Naga people in Nagaland State, **India** (Meyer-Rochow 2005).

Odontotermes yunnanensis The larvae, alates and the nests are eaten fried in **China** (Cheng and Feng 1999, Hu and Zha 2009).

Odontotermes **spp. DRC:** The adults of a species belonging to this genus are considered good for eating grilled by the Sanga people (Malaisse 1997). **India:** Some species belonging to this genus are used as a medicine for asthma by six out of nine tribes in Tamil Nadu (Solavan et al. 2004). **Kenya:** The Kitosh people, who live around Mt. Elgon, eat the termites belonging to this genus raw or roasted as an important food (Bryk 1927). **Thailand:**

People eat a species belonging to this genus (Leksawasdi 2008). **Zimbabwe:** A species belonging to this genus is eaten (Weaving 1973).

Protermes **sp.** A species belonging to this genus is considered to be edible by the Aka people in **CAR** (Bahuchet 1985).

Pseudacanthotermes militalis **Tanzania:** The alates are eaten raw (Harris 1940). **Uganda:** The alates and queens are eaten raw (Pomeroy 1976). This species is also consumed in **Angola** (Silow 1983) and **DRC** (Bequaert 1921).

Pseudacanthotermes spiniger **Angola:** This species is consumed as a foodstuff by the Mbunda people (Silow 1983). **DRC:** People of the Azande and Mangbetu tribes eat the alates roasted (Bequaert 1921). This species is also considered edible in following countries; **Uganda** (Pomeroy 1976) and **Zambia** (Silow 1983).

Pseudacanthotermes **spp.** The alates, queens, and soldiers of several species belonging to this genus are eaten raw or dried in **Tanzania** (Harris 1940).

Syntermes aculeosus **Brazil:** Native people of the Makiritare tribe eat the heads of soldier ants raw or boiled along with red pepper and casaba (Marconi et al. 2002, Paoletti et al. 2003). **Colombia:** The soldiers are used as a foodstuff by the Tukanoan people (Dufour 1987). **Venezuela:** The Ye'Kuana people living around Alto Orinoco eat the adults raw (Araujo and Beserra 2007).

Syntermes parallelus The adults are eaten in **Colombia** (Dufour 1987).

Syntermes spinosus **Colombia:** The soldiers are eaten by the Tukanoan people (Dufour 1987). **Venezuela:** The adults are consumed raw by the Yanomamo people around Alto Orinoco (Araujo and Beserra 2007).

Syntermes **pr.** *spinosus* In **Brazil,** the alates and soldiers are used as food source (Setz 1991).

Syntermes synderi The adults are edible in **Colombia** (Dufour 1987).

Syntermes tanygnathus **Colombia:** The soldiers are eaten by the Tukanoan people (Dufour 1987). **Venezuela:** The adults are consumed raw by the Yanomamo people around Alto Orinoco (Araujo and Beserra 2007).

Syntermes territus In **Venezuela,** only the soldier's heads are fried by the Yanomamo people (Paoletti and Dufour 2005).

Syntermes **spp. Brazil:** Native people of the Yekuana tribe eat termites belonging to this genus (Milton 1984, Paoletti and Dreon 2005). **Venezuela:** The soldiers of a species belonging to this genus are considered good for eating by the Guajibo (Paoletti and Dufour 2005), Piaroa (Zent 1992) and Yanomamo (Lizot 1977) people.

Termes arborum In **South America,** this species has been reported as food insects in a report of the 19th century (Hope 1842).

Termes atrox In **Indonesia**, after removing the wings the alates are roasted with flour, and baked into a type of cake. The roasted queens are especially relished (van Burg 1904).

Termes capencis This species is eaten raw or cooked in RSA (Sparrmann 1778).

Termes destructor **Equador:** Adults are fried (Cutright 1940). **Guyana:** The adults are considered to be edible (DeFoliart 2002). **Indonesia:** After removing the wings of the alates are roasted with flour, and baked into a type of cake. The roasted queens are especially relished (van Burg 1904). The alates are roasted (van Burg 1904).

Termes fatale In **Indonesia**, after removing the wings of the alates are roasted with flour, and baked into a type of cake. The roasted queens are especially relished (van Burg 1904). The alates are roasted (van Burg 1904). **Africa (southern area):** The Khoikhoi people relish this termite (Sparrmann 1778). **RSA:** This species has been reported as a favorite food of the Bosjesman people (Hope 1842).

Termes flavicole **Amazonia:** The workers are eaten raw or roasted (Wallace 1852/53). **Thailand:** The winged adults are fried (Jonjuapsong 1996).

Termes gabonensis In **DRC**, the alates, and soldiers are consumed raw in a form of paste or dried (Chinn 1945).

Termes mordax In **Indonesia**, after removing the wings the alates are roasted with flour, and baked into a type of cake. The roasted queens are especially relished (van Burg 1904).

Termes natalensis → Macrotermes natalensis

Termes smeathmanni **Africa:** This species has been reported as food insects in a report of the 19th century (Hope 1842).

Termes sumatranum In **Indonesia**, after removing the wings of the alates are roasted with flour, and baked into a type of cake. The roasted queens are especially relished (van Burg 1904).

Termes **spp.** In **Tanzania**, alates, queens and soldiers of several species belonging to this genus are eaten raw or dried (Harris 1940).

Trinervitermes sp. A species belonging to this genus is an article of food by the Luba Kasai people in **DRC** (Callewaert 1922).

Hodotermitidae (Rottenwood termites) **[3]**

Hodotermes mossambicus In **Botswana,** the alates are a food source, and are roasted by people of the Tswana (Grivetti 1979) and San tribes (Nonaka 1996).

Hodotermes **sp.** A species belonging to this genus is a foodstuff in **RSA** (Fuller 1918).

Microhodotermes viator The larvae are an article of food in **RSA** (Bodenheimer 1951).

Microhodotermes **spp.** Several species belonging to this genus are used as food in **Africa** (Logan 1992).

Zootermopsis angusticollis The **USA:** This termite is recommended as a survival food in a course of the US army training school (Luttrell 1992).

8. BLATTARIA (Cockroaches) [12]

Blaberidae [0]

Gomphadorhina **sp.** The adults are eaten roasted by the Nishi people in Arunachal Pradesh State, **India** (Singh et al. 2007).

Gyna **sp.** This species is a food item in **Cameroon** (Seignobos et al. 1996).

Cryptocercidae (Brown-headed cockroaches) [0]

Cryptocercus **sp.** The adults are consumed roasted by the Nishi people in Arunachal Pradesh State, **India** (Singh et al. 2007).

Panesthiidae (Ovoviviparous cockroaches) **[1]**

Panesthia angustipennis **Indonesia:** In Java Island, people let children eat this species when coughing persists (Roepke 1952).

Panesthia **sp.** A species belonging to this genus is a food source by people in Sabah State, **Malaysia** (Chung et al. 2002).

Blattellidae [3]

Blatella germanica (German cockroach) **China:** This species is used as an antidote, and is said to be effective against renal cancer (Zimian et al. 1997). **Mexico:** The adults are eaten. The protein content of this species is about 78% of dry weight (Pennino et al. 1991). This species is raised as food (Ramos-Elorduy 1997, Ramos-Elorduy and Pino 2002).

Cosmozosteria **sp.** A species belonging to this genus is an article of food in **Australia** (Yen 2005).

Ectobios lapponicus (dusky cockroach) **Europe:** The powder of this species is used as a medicine for whooping cough (Gordon 1996).

Eurycotis manni This species is used to cure headaches in Bahia State, **Brazil** (Costa-Neto and Pacheco 2005).

Pseudomops **sp.** The adults are eaten in **Mexico** (Ramos-Elorduy and Pino 2002). The protein content of this species is very high (Pennino et al. 1991).

Blattidae (Cockroaches) **[5]**

Blatta orientalis (oriental cockroach) **China:** This species has been used medicinally for more than 2000 years (Shennung Ben Ts'ao King: Read 1941). It is said to be effective against renal cancer (Zimian et al. 1997). **Russia:** The powder of this species was used as a diuretic (Ealand 1915). **Thailand:** The eggs of species are roasted (Bristowe 1932).

Neostylopyga rhombifolia → Stylophyga rhombifolia

Periplaneta americana (American cockroaches) **Brazil:** This species is eaten (Lenko and Papavero 1979). The water used for boiling the entire body of the cockroach is said to be good for stomachaches, asthma, etc. (Costa-Neto and de Melo 1998). **China:** This species is used as an antidote. It is said to be effective against renal cancer (Zimian et al. 1997). In some districts, people eat the larvae and the adults non-officinally (Bodenheimer 1951). **Japan:** The Japanese name is *wamon-gokiburi*. The adults are said to be good for colds, indigestion, etc. (Koizumi 1935). **Malaysia:** This species is eaten in Sabah State, and for medicinal purposes by the Chinese and Kadazandusuns (Chung et al. 2002). **Mexico:** This species is raised as food (Ramos-Elorduy 1997).

Periplaneta australasiae (Australian cockroaches) **China:** This species is eaten in some districts (Bodenheimer 1951). This species is also used as an antidote, and is said to be effective against renal cancer (Zimian et al. 1997). **Mexico:** This species is eaten in Chiapas, Tabasco and Veracruz States (Ramos-Elorduy 2009).

Periplaneta fuliginosa (smoky brown cockroach) **(PL III-1) China:** This species is said to be effective against renal cancer (Zimian et al. 1997).

Periplaneta picea → Periplaneta fuliginosa

Periplaneta **sp.** In Mexico, the larvae and adults of the cockroach belonging to this genus are eaten (Ramos-Elorduy and Pino 1990).

Polyzosteria **sp.** A species of this genus is eaten in **Indonesia** (West-Papua State) (Tommaseo Ponzetta and Paoletti 1997).

Stylophyga rhombifolia The fried eggs of this species are consumed by children in **Thailand** (Bristowe 1932).

Corydiidae [2]

Eupolyphaga sinensis (sand cockroach) **China:** This species is raised for medicinal purposes (Zimian et al. 1997). This species is used as an anodyne or for curing internal hemorrhage (Inagaki 1984, Zimian et al. 1997).

Polyphga plancyi **China:** This species is used to cure internal hemorrhage (Inagaki 1984).

Plate III. 1. *Periplaneta fuliginosa* adult. Body length: 35 mm (cf. p.48); 2. *Tenodera aridiforia* adult. Body length: max. 90 mm (cf. p.51); 3. *Pediculus humanus capitas* adult. Body length: 3 mm (cf. p.52). (Photo: Utsunomiya Teisoukasei Company Ltd., Utsunomiya, Japan); 4. *Hyaloptera pruni* adults and larvae on a reed leaf. Body length: 2 mm (cf. p.53); 5. *Graptopsaltoria nigrofascata*. a: adult. Body length: 50 mm (cf. p.56), b: Fried larvae. Body length: 33 mm (cf. p.56).

Phyllodromiidae [1]

Opisthoplatia orientalis **China:** This species is used to cure internal hemorrhage (Inagaki 1984). **Japan:** The Japanese name is *satsuma-gokiburi*. The adults are used for meningitis, colds, stomach disorders, etc. (Umemura 1943).

9. MANTODEA (Mantes) [20]

Hymenopodidae (Flower mantises) **(1)**

Pseudoharpax virescens This species is eaten in **Africa** (Seignobos et al. 1996).

Mantidae (Praying mantis) **(19)**

Epitenodera gambiense → *Epitenodera houyi*

Epitenodera houyi This species is treated as a foodstuff in **Africa** (Roulon-Doko 1998).

Hierodula bipapilla → *Hierodula patellifera*

Hierodula coarctata This species is consumed by the Ao-Naga people in Nagaland State, **India** (Meyer-Rochow 2005).

Hierodula patellifera **China:** This species is considered good for eating in China (Mao 1997). Its egg case is used medicinally as a diuretic and various other diseases and disorders (Chen 1982). **Japan:** The Japanese name is *harabiro-kamakiri*. In the Nagano Prefecture, the wings, legs and intestines of the larvae and adults are removed before cooking with sugar-soy sauce (Takagi 1929a). The roasted adults are used for treating beriberi (Miyake 1919) and for lumbago, hernia, etc. (Shiraki 1958). **North Korea:** The adults are pulverized after desiccation, or infused with sesame oil for treating hernia, constipation, lumbago, etc. in North P'yongan, North Hamgyong (Okamoto and Muramatsu 1922). **South Korea:** The adults are pulverized after desiccation, or infused with sesame oil for treating hernia, constipation, lumbago, etc. in North Hamgyong, South Gyeongsang, South Joella, Gyeonggi, Kangwon and South Chungcheong (Okamoto and Muramatsu 1922).

Hierodula saussurei In **China,** the oothecae of this species have been used medicinally more than 2000 years (Shennung Ben Ts'ao King: Read 1941). This species is used to ameliorate impotence (Zimian et al. 1997).

Hierodula sternosticta This species is eaten by the Kiriwina people in Trobriand Island, **PNG** (Meyer-Rochow 1973).

Hierodula westwoodi This species is considered good for eating in **India** (Gope and Prasad 1983).

Hierodula **spp. PNG:** Several species belonging to this genus are treated as a foodstuff (Meyer-Rochow 1973). **Thailand:** The Laotian name: *mang naap*. The eggs and adults of a green mantis of this genus were eaten around Hua Hin (Bristowe 1932).

Hoplocorypha garuana This species is considered edible in **Cameroon** (Barreteau 1999).

Mantis religiosa (praying mantis, European mantids) **Africa:** This species is considered edible in large areas of the African continent (Quin 1959). **China:** The adults are a food item (Hu and Zha 2009). The egg case is used medicinally to cure diuretic and various other diseases and disorders (Chen 1982). This species is also used to ameliorate impotence (Zimian et al. 1997). **India:** The adults and the eggs are roasted by the Nishi people in Arunachal Pradesh State (Singh et al. 2007). The Karbi and Rengma Naga people in Assam State eat the larvae and the adults as a chutney or roasted (Ronghang and Ahmed 2010). **Thailand:** People eat this species (Hanboonsong 2010).

Miomantis paykullii This species is eaten in **Africa** (Seignobos et al. 1996).

Paralempdera angustipennis This species is a food article in **China** (Mao 1997).

Paratempdera sinensis This species is a food source in **China** (Mao 1997). Its egg case is used medicinally as a diuretic and for various other diseases and disorders (Chen 1982).

Paratenodera →*Tenodera aridifolia*

Paratenodera superstitiosa In **Japan,** the egg cases are used for treating eye diseases (Shiraki 1958).

Pseudoharpax virescens This species is eaten in **Africa** (Seignobos et al. 1996).

Pseudomantis maculate → *Statilia maculata*

Sinensisa sp. In **Thailand,** the eggs and adults are fried (Chen and Wongsiri 1995).

Sphodromantis centralis This species is eaten in **Africa** (Malaisse 2005).

Statilia maculata **China:** This species is used to ameliorate impotence (Zimian et al. 1997). **Japan:** The Japanese name is *ko-kamakiri*. In the Nagano Prefecture, the wings, legs and intestines of the larvae and adults are removed before cooking with sugar-soy sauce (Takagi 1929a). Its egg case is used medicinally as a diuretic and for various other diseases and disorders (Matsumura 1918).

Tarachodes saussurei This species is eaten in **Cameroon** (Barreteau 1999).

Tenodera aridifolia (Chinese mantid) **(PL III-2) Japan:** The Japanese name is *ō-kamakiri*. The eggs are a food item in the Nagano Prefecture (Katagiri and Awatsuhara 1996). The egg cases are used officinally for pleurisy, phlegm, etc. (Koizumi 1935). The adults are used for treating convulsive fits, lumbago, etc. (Shiraki 1958). **North Korea:** The adults are used as dried powder for treating hernia, constipation, lumbago, etc. in North P'yongan, North Hamgyong (Okamoto and Muramatsu 1922). **South Korea:** The adults are used as dried powder for treating hernia, constipation, lumbago, etc. in North Hamgyong, South Gyeongsang, South Joella, Gyeonggi, Kangwon and South Chungcheong (Okamoto and Muramatsu 1922).

Tenodera aridifolia sinensis → *Tenodera sinensis*

Tenodera bravico The larvae and the adults are eaten in **China** (Hu and Zha 2009).

Tenodera capitata **China:** This species is considered edible (Mao 1997). **Japan:** The Japanese name is *ō-kamakiri*. The roasted adults were used for treating tuberculosis and other diseases (Miyake 1919). **North Korea:** The adults are pulverized after roasting, and used for treating hernia, constipation, lumbago, etc. in North P'yongan, North Hamgyong (Okamoto and Muramatsu 1922). **South Korea:** The adults are pulverized after roasting, and used for treating hernia, constipation, lumbago, etc. in North Hamgyong, South Gyeongsang, South Joella, Gyeonggi, Kangwon and South Chungcheong (Okamoto and Muramatsu 1922).

Tenodera sinensis (Chinese mantids) **China:** The larvae and adults are eaten (Hu and Zhz 2009). **South Korea:** The adults are eaten roasted (Nonaka 1991). **Thailand:** The eggs and adults are considered good for eating fried (Hanboonsong et al. 2000).

Tenodera **sp.** In **PNG,** a species belonging to this genus is treated as a foodstuff by the Kiriwina people in Trobriand Island (Meyer-Rochow 1973).

10. ANOPLURA (Lice) [4]

Haematopinidae (Mammal sucking lice) (1)

Haematopinus trichechi In **Canada**, the Inuit people eat a louse of presumably this species. This species is an ectoparasite of the walrus, and the people take the lice from the fins of walrus and eat them raw (Honda 1981).

Pediculidae (Lice) **(2)**

Pediculus humanus (louse) **Brazil:** The Brazilian name is *piojo*. The adults are consumed by the Tapirapé and Maku people (Paoletti and Dufour 2005). **Brazil-Colombia:** This species is eaten raw by the Maku people (Paoletti and Dufour 2005), **Venezuela:** This species is also used as a food by the Yanomamo (Smole 1976) and Yukpa people (Ruddle 1973).

Pediculus humanus capitis (head louse) **(PL III-3) Amazonia:** The adults of the lice are eaten raw (Pereira 1954). **Indonesia:** Women in Java Island liked to eat the head lice (van der Burg 1904). **Japan:** The Japanese name is *atama-jirami*. This species was used as a remedy for meningitis in the Okayama Prefecture in the olden times (Miyake 1919). The adults were also used for whooping cough (Koizumi 1935). **USA:** Native Americans of the Massachusett tribes ate this species (Carr 1951, Meyer-Rochow 1973). This subspecies is used as a foodstuff also in following countries; **Australia** (Lumholtz 1890, Bonwick 1898), **Mongolia** (Nishikawa 1974), **Peru** (Simmonds 1885), **PNG** (Meyer-Rochow 1973), **Ecuador** (Onore 1997 and 2005).

Pediculus humanus corporis (body louse) **Africa:** The adults of the lice are consumed raw (Kolben 1738). **Brazil:** People of the Tapirapé tribe and also some other tribes eat this species (Hitchicock 1962). **Japan:** The Japanese name is *koromo-jirami*. This subspecies was used as a remedy for meningitis in the Okayama Prefecture in the olden times (Miyake 1919). **Mexico:** The adults are eaten raw in Oaxaca State (Ramos-Elorduy et al. 1997). **Mongolia:** People eat this species (Nishikawa 1974), **USA:** Native Americans of the Massachusett tribes ate this species (Carr 1951, Sutton 1988).

Pediculus **sp.** (louse) **Brazil:** The adults of a species of a louse belonging to this genus are an article of food (Paoletti and Dufour 2005). **Colombia:** The louse is also eaten (Smole 1976, Ruddle 1973). **Venezuela:** (Smole 1976, Ruddle 1973).

11. HEMIPTERA [240]

(Homoptera) [98]

Aphididae (Aphids or plant lices) (2)

Hyalopterus pruni (mealy plum aphid) **(PL III-4)** In the **USA,** Native Americans of Yuma, Chemehuevi (Drucker 1937), Cochimi (Clavigero 1937), Kiliwa (Orcutt 1887), Washoe, Northern Paiute, Owens Valley Paiute, Panamint western Shoshone, Utah southern Paiute, western Paiute ate the honey dew of this species raw (Jones 1945, Fowler 1986).

Schlechtendalis chinensis The Japanese name is *nurude-ōmimifushi-aburamushi*. The gall made by this aphids were used as a remedy for stomach disorders, diarrhea, etc. in **Japan** (Koizumi 1935).

Asterolecaniidae (Pit scales) **(2)**

Asterolecanium bambusae **China:** This species was used for treatment of paralysis (Li 1596).

Cerococcus quercus **USA:** Some American Indian tribes eat honey dew of this species raw (Essig 1934 and 1965).

Cicadellidae (Leaf-hoppers) **(2)**

Euscelis decorates The honey dew of this species (called tamarisk manna) is eaten raw in **Iraq** (Bodenheimer 1951).

Goniagnathus decpratus → *Euscelis decorates*

Opsius jucundus The honey dew of this species (called tamarisk manna) is eaten raw in **Iraq** (Bodenheimer 1951).

Cicadidae (Cicadas) **(62)**

Afzeliada afzelii This species is a food source in **DRC** (Malaisse and Parent 1997a).

Afzeliada duplex This species is a food item in **DRC** (Malaisse and Parent 1997a).

Afzeliada **sp.** The adults of a species belonging to this genus are eaten raw or fried in **PRC** (Moussa 2004).

Andropogon gayanus The adults are used as a food item in **Africa** (van Huis 2008).

Baeturia **sp.** The Chuave people in **PNG** eat a species belonging to this genus raw or grilled on hot ashes (Meyer-Rochow 1973).

Carineta fimbriata The adults are consumed by the Quichua people in **Ecuador** (Onore 1997).

Catharsius molasses This species is a foodstuff in **Thailand** (Leksawasdi 2008).

Chremistica **sp.** The adults of a species belonging to this genus are considered edible in **Thailand** (Hanboonsong et al. 2000).

Cicada flammata The adults are fried in **China** (Chen and Feng 1999).

Cicada montezuma The adults are eaten in **Mexico** (Ramos-Elorduy and Pino 2002).

Cicada nigriventris The adults are eaten in **Mexico** (Ramos-Elorduy and Pino 2002).

Cicada orni **Italy:** The excreta produced by this species on ash tree were recognized as a kind of manna, and were used officinally (Simmonds 1877).

Cicada plebeja The larvae are considered a delicacy in **France** (Fabre 1922).

Cicada pruinosa **India:** The adults are consumed boiled or as paste. The wings are removed (Chakravorty et al. 2011). **Mexico:** The adults are treated as a foodstuff (Ramos-Elorduy and Pino 2002).

Cicada verides This species is used as food by the Ao-Naga people in Nagaland State, **India** (Meyer-Rochow 2005).

Cicada **spp. India:** A species belonging to this genus is consumed by the Ao-Naga people in Nagaland State (Meyer-Rochow 2005). **Mexico:** The adults of a species belonging to this genus is treated as a foodstuff (Ramos-Elorduy and Pino 2002).

Cosmopsaltria opalifera → Meimuna opalifera

Cosmopsaltria waine **Indonesia** (West-Papua): This cicada outbreaks occasionally, although it has a two year life cycle. People in West Papua eat it (Ramandey and van Mastrigt 2010).

Cosmopsaltria **spp. Thailand:** The adults of several species belonging to this genus are fried (Sangpradub 1982, Jonjuapsong 1996).

Cryptotympana aquilla The larvae and adults are roasted by the Nishi people in Arunachal Pradesh State, **India** (Singh et al. 2007).

Cryptotympana atrata The larvae and adults are fried in **China** (Chen and Feng 1999).

Cryptotympana facialis **China:** The larvae, adults and the exuviae were used medicinally since ancient times (Li 1596). This species is said to be effective against thyroid cancer (Zimian et al. 1997). **Japan:** The Japanese name is *kuma-zemi*. The exuviae of the last instar larvae were used as antifebrile (Miyake 1919). **North Korea:** The dried powder of the exuviae is used as antifebril of colds in North Hamgyong (Okamoto and Muramatsu 1922).

Cryptotympana intermedia → Cryptotympana facialis

Cryptotympana japonensis → Cryptotympana facialis

Cryptotympana pustulata → Cryptotympana facialis

Cyclochila virens The adults are eaten by the Galo people in Arunachal Pradesh State, **India** by roasting or as a paste. The wings are removed (Chakravorty et al. 2011).

Diceroprocta apache The adults are roasted in the **USA** (Ebeling 1986).

Diceroprocta **spp. USA:** Native Americans of Panamint, southern Paiute, and Chemehuevi eat this species (Fowler 1986).

Diceropyga **sp.** The Chuave people in **PNG** eat this species raw or grilled on hot ashes (Meyer-Rochow 1973).

Dundubia emanatura **Thailand:** The adults are roasted after removing the wings (Mungkorndin 1981).

Dundubia intemerata **China:** People of the Thai tribe living in the southern area of the country eat this species (Shu 1989). **Laos:** The Laotian name is *tua chuck-a-chun*. The adults are eaten (Bodenheimer 1951). **Thailand:** The adults are roasted after removing the wings (Mungkorndin 1981).

Dundubia intermerata → Dundubia intemerata

Dundubia jacoona The adults are eaten in Sabah State, **Malaysia** (Chung et al. 2002).

Dundubia mannifera **Thailand:** The adults are roasted after removing the wings (Mungkorndin 1981).

Dundubia spiculata The larvae and the adults are roasted by the Nishi people in Arunachal Pradesh State, **India** (Singh et al. 2007).

Dundubia **spp. Malaysia:** In Sabah State, a species called *tavir* is roasted or stir-fried with some salt and other flavorings but without oil (Chung 2010). **Thailand:** The adults of some species belonging to this genus are roasted after removing the wings (Sangpradub 1982, Jonjuapsong 1996).

Euterphosia crowfooti The adults are roasted or as a paste after removing the wings in Arunachal Pradesh State in **India** (Chakravorty et al. 2011).

Graptopsaltoria colorata → *Graptopsaltria nigrofuscata*

Graptopsaltria nigrofuscata **(PL III-5a)** The Japanese name is *abura-zemi*. This cicada together with *Oncotympana maculaticollis* is a representative of cicadas in **Japan**. The larvae and adults are fried, or roasted with soy sauce in the Nagano Prefecture (Takagi 1929d). The matured larvae come out from below the earth for molting to adults and are known to be good to eat (Mitsuhashi 2003 and unpublished data) **(PL III-5b)**. The infusion of the exuviae is used as antifebrile (Miyake 1919).

Graptopsaltria tienta **China:** This species was used medicinally since ancient times (Li 1596).

Huechys sanguinea (red ladybug) **China:** This species is used as an anodyne or as a medicine for a cough. It is also said to be effective against skin cancer (Zimian et al. 1997).

Huechys thoracica (short-winged ladybug) **China:** This species is used as an anodyne or as a medicine for a cough (Zimian et al. 1997).

Ioba horizontalis This species is eaten in **DRC** (Mbata 1995).

Ioba leopardina The adults of this species are fried in **DRC, Zambia** and **Zimbabwe** (Mbata 1995).

Ioba **sp.** The adults of a species belonging to this genus is a food item in **Malawi** (Shaxon et al. 1974).

Leptopsaltria japonica → *Tanna japonensis japonensis*

Magicicada cassini (Cassini periodical cicada, dwarf cicada) This species is one of 17-year old periodical cicada. The nymphs and adults are widely eaten even at present in the **USA** (DeFoliart 2002).

Magicicada septendecim (periodical cicada, 17-year locust) In the **USA,** Native Americans of the Cheroke tribes ate this species (Waugh 1916, Carr 1951, Howard 1886). Now-a-days, different recipes have been made (Rehert 1984).

Magicicada septendecula One of the 17-year periodical cicada a little bit smaller than *Magicicada septendecim*. It is eaten like *M. septendecim* in the **USA** (DeFoliart 2002).

Magicicada tredecassini → *Magicicada tredecim*

Magicicada tredecim (13-year periodical cicada) This species is eaten in the **USA** like *M. septendecim* (DeFoliart 2002).

Magicicada tredecula → *Magicicada tredecim*

Meimuna opalifera **Japan:** The Japanese name is *tsukutsukuboushi*. The larvae and the adults are fried or roasted in the Nagano Prefecture (Takagi 1929d). The exuviae of the last instar larvae are used for treating various diseases such as tympanitis, constipation, difficulty in urination, etc. (Umemura 1943). **Laos, Myanmar, Thailand** and **Vietnam:** The adults are considered good for eating (Yhoung-Aree and Viwatpanich 2005).

Monomatapa insignis This species is a foodstuff in **Botswana** (Roodt 1993).

Monomotapa **sp.** The adults of a species belonging to this genus are consumed as a food item in **Malawi** (Shaxon et al. 1974).

Munza furva This species is a food source in **DRC** (Malaisse and Parent 1997a).

Okanagana bella The adults are roasted and relished in the **USA** (Sutton 1988).

Okanagana cruentifera The adults are roasted and eaten in the **USA** (Sutton 1988).

Okanagodes **spp.** In **USA,** Native Americans belonging to following tribes eat some species of cicadas belonging to this genus; Washoe, northern Paiute, Owens Valley Paiute, Panamint, western Shoshone, northern Shoshone, Bannock, Utah southern Paiute, northern Ute, western Ute, southern Ute (Fowler 1986).

Oncotympana maculaticollis **China:** This species has been used medicinally since ancient times (Li 1596). **Japan:** The Japanese name is *minmin-zemi*. This cicada together with *Graptopsaltria nigrofuscata* is a representative cicada in Japan. The larvae and adults are fried, or roasted with soy sauce in the Nagano Prefecture (Takagi 1929d). The matured larvae come out from below the earth for molting to adults are known to be good to eat (Mitsuhashi 2003 and 2005). The paste made of the exuviae or dried powder of the exuviae were used for treating dizziness (Miyake 1919).

Orapa **sp.** A species belonging to this genus is a foodstuff in **Botswana** and **Malawi** (Roodt 1993).

Orientopsaltria **spp. Laos:** A species belonging to this genus (Laotian name is *chak chanh*) is relished (Boulidam 2008). **Malaysia:** In Sabah State, a species called *tengir* is roasted or stir-fried with some salt and other flavorings but without oil (Chung 2010). **Thailand:** The adults of a species belonging to this genus are roasted after removing the wings (Hanboonsong et al. 2000).

Platylomia radha **Thailand:** This species is roasted after removing the wings (Mekloy 2002).

Platylomia **spp. Malaysia:** A species belonging to this genus is eaten in Saba State (Chung et al. 2002). **Thailand:** The adults of a species belonging to this genus are roasted after removing the wings (Hanboonsong et al. 2000).

Platypedia areolata The adults of this species are roasted in the **USA** (Sutton 1988).

Platypedia lutea The adults of this species are roasted in the **USA** (Sutton 1988).

Platypleura adouma The adults of this species are eaten raw or fried in **PRC** (Moussa 2004).

Platypleura insignis The larvae of this species are used as a food item in **Myanmar** (Distant 1889–1892).

Platypleura kaempferi **China:** The adults of this species are fried (Chen and Feng 1999). This species is used as a suppressor of swelling (Zimian et al. 1997). **Japan:** The Japanese name is *niinii-zemi.* The infusion of the exuviae was used as antifebrile (Miyake 1919).

Platypleura quadraticollis The adults of this species are eaten raw or fried in **Zimbabwe** (Rice 2000).

Platypleura stridula The adults of this species are fried in **Zambia** (Mbata 1995).

Platypleura **sp.** The adults of a species belonging to this genus are eaten **Malawi** (Shaxon et al. 1974).

Pomponia imperatorial (giant cicada, empress cicada) This adults are used as a foodstuff in **Malaysia** (Essig 1947).

Pomponia linearis The larvae and the adults are roasted by the Nishi people in Arunachal Pradesh, **India** (Singh et al. 2007).

Pomponia maculaticolis → *Oncotympana maculaticolis*

Pomponia merula: This large cicada is often eaten by stir- or deep-frying in Kalimantan State, **Indonesia** (Chung 2010).

Pomponia **sp.** A species belonging to this genus is used as a food item in **Thailand** (Leksawasdi 2008).

Proarna **sp.** The Mexican name is *chicharra.* In **Mexico,** a species which is called *gints'yo* by the Hñähñu native people in Hidalgo State eat the adults (Ramos-Elorduy de Conconi et al. 1984, Ramos-Elorduy et al. 1997, Aldasoro Maya 2003).

Pycna repandar The adults are eaten roasted or as a paste after removing the wings in Arunachal Pradesh State in **India** (Chakravorty et al. 2011).

Pycna **sp.** The adults of a species belonging to this genus is considered edible in **Malawi** (Shaxon et al. 1974).

Quesada gigas The adults are considered good for eating in **Mexico** (Ramos-Elorduy and Pino 2002).

Rihana **sp.** The adults of a species belonging to this genus are roasted in **Thailand** (Sangpradub 1982, Jonjuapsong 1996).

Sadaka radiata This species is used as food in **DRC** (Malaisse and Parent 1997a).

Tanna japonensis japonensis The Japanese name is *higurashi.* The larvae and adults are fried, or roasted with soy sauce in the Nagano Prefecture, **Japan** (Takagi 1929d). The exuviae of the last instar larvae were used for treating various diseases such as tympanitis, constipation, difficulty in urination, etc. (Umemura 1943).

Terpnosia vacua The Japanese name is *haru-zemi.* In **Japan,** the exuviae of the last instar larvae are used for treating various diseases such as tympanitis, constipation, difficulty in urination, etc. (Umemura 1943).

Tettigonia antiquorum **Greece:** This species may be the cicada, which Aristotle described as good food insects (Hope 1842).

Tibicen montezuma → *Cicada montezuma*

Tibicen pruinosa → *Cicada pruinosa*

Tibisen septendecim → *Magicicada septendecim*

Ugada giovanninae The adults are eaten raw or fried in **PRC** (Moussa 2004).

Ugada limbalis **DRC:** The adults are eaten raw, or fried (Malaisse and Parent 1997a). **Zambia:** The adults are fried after removing the wings (Mbata 1995).

Ugada limbata The adults are considered good for eating in **PRC** (Moussa 2004).

Ugada limbimaculata **DRC:** The adults area eaten raw or fried (Malaisse and Parent 1997a). **PRC:** The adults are used as food (Moussa 2004).

Yezoterpnosia vacua → *Terpnpsia vacua*

Coccidae (Soft scales) **(2)**

Ceroplastes sinensis → *Ericerus pela*

Coccus (*Llavea*) *axin* **Mexico:** The boiled larvae and adults are used to relieve the toxic effects of poisonous fungi. They are also used as a binding medicine (Ramos-Elorduy de Conconi and Pino 1988).

Ericerus pela (white wax scale) **China:** People fry the larvae and adults or dip the eggs in an alcoholic drink (Chen and Feng 1999). This species is used as a styptic or an anodyne (Zimian et al. 1997). **Japan:** The wax secreted by this insect was used as an anodyne, or as a remedy for tuberculosis, stomach disorders, etc. (Umemura 1943, Shiraki 1958).

Dactylopiidae (Cochineal insects) (4)

Dactylopius coccus (cochineal insect) (**PL IV-1**) **Italy** and **Morocco:** A type of dye (carmine) is extracted from the larvae and adults of this species, and is frequently used for coloring milk products (Mitsuhashi 2003, Benhalima et al. 2003). **Mexico:** The Mexican name is *cochinilla grana*. This species is eaten in the Hidalgo, Morelos and Oaxaca States (Ramos-Elorduy 2009), and raised as food (Ramos-Elorduy 1997). The boiled larvae and adults are used to relieve the toxic effects of poisonous fungi. They are also used as a medicine for caries (Ramos-Elorduy de Conconi and Pino 1988).

Dactylopius confusus The Mexican name is *cochinilla grana*. This species is eaten in the Hidalgo, Morelos and Oaxaca States (Ramos-Elorduy 2009), and raised as food for humans in **Mexico** (Ramos-Elorduy 1997). The boiled larvae and adults are used to relieve the toxic effects of poisonous fungi. They are also used as a medicine for caries (Ramos-Elorduy de conconi and Pino 1988).

Dactylopius indicus This species is eaten in the Hidalgo, Morelos and Oaxaca States (Ramos-Elorduy 2009), and raised as food for humans in **Mexico** (Ramos-Elorduy 1997). The boiled larvae and adults are used to relieve the toxic effects of poisonous fungi. They are also used as a medicine for caries (Ramos-Elorduy de conconi and Pino 1988).

Dactylopius tomentosus The Mexican name is *cochinilla grana*. The larvae and adults are used in **Mexico:** (Ramos-Elorduy et al. 1997). This species is raised as food (Ramos-Elorduy 1997). The boiled larvae and adults are used to relieve the toxic effects of poisonous fungi. They are also used as a medicine for caries (Ramos-Elorduy de conconi and Pino 1988).

Eriosomatidae (4)

Melaphis chinensis **China:** This species was used as an antidote (Li 1596). **Japan:** The Japanese name is *nurudeshiro-aburamushi*. The gall made by this insect is used as a remedy for stomach ulcer, and in olden times for treating scrofula (Hayashi 1903).

Nurudae rosea (rosy gall aphid) This species is said to be effective against hepatic cancer in **China** (Zimian et al. 1997).

Nurudae shiraii (sumac gall aphid) **China:** This species is said to be effective against hepatic cancer (Zimian et al. 1997).

Nurudae sinica (Chinese gall aphid) This species is said to be effective against hepatic cancer in **China** (Zimian et al. 1997).

Flatidae (Flatid planthoppers) **(3)**

Flata limbata **China:** The waxy substance secreted by this species was used for treating heart diseases (Ealand 1915).

Lawana imitata The larvae and adults are eaten fried in **China** (Chen and Feng 1999).

Phremnia rubra → *Phromnia rubra*

Phromnia rubra This species is widely distributed in the western and southern areas of **Madagascar**. The larvae secrete white sweet droplets, which sometimes accumulate as the size of a fist on the branch of Combretaceae or on the ground. The Sakalava, Bara and Mahafaly people are fond of this sweet substance called *tantely sakondry* (Decary 1937).

Fulgoridae (Lantern flies) **(2)**

Lycorma delicatula (Chinese blistering cicada) In **China,** this species was used medicinally more than 2000 years (Shennung Ben Ts'ao King: Read 1941), and is still used as a suppressor of swelling (Zimian et al. 1997).

Pyrops madagascariensis → *Zana tenebrosa*

Pyrops tenebrosa → *Zana tenebrosa*

Zana tenebrosa This species is eaten in **Madagascar** (Decary 1937).

Lacciferidae (Lac insects) **(1)**

Laccifer lacca (lac insect) **China:** This species was used medicinally since the third century (Shennung Ben Ts'ao King: Read 1941), and is still used as a suppressor of swelling (Zimian et al. 1997). **Thailand:** This species is eaten in Phu Wiang Valley (DeFoliart 2002).

Membracidae (Tree-hoppers) **(5)**

Anthiante expansa The Mexican name is *torito*. The larvae and adults are a foodstuff in **Mexico** (Ramos-Elorduy et al. 1998).

Anthiante expensa → *Anthiante expansa*

Darthula hardwicki The larvae and adults are fried in **China** (Chen and Feng 1999).

Hoplophorion monograma The Mexican name is *periquito*. The larvae and adults are eaten in the Michoacanm, Guerrero States and Mexico FD, **Mexico** (Ramos-Elorduy de Conconi et al. 1984, Ramos-Elorduy et al. 1997).

Metcalfiella monograma → *Hoplophorion monograma*

Umbonia reclinata The Mexican name is *torito*. The larvae and adults are food item in Puebla State, **Mexico** (Ramos-Elorduy de Conconi et al. 1984, Ramos-Elorduy et al. 1997).

Umbonia spinosa **Brazil:** The adults are used as food (Wallace 1953). **Colombia:** This species is eaten raw or toasted by the Tukanoan people (Paoletti and Dufour 2005). **Ecuador:** The larvae and adults are used as foodstuff by the Quichua people (Onore 1997 and 2005).

Umbonia sp. The larvae and adults of a species belonging to this genus are preferred insects as food in the Morelos and Guerrero States, **Mexico** (Ramos-Elorduy de Conconi et al. 1984).

Pseudococcidae (Mealy bugs) (6)

Apiomorpha pomiformis The galls containing the larvae of this insect are eaten raw in **Australia** (Cleland 1966).

Cystococcus echiniformis In **Australia**, the galls containing the larvae and adults of this insect are eaten raw (Menzel and D'Aluisio 1998).

Cystococcus pomiformis: In **Australia,** the large galls measured 7.5 cm in diameter containing the larvae of this insect are called apple of bloodwood, and consumed raw (Sweeney 1947).

Cystococcus **sp. Australia:** The galls containing the larvae and adults of this insect are edible (James 1983).

Naiacoccus serpentines → *Najacoccus serpentinus*

Najacoccus serpentinus In **Iran** and **Iraq,** the honey dew excreted by this species is eaten raw. It is called tamarisk manna (Leibowitz 1943, Bodenheimer 1951).

Phenacoccus prunicola In **China,** people fry the larvae and adults or dip the eggs in an alcoholic drink (Chen and Feng 1999).

Trabutina mannipara **Iraq:** The Bedouin people eat the honey dew excreted by this species raw. It is called *tamarisk* manna (Bodenheimer 1951). The honey dew of this species is relished in **Israel** and **Iraq** (Bodenheimer 1951, Leibowitz 1943).

Trabutina serpentines → *Najacoccus serpentinus*

Trabutina **sp.** The honey dew excreted by this species is used as a sweet material in **Iran.** It is called *tamarisk* manna (Bodenheimer 1951).

Psyllidae (Jumping plantlice) **(3)**

Arytaira mopane The honey dew of this species is an important food in **Botswana** (Sekhwela 1989) and **Zimbabwe** (Weaving 1973).

Austrotachardia acacia In **Australia**, aborigines eat the honey dew of this species raw (Cleland 1966). This plantlice suck the sap of *Acacia aneura*, and excrete red honey dew, which is called red mulga lerp (Low 1989).

Chermes sp. In **Lapland**, the gall made by this species is eaten (Linné 1811).

Eucalyptolyma sp. In **Australia**, the honey dew excreted by this insect is called lerp sugar, and is eaten (DeFoliart 2002).

Glycaspis spp. In **Australia,** conical shaped honey dew produced by immature stages of psillids belonging to this genus is called lerps. Aboriginal people eat them (Yen 2005) **(PL IV-2)**.

Psylla eucalypti → *Spondyliaspis eucalypti*

Psylla sp. The honey dew of a species belonging to this genus is preferred foodstuff in the southern area of **Africa** (Livingstone 1857).

Spondyliaspis eucalypti In **Australia**, aborigines eat the honey dew of this species raw (Bodenheimer 1951).

(Heteroptera) [141]

Alydidae (2)

Leptocorisa acuta **India:** The adults are roasted by the Nishi people in Arunachal Pradesh State (Singh et al. 2007). **Indonesia:** The adults are used to prepare "sambal" (DeFoliart 2002).

Leptocorisa oratorius **Malaysia:** The vernacular name is *pesisang*. In Sabah State, elderly Kadazandusun villagers mash the bugs with chili and salt, and cook them in hollow bamboo stems. This dish is served as a condiment (Chung 2010). **Indonesia:** In Kalimantan State, this species is eaten raw (Chung 2010).

Aphelochiridae (1)

Aphelocheirus vittatus The Japanese name is *nabebutamushi*. In **Japan,** this species is a member of *zazamushi* (see Tricoptera-Leptoceridae-*Parastenopsyche sauteri*). This species is cooked with soy sauce and sugar in the Nagano Prefecture (Torii 1957).

Belostomatidae (13)

Abedus dilatatus The Mexican name is *cucarachón*. The larvae and adults are used as a foodstuff in **Mexico** (Ramos-Elorduy et al. 1998).

Plate IV. 1. Mass rearing of *Dactylopius coccus* on a cactus leaf in Peru. Body length of large larvae: 7 mm (cf. p.60); 2. Lerp manna secreted by *Glycaspis* sp. on a leaf of *Eucalyptus* tree (Photo: A.L. Yen, Victoria, Australia) (cf. p.63); 3. Water boatmen sold at markets in Mexico City, Mexico. Insert: Close-up of adults. Body length: 4–6 mm (cf. p.68); 4. *Lethocerus indicus* adult. Body length: ♂ 70 mm, ♀ 80 mm (cf. p.66); 5. Bugs, *Euchosternum delegorguei*, sold at a market in Republic of South Africa (cf. p.74) (Photo: K. Nonaka, Tokyo, Japan); 6. Roasting of skewered *Tessaratoma pappilosa* above a charcoal burner in Vientiane, Laos. Insert: An adult. Body length: 20 mm (cf. p.77).

Abedus ovatus The Mexican name is *cucarachon de agua.* The larvae and adults are a food source in **Mexico** (FD) (Ramos-Elorduy de Conconi et al. 1984).

Abedus **sp.** The Mexican name is *cosha.* The larvae and adults of a species belonging to this genus are treated as a foodstuff in **Mexico** (Ramos-Elorduy et al. 1998).

Belastoma → *Lethocerus*

Diplonychus japonicas The Japanese name is *kooimushi.* In the Nagano Prefecture, **Japan,** the adults are fried with oil, and then maybe cooked with sugar-soy sauce (Takagi 1929f). The eggs are eaten raw, or cooked with predaceous diving beetles (Coleoptera-Dytiscidae) (Mukaiyama 1987).

Diplonychus rusticus This species is considered edible in **Thailand** (Chen and Wongsiri 1995).

Diplonychus **sp.** A species belonging to this genus is used as food in **Thailand** (Hanboonsong 2010).

Kirkaldgia degrollei The larvae and adults are considered good for eating in **China** (Hu and Zha 2009).

Kirkaldia deyrolei → *Lethocerus deyrolei*

Lethocerus americanus (giant water bug) The adults are eaten by western native Americans in the **USA** (Essig 1949).

Lethocerus cordofanus The adults are roasted in **Cameroon** (Malaisse and Parent 1997a, De Colombel 2003).

Lethocerus deyrolei The Japanese name is *tagame.* In olden times, people of several prefectures in **Japan** such as, Chiba, Okayama, Kyoto and Shizuoka roasted the eggs and put a touch of soy sauce on them. The adults were also used for treating convulsive fits (Miyake 1919). In the Tochigi Prefecture, the adults were mashed with miso paste and roasted (Hori 2006). Recently the population of this bug has decreased a great deal, and the species has became endangered.

Lethocerus europdeum **Thailand:** This species is an important food (Leksawasdi 2008).

Lethocerus indicus (giant water bug) **(PL IV-4) China:** The adults are boiled with salt, or fried (Hoffmann 1947). **India:** This species is considered good for eating by people of the Nishi and Galo tribes in Arunachal Pradesh State. The people eat the adults boiled or fried (Chakratorty et al. 2011). The Karbi and Rengma Naga people in Assam State eat the larvae and adults as a chutney (Ronghang and Ahmed 2010). This species is also eaten by the Ao-Naga people in Nagaland State (Meyer-Rochow 2005). **Laos:** The Laotian name is *meang da.* The adults are roasted with salt to make a spicy dish (Mitamura 1995, Yhoung-Aree and Viwatpanich 2005). **Myanmar:** The adult insects are placed on hot coals, and the cooked insides are eaten just as one would the soft parts of the limbs of lobsters and crabs (DeFoliart 2002). **Singapore:** A specially flavored salt (*kwai fa shim im*) is sold with *L. indicus* adults. The salt is fragrant and probably has henna flowers added to it (Hoffmann 1947). **Sri Lanka:** This species is eaten (Nandasena et al. 2010). **Thailand:** The Thai name is *malang da na*. It is a very popular food insect. The adults are fried, or used as a spice material (Mungkorndin 1981). **Vietnam:** The eggs and adults are steamed, roasted, fried, or used to extract scent material from scent glands (Tiêu 1928).

Lepthocerus micantulum The Yekuana people in Alto Orinoco area, **Venezuela**, eat the adults (Araujo and Beserra 2007).

Lethocerus **spp. India:** A species belonging to this genus is eaten roasted (Pathak and Rao 2000). **Mexico:** The larvae and adults of a species, whose Mexican name is *cucarachón,* are a food item (Ramos-Elorduy et al. 1998). Several species are raised as food for humans (Ramos-Elorduy 1997). **PRC:** The adults of several species belonging to this genus are edible (Moussa 2004). **Venezuela:** A species belonging to this genus is edible by the Ye'kuana people (Ramos-Elorduy de Conconi et al. 1984, Paoletti and Dufour 2005).

Lohita grandis (giant red bug) This species is used as foodstuff by the Ao-Naga people in Nagaland State, **India** (Meyer-Rochow 2005).

Sphaerodema rustica This species is used as a food fried or boiled in **China** (Chen and Feng 1999). **Thailand:** This species is treated as a food item (Leksawasdi 2008).

Cimicidae (Bed bugs) **(1)**

Cimex lectularis **Rome:** Around AD 0, people ate the bed bugs to cure bite wounds by snakes (Plinius AD 77–79).

Coreidae (Coreid bugs) **(14)**

Acantocephala declivis The larvae and the adults are eaten in **Mexico** (Ramos-Elorduy de Conconi 1982).

Acantocephala luctuosa The larvae and adults are eaten in **Mexico** (DeFoliart 2002).

Acantocephala **sp.** The larvae and adults of a species belonging to this genus are treated as a foodstuff in **Mexico** (Ramos-Elorduy and Pino 2002).

Anoplocnemis curvipes People suck the body fluid of the larvae and the adults in **Cameroon** (De Colombel 2003).

Anoplocnemis phasianus (leaf-footed bug) People eat this species in **Thailand** (Chen and Wongsiri 1995).

Carlisis wahlbergi → *Pentascelis wahlbergi*

Dalader acuticosta The adults are fried by the Nishi people of Arunachal Pradesh State, **India** (Chakravorty et al. 2011).

Dicranocephalus wallichi bowringi This species is considered edible in **China** (Chen and Feng 1999).

Homoeocerus sp. People eat a species belonging to this genus in **Thailand** (Hanboonsong et al. 2000).

Mamurius mopsus This species is eaten in **Mexico** (Ramos-Elorduy et al. 1985).

Mictis tenebrosa **China:** The larvae and adults are fried (Chen and Feng 1999). **India:** The adults are consumed fried or as a raw paste by the Nishi people in Arunachal Pradesh State (Chakravorty et al. 2011).

Mictis sp. A species belonging to this genus is considered edible by the Chauve people in **PNG** (Meyer-Rochow 1973).

Onchorochira nigrorufa This species is used as food in **Thailand** (Leksawasdi 2008).

Pachilis gigas (giant mesquite bug) The Mexican name is *xamues*. In **Mexico,** native people of Hñähñu tribe eat the larvae and adults (Ramos-Elorduy de Conconi et al. 1984, Ramos-Elorduy et al. 1997, Aldasoro Maya 2003). The larvae are collected and kept on trees near the house in many states (Ramos-Elorduy 2009). This species is also edible in the Queretaro, Guerrero, Hidalgo, San. Luis Potosi States (Ramos-Elorduy de Conconi et al. 1984). The oil extracted from the bug is used for treating scrofula, stomach disorders, etc. (Ramos-Elorduy de Conconi and Pino 1988).

Pe(n)tascelis remipes The adults are eaten in **Zimbabwe** (Chavanduka 1975).

Pe(n)tascelis wahlbergi The adults are edible in **Zimbabwe** (Chavanduka 1975).

Prionolomia sp. A species belonging to this genus is considered edible in **Thailand** (Chen and Wongsiri 1995).

Sephina vinula The larvae and adults are edible in **Mexico** (Ramos-Elorduy et al. 1985).

Stenocoris varicornis This species is eaten in **Indonesia** (Java) (van der Burg 1904).

Thasus gigas → *Pachilis gigas*

Corixidae (Water-boatmen) **(13)**

Ahauhtlea azteca → *Krizousacorixa azteca*

Corisella edulis The Mexican name is *ahuahutle*, or *axayacatl*. The eggs, larvae and the adults are considered edible in **Mexico** (Ramos-Elorduy et al. 1998, DeFoliart 2002). This species is raised as food (Ramos-Elorduy 1997).

Corisella mercenaria The Mexican name is *ahuahutle*, or *axayacatl*. In **Mexico,** the eggs, larvae, and the adults are used as food in the Guanajuato, and DF State (Ramos-Elorduy de Conconi et al. 1984), and can be raised as food for humans **(PL IV-3)** (Ancona 1933, Ramos-Elorduy 1997, Ramos-Elorduy and Pino 2002).

Corisella texcocana The Mexican name is *ahuahutle*, or *axayacatl*. In **Mexico,** the eggs, larvae and the adults are eaten in the Guanajuato, Michoacan and DF States (PL IV-3) (Ancona 1933, Ramos-Elorduy de Conconi et al. 1984).

Corisella **spp. Mexico:** The eggs and adults of several species belonging to this genus are eaten, and are raised as foods (Ramos-Elorduy 1997, Ramos-Elorduy et al. 1997).

Corixa esculenta The eggs are eaten in **Egypt** (Motschoulsky 1856). This species is also eaten in many regions of Africa (Ramos-Elorduy de Conconi 1987).

Corixa femorale **Mexico:** The eggs are considered edible (Ealand 1915).

Corixa mercenaria → *Corisella mercenaria*

Graptocorixa abdominalis The Mexican name is *ahuahutle*, or *axayacatl*. In **Mexico,** this species is preferred edible insect by the Mazahua, Otomi and Nahus people in DF State (Ramos-Elorduy 2009), and raised as food (Ramos-Elorduy 1997, Ramos-Elorduy 2006).

Graptocorixa bimaculata The eggs, larvae and the adults are eaten by the Mazahua, Otomi and Nahus people in DF State, **Mexico** (Ramos-Elorduy 2009). This species is raised as food for humans (Ramos-Elorduy 1997).

Graptocorixa **sp.** The eggs, larvae and the adults of a species belonging to this genus is good for eating in **Mexico** (Ramos-Elorduy and Pino 2002).

Hesperocorixa distanti distanti **Japan:** The Japanese name is *mizumushi*. This species is a member of *zazamushi* (see Tricoptera-Leptoceridae-*Parastenopsyche sauteri*) (Nagano Fishery Experiment Station, Suwa 1985).

Hesperocorixa laevignata **Mexico:** This species is raised as food (Ramos-Elorduy 1997).

Krizousacorixa azteca The Mexican name is *ahuahutle*, or *axayacatl*. In **Mexico**, the eggs, larvae and the adults are considered a delicacy in the Guanajuato, Michoacan and DF States (Ancona 1933, Ramos-Elorduy de Conconi et al. 1984). This species is raised as food (Ramos-Elorduy 1997).

Krizousacorixa femorata The Mexican name is *ahuahutle*, or *axayacatl*. In **Mexico**, the eggs, larvae and the adults are eaten in the Guanajuato,

Michoacan and DF States (Ancona 1933, Ramos-Elorduy de Conconi et al. 1984). This species can be raised as food for humans (Ramos-Elorduy 1997).

Krizousacorixa **sp.** The eggs and adults of a species belonging to this genus are used as food in **Mexico** (Ramos-Elorduy et al. 1997).

Micromecta quadriseta The larvae and adults are a foodstuff in **China** (Hu and Zha 2009).

Sigara substriata The larvae and adults are used as food in **China** (Hu and Zha 2009).

Gerridae (Water-striders) **(2)**

Cylindrostethus scrutator People eat this species in **Thailand** (Hanboonsong et al. 2000).

Gerris spinole This species is eaten by the Ao-Naga people in Nagaland State, **India** (Meyer-Rochow 2005).

Geris **sp.** A species belonging to this genus is used as a foodstuff by the Ao-Naga people in Nagaland State, **India** (Meyer-Rochow 2005).

Naucoridae (Creeping water-bugs) **(4)**

Ambrysus stali The Yekuana people in Alto Orinoco area, **Venezuela**, eat the adults (Araujo and Beserra 2007).

Ambrysus **sp.** A species belonging to this genus is roasted by the Yekuana people in **Venezuela** (Paoletti and Dufour 2005).

Limnocoris **cf.** *minutes* The Yekuana people in Alto Orinoco area, **Venezuela**, eat the adults (Araujo and Beserra 2007).

Sphaerodema molestum The adults are roasted in **Thailand** (Bristowe 1932).

Sphaerodema rustica **China:** This species is fried (Chen and Feng 1999). **Thailand:** The adults are roasted (Bristowe 1932).

Nepidae (Water-scorpions) **(6)**

Laccotrephes flavovenosa **Japan:** The Japanese name is *taikōchi*. The adults were said to be effective for treating nervous breakdowns (Shiraki 1958). **South Korea:** The infusion of the adults is used for nervous prostration in South Joella Province (Okamoto and Muramatsu 1922).

Laccotrephes griseus The Laotian name is *mang dah*. The adults are skewered and roasted in **Thailand** (Bristowe 1932).

Laccotrephes ruber People eat the adults of this species in **Thailand** (Hanboonsong et al. 2000).

Laccotrephes sp. A species belonging to this genus is eaten steamed, or roasted by the Meeteis people in Nagaland State, **India** (Meyer-Rochow 2005). It is also used as a spice material.

Nepa spp. **Madagascar:** Several species belonging to this genus are used as food (Decary 1937, Paulian 1943). **Thailand:** Several species belonging to this genus are used as food (Sangpradub 1982).

Ranatra chinensis **Japan:** The Japanese name is *mizu-kamakiri*. The adults are cooked in the Nagano Prefecture (Mukaiyama 1987). **Laos:** This species is also eaten (Nonaka 1999a). **Thailand:** This species is consumed as food (Leksawasdi 2008).

Ranatra longipes thai People eat this species in **Thailand** (Hanboonsong et al. 2000).

Ranatra variipes People eat this species in **Thailand** (Hanboonsong et al. 2000).

Ranatra sp. A species belonging to this genus is fried in oil by the Meeteis people in Nagaland State, **India** (Meyer-Rochow 2005).

Sphaerocoris sp. The local name of this species in **Malawi** is *nsensenya*, and is prepared by washing and frying with little salt until brown (Shaxon et al. 1985).

Notonectidae (Backswimmers) **(8)**

Anisops barbatus People eat this species in **Thailand** (Hanboonsong et al. 2000).

Anisops bouvieri People eat this species in **Thailand** (Hanboonsong et al. 2000).

Anisops fieberi The larvae and adults are eaten in **China** (Hu and Zha 2009).

Anisops spp. **India:** A species belonging to this genus is fried in oil by the Meeteis people in Nagaland State (Meyer-Rochow 2005). **Laos, Myanmar, Thailand** and **Vietnam:** Some species belonging to this genus are eaten (Yhoung-Aree and Viwatpanich 2005).

Enithares sinica The larvae and adults are used as a food item in **China** (Hu and Zha 2009).

Notonecta chinensis The larvae and adults are edible in **China** (Hu and Zha 2009).

Notonecta fasciata **Mexico:** This species is raised as food (Ramos-Elorduy 1997).

Notonecta undulata **Thailand:** This species is eaten (Hanboonsong et al. 2000).

Notonecta unifasciata The Mexican name is *ahuahutle,* or *axayacatl.* In **Mexico,** the eggs, larvae and the adults are relished in Guanajuato, Michoacan and Mexico FD States (Ealand 1915, Ramos-Elorduy de Conconi et al. 1984).

Notonecta **sp. Mexico:** The eggs, larvae and the adults of a species belonging to this genus are used as a foodstuff (Ramos-Elorduy and Pino 2002). **Myanmar:** People mash the bugs in a mortar, and use it as a shrimp substitute to give "body" to gravies, soups and other dishes (DeFoliart 2002).

Pentatomidae (Stink bugs) **(75)**

Acantocephala lucuosa The larvae and adults are consumed as food in **Mexico** (Ramos-Elorduy and Pino 1990).

Acantocephala decrivis **Mexico:** The oil extracted from the bug is used for treating scrofula, stomach disorders, etc. (Ramos-Elorduy de conconi and Pino 1988).

Acrosternum millieri **Cameroon:** The Mofu people use the adults to extract an aromatic substance for foods by dipping them in oil (Seignobos et al. 1996).

Agonoscelis pubescens **Sudan:** The Gaaliën people in the Nuba mountains in Kurdufan, press the hibernating adults, and extract oil, which is used for cooking as well as to treat scab disease of camels (van Huis 1996 and 2003).

Agonoscelis versicolor In **Sudan**, the adults are used to extract an aromatic substance for foods by dipping them in oil (van Huis 2008).

Alcaerrhynchus grndis The adults are consumed fried or boiled with vegetables by the Nishi people in Arunachal Pradesh State, **India** (Chakravorty et al. 2011).

Amissus testaceus This species is used as a food item in **Thailand** (Leksawasdi 2008).

Anoplocnemis curvipes The adults are eaten raw in **Cameroon** (De Colombel 2003).

Arcana tenebrosa → *Brochymena tenebrosa*

Aspongopus chinensis **China:** This species is used as a food source (Mao 1997), and is used as a cordial (Li 1596).

Aspongopus nepalensis In Assam, **India**, this bug is sought after by people as food. They pounded the bug and mix it with rice (Distant 1902).

Aspongopus viduatus The adults are eaten raw in **Cameroon** (De Colombel 2003).

Atizies sufultus **Mexico:** This species is eaten in Guerrero State (Ramos-Elorduy de Conconi et al. 1984).

Atizies taxcoensis → *Euschistus taxcoensis*

Bagrada picta (painted bug) **India:** This species is consumed as food by the Ao-Naga people in Nagaland State (Meyer-Rochow 2005). **Sri Lanka:** This species is eaten (Nandacena et al. 2010).

Banasa subrufescens The larvae and adults are used as a foodstuff in **Mexico** (Ramos-Elorduy 2003).

Banasa **sp.** A species belonging to this genus is consumed as a foodstuff in **Mexico** (Ramos-Elorduy 2003).

Basicrryptus **sp.** The adults are used as a spice in **Cameroon** (De Colombel 2003).

Brochymena tenebrosa This species is a food source in **Mexico** (Ramos-Elorduy 2003).

Brochymena **sp.** The larvae and adults are considered as a food item in **Mexico** (Ramos-Elorduy 2003).

Carbula pedalis **Cameroon:** The Mofu people use the adults to get aromatic substance by extraction with oil (Seignobos et al. 1996).

Chlorocoris distinctus The larvae and adults are eaten in **Mexico** (Ramos-Elorduy 2003).

Chlorocoris irroratus The larvae and adults are consumed as food in **Mexico** (Ramos-Elorduy 2003).

Chlorocoris **sp.** The larvae and adults are used as a food item in **Mexico** (Ramos-Elorduy 2003).

Coridicus chinensis The larvae and adults are eaten in **China** (Hu and Zha 2009).

Coridius chinensis → *Aspongopus chinensis*

Coridius nepalensis → *Aspongopus nepalensis*

Cyclopelta parva (dadap bug) The larvae and adults are fried in **China** (Chen and Feng 1999). This species is said to be effective against esophageal cancer (Zimian et al. 1997).

Cyclopelta subhimalayensis The adults are consumed by people of the Miris, Mishmas, Abors, and some Nagas tribes in Assam district, **India** (Strickland 1932).

Dendrocerus sufultus → *Euchistus sufultus*

Diploxys cordofana **Cameroon:** The Mofu people use the adults to get aromatic substance by extraction with oil (Seignobos et al. 1996).

Diploxys **sp.** The adults are used as a spice in **Cameroon** (De Colombel 2003).

Dolycoris indicus (bamboo bug) This species is eaten by the Ao-Naga People in Nagaland State, **India** (Meyer-Rochow 2005).

Edessa championi **Mexico:** The adults are considered edible (Ramos-Elorduy 2003).

Edessa conspersa The Mexican name is *jumil*. The adults are eaten in **Mexico** FD (Ramos-Elorduy de Conconi et al. 1984). This species is sold at the markets of Ozumba in Mexico city (Ramos-Elorduy 2003).

Edessa cordifera The Mexican name is *jumil*. The adults are eaten in **Mexico**. This species had been sold at the markets of Taxco city in Guerrero State (Ramos-Elorduy 2003).

Edessa discors The adults are relished in **Mexico** (Ramos-Elorduy 2003).

Edessa fuscidorsata The adults are considered good for eating in **Mexico** (Ramos-Elorduy 2003).

Edessa helix The adults are consumed as food in **Mexico** (Ramos-Elorduy 2003).

Edessa indigena The adults are good for eating in **Mexico** (Ramos-Elorduy 2003).

Edessa lepida The adults are relished in **Mexico** (Ramos-Elorduy 2003).

Edessa mexicana The Mexican name is *jumil*. In **Mexico**, the adults are eaten in Morelos State (Ramos-Elorduy de Conconi et al. 1984). This species has been sold at the markets of Ozumba in Mexico FD (Ramos-Elorduy 2003).

Edessa montezumae The Mexican name is *jumil*. The larvae and adults are consumed in **Mexico** (Ramos-Elorduy et al. 1998).

Edessa petersii The Mexican name is *jumil*. The larvae and adults are a food source in **Mexico** (Ramos-Elorduy et al. 1998). The oil extracted from the bug is used for treating scrofula, stomach disorders, etc. (Ramos-Elorduy de Conconi and Pino 1988).

Edessa reticulat The adults are edible in **Mexico** (Ramos-Elorduy 2003).

Edessa rufomarginatus The larvae and adults are treated as a food item in **Mexico** (Ramos-Elorduy and Pino 2002).

Edessa **sp.** The Mexican name is *jumil*. The larvae and adults are used as foodstuff in **Mexico** (Ramos-Elorduy et al. 1998).

Encosternum delegorguei → *Euchosternum delegorguei*

Erthesina fullo **China:** The larvae and adults are eaten in China (Hu and Zha 2009). **India:** The adults are used as a food item by the Naga people in Assam district (Hoffmann 1947).

Euchosternum delegorguei **RSA:** The adults are consumed raw, roasted or cooked **(PL IV-5)** (van Huis 2005). **Zimbabwe:** The adults are consumed raw, roasted or cooked. The Mapulana people relish this species, and eat them raw or cooked (Faure 1944, Chavanduka 1975).

Euchosternum pallidus The adults are eaten raw, roasted or cooked in **Zimbabwe** (van Huis 2005).

Eurostus validus In **China**, the larvae and the adults are fried (Chen and Feng 1999).

Euschistus bifibulus The larvae and adults are used as food in **Mexico** (Ramos-Elorduy 2003).

Euschistus biformis The larvae and adults are food item in **Mexico** (Ramos-Elorduy 2003).

Euschistus comptus The larvae and adults are consumed in **Mexico** (Ramos-Elorduy 2003).

Euschistus crenator orbiculator In **Mexico**, the larvae and adults are eaten in the Morelos, Hidalgo, Veracruz, Guerrero States and Mexico FD (Ramos-Elorduy 2003, Ramos-Elorduy de Conconi et al. 1984). This species is also used to prepare a type of sauce (Ancona 1933). The oil extracted from the bug is used for treating scrofula, stomach disorders, etc. (Ramos-Elorduy de Conconi and Pino 1988).

Euschistus egglestoni **Mexico:** The larvae and adults are eaten (Ramos-Elorduy 2003). The oil extracted from the bug is used for treating scrofula, stomach disorders, etc. (Ramos-Elorduy de Conconi and Pino 1988).

Euschistus integer The larvae and adults are consumed in **Mexico** (Ramos-Elorduy 2003).

Euschistus lineatus In **Mexico**, the larvae and adults are used as a foodstuff in the Morelos, Hidalgo, Veracruz, Guerrero States and Mexico FD (Ramos-Elorduy de Conconi et al. 1984). The oil extracted from the bug is used for treating scrofula, stomach disorders, etc. (Ramos-Elorduy de Conconi and Pino 1988).

Euschistus rugifer The larvae and adults are food source in **Mexico** (Ramos-Elorduy 2003).

Euschistus schaffneri The larvae and adults are used as food item in **Mexico** (Ramos-Elorduy 2003).

Euschistus spurculus The larvae and adults are fit for eating in Mexico (Ramos-Elorduy 2003).

Euschistus stali The larvae and adults are an article of food in **Mexico** (Ramos-Elorduy 2003).

Euschistus strennus → *Euschistus strenuus*

Euschistus strenuus The Mexican name is *jumil*. **Mexico:** The larvae and adults are eaten in the Morelos, Hidalgo, Veracruz, Guerrero States and

Mexico FD (Ramos-Elorduy de Conconi et al. 1984). The oil extracted from the bug is used as a remedy for scrofula, stomach disorders, etc. (Ramos-Elorduy de Conconi and Pino 1988). This species had been sold at the markets of Cuernavaca in Morelos State (Ramos-Elorduy 2003).

Euschistus sufultus The larvae and adults are eaten in **Mexico** (Ramos-Elorduy de Conconi et al. 1984).

Euschistus sulcacitus **Mexico:** The larvae and adults are used as food (Ramos-Elorduy 2003). This species had been sold at the markets of Guautla in Morelos State (Ramos-Elorduy 2003).

Euschistus taxcoensis In **Mexico,** native people of the Tlahuica tribe eat the larvae and adults (Menzel and D'Aluiso 1998, Ramos-Elorduy 2003). The adults are eaten during the bug festival in Cerro El Huizteco Mountain (Ramos-Elorduy 2003). The oil extracted from the bug is used for treating scrofula, stomach disorders, etc. (Ramos-Elorduy de Conconi and Pino 1988).

Euschistus zopilotensis → *Euschistus strenuus*

Euschistus **sp.** The larvae and adults of a species belonging to this genus are consumed in **Mexico** (Ramos-Elorduy et al. 1998).

Eusthenes curpreus In **China,** the larvae and adults are fried (Chen and Feng 1999).

Eusthenes saevus In **China,** the larvae and adults are fried (Chen and Feng 1999).

Gonielytrum circuliventrie → *Euchosternum delegorguei*

Haplosterna delegorguei → *Euchosternum delegorguei*

Halyomorpha picus The adults are eaten raw as a chutney in Arunachal Pradesh State, **India** (Chakravorty et al. 2011).

Mormidea (Mormidea) notulata The adults are relished in **Mexico** (Acuña et al. 2011).

Mormidea **sp.** The larvae and adults are consumed in **Mexico** (Ramos-Elorduy 2003).

Moromorpha tetra The larvae and adults are eaten in **Mexico** (Ramos-Elorduy 2003).

Natalicola circuliventris → *Euchosternum delegorguei*

Natalicola delegorguei → *Euchosternum delegorguei*

Natalicola pallidus → *Euchosternum pallidus*

Nezara antennata **China:** This species is used for treating stomach aches, disorders of the kidney or spleen (Nanba 1980).

Nezara (=*Acrostemum*) *majuscule* (=*Chinavia montivaga*) The larvae and adults are used as a foodstuff in **Mexico** (Ramos-Elorduy 2003).

Nezara robusta The adults are relished in **Malawi** (Shaxon et al. 1974).

Nezara viridula (southern green stink bugs) **Cameroon:** The adults are eaten raw (De Colombel 2003). **China:** This species is used as a suppressor of swelling (Zimian et al. 1997). **India:** The adults are consumed by the Nishi people of Arunachal Pradesh State. People eat the bug raw or as a chutney (Chakravorty et al. 2011). **Indonesia:** In West Papua State, children in Pass Valley eat this species raw or roasted (Ramandey and van Mastrigt 2010). In Kalimantan State, this species is often eaten raw (Chung 2010). **Japan:** The Japanese name is *minamiao-kamemushi*. The adults were used for treating stomach disorders (Shiraki 1958). **Malaysia:** In Sabah State, this species is called *tangkayomot*, and is relished by the villagers (Chung 2010). **Mexico:** The larvae and adults are consumed (Ramos-Elorduy 2003).

Ochrophora montana **India:** The larvae and adults are eaten raw or fried. This bug contains fat about 20% of its wet weight, and people extract oil for cooking from it (Sachan et al. 1987).

Oebalus mexicana The larvae and adults are used as a food item in **Mexico** (Ramos-Elorduy 2003).

Oebalus pugnax The larvae and adults are consumed in **Mexico** (Ramos-Elorduy 2003).

Padaeus trivittatus The larvae and adults are food source in **Mexico** (Ramos-Elorduy 2003).

Padaeus viduus The larvae and adults are an article of food in **Mexico** (Ramos-Elorduy 2003).

Pellaea stictica The larvae and adults are good for eating in **Mexico** (Ramos-Elorduy 2003).

Pharypia fasciata The larvae and adults are eaten in **Mexico** (Ramos-Elorduy and Pino 1989).

Proxys puntalatus The larvae and adults are a food item in **Mexico** (Ramos-Elorduy 2003).

Proxys **sp.** The larvae and adults are treated as foodstuff in **Mexico** (Ramos-Elorduy 2003). A species belonging to this genus is eaten in **Thailand** (Hanboonsong et al. 2000).

Ptilarmus fasciata → *Pharypia fasciata*

Solubea mexicana → *Oebalus mexicana*

Tessaratoma javanica The larvae and adults are fried in **Thailand** (Chen and Wongsiri 1995).

Tessaratoma papillosa (litchi stink bag) **China:** The larvae and adults are salted and heated (Luo 1990). In the Guangxi, Guangdong and Fujian regions, people eat the bugs by wrapping them in cabbage leaves, then instantly cooking them on hot ashes, after removing the heads, wings, legs and viscera (Zhi-Yi 2005). This species is used as a suppressor of swelling (Zimian et al. 1997). **Laos:** This species is sold in markets around Vientiane, and roasted (Mitsuhashi unpubl. observation, 2012) **(PL IV-6)**. **Thailand:** The larvae and adults are fried (Hanboonsong et al. 2000).

Tessaratoma quadrata **Laos:** The Laotian name is *meang kieng*. This species is eaten (Boulidam 2008). **India:** The adults are consumed by the Galo people in Arunachal Pradesh State. People make a paste (chutney) from the adults after removing their wings (Chakravorty et al. 2011).

Pyrrhocoridae (Red bugs) (1)

Antilochus coqueberti In **India**, the adults are eaten by the Nishi people of Arunachal Pradesh State, fried or boiled with vegetables (Chakravorty et al. 2011).

Rhopalidae (Scentless plant bugs) (1)

Leptocoris acuta This species is a food item in **Indonesia** (Java) (van der Burg 1904).

Scutelleridae (Shield-backed bugs) (0)

Sphaerocoris **sp.** People eat a species belonging to this genus in **Malawi** (DeFoliart 2002).

12. COLEOPTERA (Beetles) [578]

Anobiidae (Death watch beetles, Drugstore beetles) (2)

Anthrenus **sp.** The adults are eaten by the Nishi people in Arunachal Pradesh State in **India** (Singh et al. 2007).

Lasioderma serricorne (cigarette beetle, tobacco beetle) The larvae are considered as food in **China** (Hu and Zha 2009).

Stegobium paniceum (biscuit beetle) **Greece:** This beetle was found in a type of pea excavated from the Akrotiri Ruin, and was thought to be eaten by ancient people (Panagiotakopulu and Buchland 1991).

Xestobium **sp.** The larvae and adults of a species belonging to this genus are roasted or boiled by the Nishi people in Arunachal Pradesh State, **India** (Singh et al. 2007).

Bostrichidae (False powderpost beetles) **(1)**

Rhizopertha dominica (lesser grain borer) **Greece:** This beetle was found in a type of pea excavated from the Akrotiri Ruin, and was thought to be eaten by ancient people (Panagiotakopulu and Buchland 1991).

Bruchidae (Pulse beetles, Bean weevils) **(5)**

Algarobius **spp.** The Pima and other tribes of Indians in **North America,** eat the larvae and pupae of several species belonging to this genus. The larvae infested most of the mesquite seeds, and people found the insects as an agreeable ingredient of the flour made from the beans (Sutton 1988).

Bruchus pisorum (pea weevil, pea beetle) The larvae and adults are an article of food in **China** (Hu and Zha 2009).

Bruchus rufimanus (broad bean weevil, broad bean beetle) The larvae and adults are considered as foodstuff in **China** (Hu and Zha 2009).

Bruchus **spp.** → *Algarobius* **spp.**

Caryoborus serripes In **Amazonia,** people of the Yukpa, Yanomamo, Matsigenka, Araweté and Jotï are known to harvest the larvae from palm seeds for consumption (Choo 2008).

Caryobruchus **sp. Brazil:** The larvae of a species belonging to this genus are eaten raw or fried by the Surui people (Costa Neto and Ramos-Elorduy 2006). **Colombia:** The succulent, greasy larvae are particularly relished by the Yukpa people, and are the most frequently consumed coleopterans. People eat the larvae raw or roasted (Ruddle 1973). **Venezuela:** The larvae of a species called *eteme* is eaten raw or roasted by the Yukpa people (Ruddle 1973).

Mimosestes **spp.** The Pima and other tribes of Indians in **North America,** eat the larvae and pupae of several species belonging to this genus. The larvae infest most of mesquite seeds, and people consider the insects as a suitable ingredient of the flour made from the beans (Sutton 1988).

Nelturmius **spp.** → *Algarobius* **spp.**

Pachymerus cardo (palm kernel borer) **Amazonia:** People of the Yukpa, Yanomamo, Matsigenka, Araweté and Jotï are known to harvest the larvae from palm seeds for consumption (Choo 2008). **Brazil:** The larvae are eaten raw or fried by the Surui people (Coimbra 1984).

Pachymerus nucleorum In **Brazil,** the larvae are considered edible by the Timbira, Araweté, Surui people (Costa Neto and Ramos-Elorduy 2006).

Pachymerus **sp.** In **Brazil,** the larvae are consumed (Costa Neto and Ramos-Elorduy 2006).

Buprestidae (Metallic wood-borers) **(21)**

Buprestis **sp. China:** People of the Thai tribe living in the southern area eat a species belonging to this genus by roasting or frying (Shu 1989). **Thailand:** The adults of a species belonging to this genus, which is known as *ma-langtap* or *ma-lang khap*, are roasted or fried. Before eating the head, legs, and wings are removed (Vara-asavapati et al. 1975, Leksawasdi and Jirada 1983).

Catoxamtha **sp.** A species belonging to this genus is eaten in **Thailand** (Leksawasdi 2008).

Chalcophora japonica The Japanese name is *uba-tamamushi*. The roasted larvae were used for treating convulsive fits in **Japan** (Miyake 1919).

Chalcophora yunnana In **China**, the larvae are fried with a touch of spice (Chen and Feng 1999).

Chalcophora **sp.** The larvae of a species belonging to this genus are a food source in **Mexico** (Ramos-Elorduy and Pino 1990).

Chrysobothris basalis **Mexico:** The larvae and adults are roasted and then pulverized. Drinking this is effective against diseases of the urinary organs and reproductive organs (Ramos-Elorduy de conconi and Pino 1988).

Chrysobothris fatalis The larvae are eaten in **Angola** (Wellman 1908).

Chrysobothris femorata (flatheaded appletree borer) The larvae and adults are consumed in **Thailand** (DeFoliart 2002).

Chrysobothris **sp. Sri Lanka:** A species belonging to this genus is a food source (Nandasena et al. 2010).

Chrysochroa **spp.** Some species, whose vernacular name is *lingaung*, belonging to this genus are considered edible in Sabah State, **Malaysia** (Chung et al. 2002).

Coraebus sidae The larvae are a foodstuff fried in **China** (Chen and Feng 1999).

Coraebus sauteri The larvae are fried in **China** (Chen and Feng 1999).

Euchroma gigantea (ceiba borer beetle) **Brazil:** The larvae are a food item by the Kayapo people (Posey 1987). **Colombia:** The larvae and adults are eaten by the Tatuyo people (Dufour 1987). **Mexico:** The larvae are edible (Ramos-Elorduy and Pino 2002).

Psiloptera wellmani The larvae are eaten in **Angola** (Wellman 1908).

Sphenoptera kozlovi The larvae are fried in **China** (Chen and Feng 1999).

Steraspis amplipennis The larvae are eaten in **Angola** (Wellman 1908).

Steraspis speciosa The larvae are consumed in **Cameroon** (De Colombel 2003).

Steraspis **sp.** A species belonging to this genus is considered edible in **Cameroon** (De Colombel 2003).

Sternocera acquisignata → *Sternocera aequisignata*

Sternocera aequisignata **China:** People of the Thai tribe eat this species by roasting or frying (Shu 1989). **Thailand:** The Laotian name is *mang khup*. People of the Laotian tribe roast the adults, remove the head, wings, and legs, and then eat it (Bristowe 1932, Hanboonsong et al. 2000).

Sternocera castairea **subsp.** *irregularis* This subspecies is eaten in **Cameroon** (De Colombel 2003).

Sternocera equisignata → *Sternocera aequisignata*

Sternocera feldspathica The larvae are a food item in **Angola** (Wellman 1908).

Sternocera funebris This species is a food source in **Zimbabwe** (Chavanduka 1975).

Sternocerca interrupta **subsp.** *immaculata* This subspecies is a food item in **Cameroon** (De Colombel 2003).

Sternocera orissa This species is eaten in **Botswana** (Nonaka 1996), **RSA** (Quin 1959), and **Zimbabwe** (Chavanduka 1975).

Sternocera ruficornis People eat this species in **Thailand** (Hanboonsong et al. 2000).

Sternocera sternicornis This species is used as food in Arunachal Pradesh State, **India** (Singh et al. 2007).

Sternocera **sp.** In **India**, the adults of a species, whose Nishi name is *jorjo punyo,* belonging to this genus are consumed by the Nishi people boiled or smoked (Chakravorty et al. 2011).

Thrincopyge alacris **Mexico:** The larvae and adults are roasted and then pulverized. Drinking this is effective against diseases of the urinary organs and reproductive organs (Ramos-Elorduy de Conconi and Pino 1988).

Carabidae (Ground beetles) **(1)**

Anthia alveolata **RSA:** The Thonga people eat this species (Junod 1962).

Euryscaphus **sp.** A species belonging to this genus is eaten by the Walbiri people in **Australia** (Meyer-Rochow and Changkija 1997).

Scarites **sp.** The larvae of a species belonging to this genus are fried in **Angola** (Bergier 1941) and **Madagascar** (Decary 1937, Paulian 1943).

Cerambycidae (Long-horned beetles) **(112)**

Acanthopus serraticornis This species is a food article in Arunachal Pradesh State, **India** (Singh et al. 2007).

Acanthophorus capensis The larvae are treated as a foodstuff cooked or fried in **Zambia** (Mbata 1995).

Acanthophorus confinis The larvae are edible cooked or fried in **Zambia** (Mbata 1995).

Acanthophorus maculatus The larvae are cooked or fried in **Zambia** (Mbata 1995).

Acrocinus longimanus (harlequin beetle) **Colombia:** Native people of the Tatuyo tribe eat the larvae and adults (Dufour 1987). **Mexico:** The larvae, pupae, and the adults are consumed (Ramos-Elorduy and Pino 2002).

Aeolesthes sp. A species belonging to this genus is a food source in **Thailand** (Hanboonsong et al. 2000).

Agrianome spinicollis This species is used as a food item in **Australia** (Reim 1962).

Analeptes trifasciata This species is considered edible in south western **Nigeria** (Womeni et al. 2009).

Ancylonotus tribulus The larvae are used as a food in **Gabon, Senegal** (Netolitzky 1918/1920) and western **Africa** (Bergier 1941).

Anoplophora chinensis The larvae and the adults are eaten in **China** (Hu and Zha 2009).

Anoplophora glabripennis → Anoplophora nobilis

Anoplophora malasiaca (white-spotted longicorn beetle) **Japan:** The Japanese name is *gomadara-kamikiri*. In the Nagano and Shimane Prefectures, the larvae are skewered and roasted. People eat them by adding soy sauce or miso paste on them. People also eat the larvae raw, just roasted or fried (Miyake 1919, Takagi 1929e). The larvae raw or roasted were medicinally used for treating various diseases such as heart ailments, convulsive fits, fever, malaria tuberculosis, etc. (Miyake 1919, Umemura 1943, Shiraki 1958). **China:** The larvae are eaten (Bodenheimer 1951).

Anoplophora nobilis The larvae are eaten raw, roasted or fried in **China** (Chen and Feng 1999).

Anoplophora sp. The larvae and adults of a species belonging to this genus are roasted by the Nishi people in Arunachal Pradesh State, **India** (Singh et al. 2007).

Aplagiognathus costata In **New Caledonia**, the Kanak people relish the larvae and pupae (Simmonds 1885).

Aplagiognathus spinosus The Mexican name is *gusano de elite podrido*. In **Mexico,** native people belonging to the Chichimeca, Chinanteco, Mazahua, Mestizo, Mixe, Mixteco, Náhua, Otomi, Popolaca, Tarasco, Tlapaneca, Tzetzale, Tzottzile, Zapoteco, and Zoque tribes eat the larvae, pupae and the adults roasted (Ramos-Elorduy de Conconi et al. 1984, Ramos-Elorduy and Pino 1989).

Aplagiognathus **sp.** The Mexican name is *gusano del sauce*. The larvae are eaten in **Mexico** (Ramos-Elorduy and Pino 1990).

Aplosonyx chalybseus This species is consumed as a food item in Arunachal Pradesh State, **India** (Singh et al. 2007).

Appectrogastra flavipilis This species is used as a food item in **Australia** (Reim 1962).

Apriona germari japonica (mulberry longhorn beetle, mulberry borer) **China:** The larvae and adults are a foodstuff (Hu and Zha 2009). The larvae are eaten raw, roasted or fried (Chen and Feng 1999). This species is used to promote blood circulation (Zimian et al. 1997). This species was used for treating heart diseases and brain fever (Li 1596). **Japan:** The Japanese name is *kuwa-kamikiri*. People eat them by adding soy sauce or miso paste on them in the Gunma Prefecture (Miyake 1919). The skewered and roasted larvae or charred larvae were used for treating various diseases including colds, convulsive fits, etc. (Miyake 1919). The larvae were also used for treating meningitis, fever, gonorrhea, etc. (Shiraki 1958). **North Korea:** The larvae are fried with sesame oil for treating meningitis, fever, gonorrhea in the Hwanghae, North P'yongan, South P'yongan and North Hamgyong Provinces (Okamoto and Muramatsu 1922). **South Korea:** The larvae are consumed fried with sesame oil to cure meningitis, fever, gonorrhea in the North Chungcheong, South Chungcheong, South Joella, North Gyeongsang Provinces (Okamoto and Muramatsu 1922). **Thailand:** People eat this species roasted (Hanboonsong et al. 2000). **Vietnam:** The larvae are used as a food item roasted (Tiêu 1928).

Apriona rugicolis → *Apriona japonica*

Arhopalus **sp.** The larvae are an article of food in **Mexico** (Ramos-Elorduy and Pino 1990).

Aristobia approximator People eat this species in **Thailand** (Hanboonsong et al. 2010).

Aristobia **sp.** In **India**, the Nishi people eat the adults of a species belonging to this genus by roasting, smoking, or boiling, after removing the wings (Chakravorty et al. 2011).

Aromia bungii The larvae are eaten raw, roasted or fried in **China** (Chen and Feng 1999).

Aromia moschota ambrosiaca The Japanese name is *kubiaka-kamikiri*. **Japan:** The larvae raw, roasted or pulverized were used medicinally for treating convulsive fits in the Ehime Prefecture (Miyake 1919).

Arophalus afin *rusticus* The Mexican name is *gusano del palo.* **Mexico:** Native people belonging to the Chichimeca, Chinanteco, Mazahua, Mestizo, Mixe, Mixteco, Náhua, Otomi, Popolaca, Tarasco, Tlapaneca, Tzetzale, Tzottzile, Zapoteco, and Zoque tribes eat the larvae and pupae (Ramos-Elorduy and Pino 1989).

Arophalus rusticus montanus The larvae and pupae are eaten in **Mexico** (Ramos-Elorduy and Pino 2004).

Baladeva walkeri → *Dorysthenes walkeri*

Bardistus cibarius In **Australia,** aborigines living in the western part of the country eat the larvae (Waterhouse 1991).

Batocera albofasciata **Indonesia:** This species is used as a foodstuff (Netolizky 1918/1920). **Philippines:** The larvae are considered edible (Gibbs et al. 1912). **Sri Lanka:** This species is treated as foodstuff (Netolizky 1918/1920).

Batocera horsfieldi **China:** This species is used to promote blood circulation (Zimian et al. 1997).

Batocera lineolata (white-striped longicorn beetle) **(PL V-1)** The Japanese name is *shirosuji-kamikiri.* In **Japan,** the larvae are skewered and roasted. People eat them by adding soy sauce or miso paste on them in the Tochigi, Gunma, Shimane and Yamaguchi Prefectures. The larvae raw, roasted or dried were used for treating various diseases such as diphtheria, stomach disorders, tuberculosis, etc. (Miyake 1919).

Batocera numitor The larvae are foodstuff in the **Philippines** (Gibbs et al. 1912).

Batocera parryi This species is an article of food in **India** (Singh et al. 2007).

Batocera roylei In **India,** the larvae and adults are used as food by the Nishi people who smoked, roasted or boiled the beetles, and removed the wings before eating (Chakravorty et al. 2011).

Batocera rubra This species is eaten by the Ao-Naga people in Nagaland State, **India** (Meyer-Rochow 2005).

Batocera rubus **Indonesia** and **Sri Lanka:** This species is used as a food item (Netolizky 1918 and 1920). **Laos** and **Thailand:** This species is eaten (Yhoung-Aree and Viwatpanich 2005).

Batocera rufomaculata **India:** The larvae are eaten fried or as a chutney by the Karbi and Rengma Naga people in Assam State (Ronghang and Ahmed

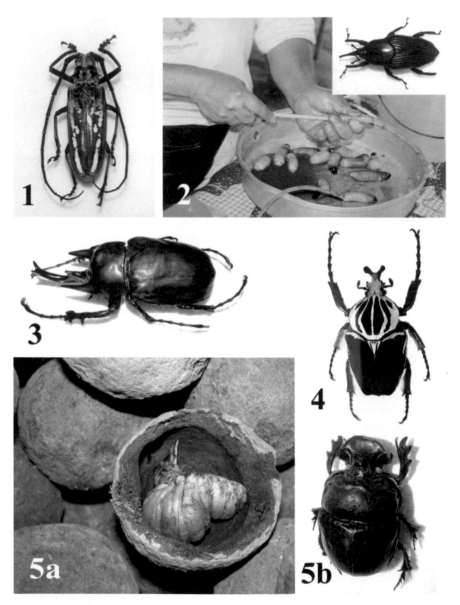

Plate V. 1. *Batocera lineolate* adult. Body length: 52 mm (cf. p.83); 2. Skewering the larvae of *Rhynchophorus palmarum* for barbeque at Iquitos, Peru. Insert: An adult. Body length: 38 mm (cf. p.97); 3. *Megasoma actaeon* adult. Body length: ♂ 145 mm, ♀ 70 mm (cf. p.100); 4. *Goliathus goliathus* adult. Body length: 120 mm. (cf. p.120); 5. *Heliocopris dominus*. a: dung balls in which larvae grow. b: an adult. Body length: 60 mm (cf. p.121).

2010). **Madagascar:** This species is fried with butter alone or together with garlic and parsley (Dufour 1987).

Batocera wallecei **Indonesia** (West Papua State): Children eat the abdomens of the adults by roasting them on fire embers, and then removing legs and elytra (Ramandey and van Mastrigt 2010).

Batocera **spp. Indonesia** (West Papua State): Several species belonging to this genus are also consumed (Tommaseo Ponzetta and Paoletti 1997). **Malaysia:** A species belonging to this genus is used as a food item in Sabah State. The adults are usually roasted after removing the hard and spiny parts (Chung et al. 2002) **PNG:** The Chuave people eat the larvae of some species belonging to this genus cooked (Mercer 1993, Meyer-Rochow 1973).

Callipogon barbatus The Mexican name is *gusano de los palos*. In **Mexico** native people of the Chontale, Huasteco, Huave, Lacandone, Maya, Mestizo, Náhua, Tarasco, Tlapaneca, Totonaca, and Zapoteco tribes eat the larvae, pupae and the adults (Ramos-Elorduy et al. 1997, Ramos-Elorduy and Pino 1989).

Cerambyx cerdo The larvae are used as food in **Morocco** (Benhalima et al. 2003).

Cerambyx heros The larvae are consumed in **France** (Fabre 1924).

Cerambyx **sp.** The larvae, pupae and the adults are used as foodstuff in the Michoacan and Guerrero States, **Mexico** (Ramos-Elorduy de Conconi et al. 1984).

Ceroplesis burgeoni This species is an article of food in southern **Africa** (Malaisse 1997).

Chloridolum thaliodes The Japanese name is *ōao-kamikiri*. In **Japan**, the larvae raw, roasted or pulverized were used for treating convulsive fits in the Ehime Prefecture (Miyake 1919).

Coelosterna scabrator This species is a food source by the Ao-Naga people in Nagaland State, **India** (Meyer-Rochow 2005).

Coelosterna **sp.** A species belonging to this genus is eaten by the Ao-Naga people in Nagaland State, **India** (Meyer-Rochow 2005).

Cnemoplites edulis This species is a food item in **Australia** (DeFoliart 2002).

Cnemoplites flavipilis This species is a food item in **Australia** (DeFoliart 2002).

Cyrtognathus forficatus The larvae are fried in **Morocco** (Benhalima et al. 2003).

Derobrachus procerus The larvae are used as food in **Mexico** (Ramos-Elorduy et al. 1997).

Derobrachus sp. In **Mexico,** the larvae, pupae, and the adults of a species, whose Mexican name is *gusano del palo*, belonging to this genus are considered as food item (Ramos-Elorduy and Pino 2002).

Diastocera wallichi This species is consumed by people in Arunachal Pradesh State, **India** (Singh et al. 2007).

Dihamnus cervinus The larvae are eaten fried or as a chutney by the Karbi and Rengma Naga people in Assam State, **India** (Ronghang and Ahmed 2010).

Dihamnus **spp.** **PNG:** The Chuave people eat the larvae of some species belonging to this genus (Meyer-Rochow 1973). **Indonesia** (West Papua State): Several species belonging to this genus are considered edible (Tommaseo Ponzetta and Paoletti 1997).

Dorysthenes buqueti **India:** This species is a food item in Arunachal Pradesh State (Singh et al. 2007). **Thailand:** People eat the adults (Hanboonsong et al. 2000).

Dorysthenus forficatus The larvae are fried in **Morocco** (Ghesquiere 1947).

Dorysthenes granulosus People eat the adults in **Thailand** (Hanboonsong et al. 2000).

Dorysthenes montanus This species is a food source in Arunachal Pradesh State in **India** (Singh et al. 2007).

Dorysthenes walkeri People eat the adults in **Thailand** (Utsunomiya and Masumoto 1999).

Dorysthenes sp. The adults of a species belonging to this genus, is consumed in **Thailand** (Hanboonsong et al. 2000).

Eburia stigmatica The larvae and pupae are used as food in **Mexico** (Ramos-Elorduy and Pino 2002).

Ergates spiculatus (pine sawyer beetle) The larvae are eaten in the **USA** (Essig 1934 and 1965).

Eupogonius tenuicornis The Japanese name is *enokino-kamikiri*. In **Japan**, the grilled larvae were said to have a vermicidal effect in the Akita Prefecture (Miyake 1919).

Eurynassa australis In **Australia** aborigines living in the Queensland State eat the larvae raw or roasted on hot ashes. The people also roast the adults after removing the elytra, and then eat the abdomen (Cambell 1926, Waterhouse 1991).

Eurynassa odewahni In **Australia,** aborigines, especially women, living in the northern part of the country are very fond of eating the larvae of this species raw (Lumholtz 1890).

Glenea (Stiroglenis) obese This species is consumed by people of Arunachal Pradesh State, **India** (Singh et al. 2007).

Haplocerambyx severus The larvae are used as foodstuff in **PNG** (Mercer 1993).

Hevorodon maxillosum → *Stenodontes* cer. *maxillosus*

Hoplocerambyx spinicornis **India:** The adults are used as a food item in Arunachal Pradesh State (Singh et al. 2007). **Malaysia** (Sarawak): The larvae are eaten raw or roasted (Mercer 1993). **Thailand:** This species is consumed (Leksawasdi 2008).

Lagocheirus rogersii In **Mexico,** native people belonging to the Chichimeca, Chinanteco, Mazahua, Mestizo, Mixe, Mixteco, Náhua, Otomi, Popolaca, Tarasco, Tlapaneca, Tzetzale, Tzottzile, Zapoteco, and Zoque tribes eat the larvae, pupae, and the adults (Ramos-Elorduy and Pino 1989, Ramos-Elorduy 2006).

Lamia 8-maculata **India:** People ate this species (Hope 1842).

Lamia rubus **India:** People ate this species (Hope 1842).

Lapnosternus buqueti → *Dorysthenes buqueti*

Macrodontia cerrocornis → *Macrodontia cervicornis*

Macrodontia cervicornis **Brazil:** The roasted larvae are consumed (Netolitzky 1918/1920). **Colombia:** The roasted larvae are eaten (Netolitzky 1918/1920). **Ecuador:** The larvae are considered edible (Onore 2005). **Guyana:** The roasted larvae are eaten (Brygoo 1946). **Paraguay:** The larvae are eaten raw or roasted by the Ache people (Hawkes et al. 1982, Hurtado et al. 1985). **West Indies:** The larvae are also used as foodstuff (Donovan 1842).

Macrotoma crenata This species is consumed by people of Arunachal Pradesh State, **India** (Singh et al. 2007).

Macrotoma edulis The larvae constitute part of a meal in **Sao Tome and Principe** (Netolitzky 1918/1920).

Macrotoma fisheri This species is considered good for eating in **Thailand** (Hanboonsong et al. 2000).

Macrotoma natala The larvae are eaten in **Botswana** (Roodt 1993).

Macrotoma pascoei → *Macrotoma fisheri*

Macrotoma **sp. Indonesia** (West Papua State): A species belonging to this genus is considered as a food item (Ramandey and van Mastrigt 2010). **Malaysia:** A species belonging to this genus is used as a food item in Sabah State. The adults are usually roasted after removing the hard and spiny parts (Chung et al. 2002).

Malambyx japonicus (=*M. raddei*) The Japanese name is *yama-kamikiri.* **Japan:** People skewer the larvae and roast it, and then add soy sauce or miso paste to them in the Gunma and Nagano Prefectures (Miyake 1919, Katagiri and Awatsuhara 1996). The larvae raw, roasted, cooked or charred were used medicinally for treating convulsive fits (Miyake 1919).

Mallodon costata → Aplagiognathuscostata

Mallodon downesii → Stenodontes downesi

Mallodon molarius → Stenodontes maxillosus

Mallodon spinosus → Aplagiognathusspinosus

Megopis mutica This species is eaten fried with butter alone or together with garlic and parsley in **Madagascar** (Dufour 1987).

Melanauster chinensis → Anoplophora malasiaca

Mnemopulis edulis The larvae are considered edible in **Australia** (Reim 1962).

Monochamus maculosus (spotted pine sawyer) The larvae are used as a food item in the **USA** (Essig 1934 and 1965).

Monochamus scutellatus The larvae are an article of food in the **USA** (Essig 1934 and 1965).

Monochamus versteegi The adult beetles are consumed by the Nishi people in **India.** They smoke, roast or boil the beetle, and eat them after removing the wings (Pathak and Rao 2000, Chakravorty et al. 2011).

Neocerambyx paris **India:** This species is considered as a foodstuff by the Ao-Naga people in Nagaland State (Meyer-Rochow 2005). **Sri Lanka:** This species is eaten (Nandacena et al. 2010).

Neoclytus conjunctus (western ash borer) The larvae and adults are used as a food item in the **USA** (Essig 1934 and 1965, Roust 1967).

Neoplocaederus → Plocaederus

Nothopleurus cer. *maxillosus → Stenodontes* cer. *Maxillosus*

Nothopleurus maxillosus → Stenodontes cer. *Maxillosus*

Nupserha fricator This species is eaten by people of Arunachal Pradesh State in **India** (Singh et al. 2007).

Oberea japonica (apple longicorn beetle) **Japan:** The Japanese name is *ringo-kamikiri.* The larvae are skewered and roasted, and then eaten by adding soy sauce in the Nagano Prefecture (Takagi 1929).

Omachanta gigas → Petrognatha gigas

Oncideres **sp.** The larvae and adults of a species belonging to this genus are considered as food by the Awa and Saraguro people in **Ecuador** (Onore 1997).

Oplatocera **sp.** Adults of a species belonging to this genus are relished by the Nishi people of **India.** They eat the beetle by smoking, roasting or boiling. They remove the heads and other appendages (Chakravorty et al. 2011).

Osphryon **sp.** A species belonging to this genus is considered edible in **Indonesia** (West Papua State) (Ramandey and van Mastrigt 2010).

Pachyterla dimidiata This species is used as foodstuff in **Thailand** (Leksawasdi 2008).

Paroplites australis **Australia:** Aborigines eat the larvae roasted (Lumholtz 1890).

Petrognatha gigas The larvae are also used as food in **Gabon** (Bergier 1941) and in **Senegal** (Netolitzky 1918/1920).

Pexteuso atys The larvae are fried in **Ecuador** (Onore 1997).

Plocaederus frenatus The larvae are fried in **CAR** and in **RSA** (Bergier 1941).

Plocaederus obesus People eat this species in **Thailand** (Hanboonsong et al. 2000).

Plocaederus rufinornis People eat this species in **Thailand** (Hanboonsong et al. 2000).

Plocaederus **sp.** A species belonging to this genus is eaten in **Thailand** (Jamjanya et al. 2001).

Polyrhaphis **sp.** The Mexican name is *gusano del palo*. The larvae are relished in **Mexico** (Ramos-Elorduy et al. 1998).

Prionoplus reticularis The Maori name is *huhu beetle*. The larvae and pupae are used as food in **New Zealand.** The larvae are found in a variety of rotting timbers (Meyer-Rochow and Changkija 1997).

Prionus californicus The larvae are considered good for eating in the **USA** (Essig 1934 and 1965). This species seems to have been eaten by the aboriginal people in Nevada State, as it was found in the coprolites excavated from the Lovelock Cave (BC 1,200) in Nevada State (Roust 1967).

Prionus coriarius **Roma Antiqua:** The larvae of this long horn beetle were relished by the Romans (Hope 1842).

Prionus damicornis → *Stenodontes damicornis*

Prionus insularis (serrate longicorn beetle) The Japanese name is *nokogiri-kamikiri*. In **Japan,** the roasted larvae were used to cure convulsive fits (Miyake 1919).

Prosopocera (*Prosopocera*) **sp.** The larvae, pupae and the adults of a species belonging to this genus are used as food in **Mexico** (Ramos-Elorduy and Pino 2002).

Psacothea hilaris The larvae and adults are eaten in **China** (Hu and Zha 2009).

Psalidognathus atys The larvae are consumed in **Ecuador** (Onore 2005).

Psalidognathus cacicus The larvae are fried in **Ecuador** (Onore 1997 and 2005).

Psalidognathus erithrocerus The larvae are eaten, or used to make soup in **Ecuador** (Onore 2005).

Psalidognathus modestus The larvae are used to make soup in **Ecuador** (Onore 1997 and 2005).

Pseudonemophas versteegii → *Monochamus versteegi*

Purpricenus temminekii **China:** This species was used for treating navel diseases (Li 1596).

Pycnopsis brachyptera This species is treated as foodstuff in **DRC** (Malaisse and Parent 1997a).

Rhagium lineatum The larvae are consumed in the **USA** (Essig 1934 and 1965).

Rhaphipodus **sp.** A species called *ngatit* in the local language is consumed in Sabah State, **Malaysia**. The adults are usually roasted after removing the hard and spiny parts (Chung et al. 2002).

Rosenbergia mandibularis In **Indonesia** (West Papua State), the Maribu people in Sentani, Jayapura, eat this species after roasting (Ramandey and van Mastrigt 2010).

Stenygrinum 4-notatum The Japanese name is *yotsuboshi-kamikiri*. In **Japan,** the roasted larvae were used medicinally for treating convulsive fits in the Ehime Prefecture (Miyake 1919).

Stenodontes damicornis **Brazil:** The larvae are eaten (Ealand 1915). **Guyana:** The larvae are roasted (Brygoo 1946). **Peru:** Native people of Yameo eat the larvae (Ealand 1915, Steward and Métraux 1963). **Suriname:** The larvae were considered an exquisite dish (Hope 1842). **West Indies:** The larvae are roasted (Donovan 1842).

Stenodontes (*=Mallodon*) *downesi* **Mozambique:** The larvae are fried (Berensberg 1907). **RSA:** The Kaffir people relish this species (Berensberg 1907). This species is also consumed in **CAR** (Bergier 1941).

Stenodontes **cer.** *maxillosus* In **Mexico,** native people of the Chontale, Huave, Huasteco, Lacandone, Maya, Mestizo, Náhua, Tarasco, Tlapaneca,

Totonaca, and Zapoteco tribes eat the larvae and pupae (Ramos-Elorduy de Conconi et al. 1984, Ramos-Elorduy and Pino 1989).

Stenodontes cer. *molaria* The Mexican name is *gusano de los palos.* The larvae are used as food in **Mexico** (Ramos-Elorduy et al. 1998).

Sternotomis itzingeri katangensis This species is considered edible in **DRC** (Malaisse and Parent 1997a).

Stromatium barbatum This species is a food source by people in Arunachal Pradesh State in **India** (Singh et al. 2007).

Stromatium longicorne The larvae are eaten raw, fried or roasted in **China** (Chen and Feng 1999).

Threnetica lacrymans This species is used as food in **Thailand** (Yhoung-Aree and Viwatpanich 2005, Leksawasdi 2008).

Thyestilla (=*Thyestes*) *gebleri* (hemp longicorn beetle) **Japan:** The Japanese name is *asa-kamikiri.* The roasted larvae were used for treating convulsive fits (Miyake 1919). **South Korea:** The infusion of the larvae is used for treating meningitis of children in North Gyeongsang Province (Okamoto and Muramatsu 1922).

Trichoderes pini The Mexican name is *gusano del pino.* In **Mexico,** native people belonging to the Chichimeca, Chinanteco, Mazahua, Mestizo, Mixe, Mixteco, Náhua, Otomi, Popolaca, Tarasco, Tlapaneca, Tzetzale, Tzottzile, Zapoteco, and Zoque tribes eat the larvae and pupae (Ramos-Elorduy de Conconi et al. 1984, Ramos-Elorduy and Pino 1989).

Xixuthrus sp. A species belonging to this genus is eaten in **Indonesia** (West Papua State) (Ramandey and van Mastrigt 2010).

Xylorhiza sp. The boiled or fried larvae of a species belonging to this genus are used as food by people of the Nishi and Galo tribes in **India** (Chakravorty et al. 2011).

Xylotrechus chinensis (tiger longicorn beetle) The Japanese name is *torahu-kamikiri.* In **Japan,** the larvae are skewered and roasted. People eat them by adding soy sauce or miso paste on them in the Nagano Prefecture (Takagi 1929e).

Xylotrechus nauticus The larvae are considered edible in the **USA** (Essig 1934 and 1965).

Xylotrechus quadripes This species is eaten by people in Arunachal Pradesh State in **India** (Singh et al. 2007).

Xylotrechus smei This species is considered good for eating by people in Arunachal Pradesh in **India** (Singh et al. 2007).

Xysterocera globosa **India:** This species is consumed by the Ao-Naga people in Nagaland State (Meyer-Rochow 2005). **Sri Lanka:** This species is treated as a food item (Nandacena et al. 2010). **Thailand:** This species is used as a food source (Leksawasdi 2008).

Xystrocera **sp.** A species belonging to this genus is consumed by the Ao-Naga people in Nagaland State, **India** (Meyer-Rochow 2005).

Zographus aulicus This species is eaten in **DRC** (Malaisse 1997a).

Chrysomelidae (Leaf beetles) (7)

Aplosonyx chalybaeus This species is relished by people of Arunachal Pradesh State in **India** (Singh et al. 2007).

Aplosonyx albicornis This species is considered edible in Sabah State, **Malaysia.** The adults are rather soft, and are either boiled, simmered until dry or stir-fried without oil (Chung et al. 2002).

Bruchus rufipes **Greece:** This beetle was found in a sort of pea excavated from the Akrotiri Ruin, and was thought to be eaten by ancient people (Panagiotakopulu and Buchland 1991).

Leptinotarsa decemlineata (Colorado potato beetle) The larvae are used as food in Oaxaca State, **Mexico** (Ramos-Elorduy de Conconi et al. 1984).

Sagra femorata This species is a food source in **Laos, Myanmar, Thailand** and **Vietnam** (Yhoung-Aree and Viwatpanich 2005).

Sagra femorata purpurea The larvae are lightly or deeply fried in **China** (Chen and Feng 1999).

Sagra **sp.** A species belonging to this genus is used as food in **Thailand** (Leksawasdi 2008).

Speciomerus giganteus In **Amazonia** people of the Yukpa, Yanomamo, Matsigenka, Araweté and Jotï are known to harvest the larvae from palm seeds for consumption (Choo 2008).

Coccinellidae (Lady beetles) (1)

Harmonia axyridis **Japan:** The Japanese name is *nami-tentō*. The adults were used for treating syphilis (Shiraki 1958). **South Korea:** The dried powder of the adults is used for treating syphilis in the Kangwon Province (Okamoto and Muramatsu 1922).

Ptychanatis axyridis → *Harmonia axyridis*

Cicindelidae (Tiger beetles) (4)

Cicindela chinensis **North Korea:** The powder of the roasted adults is eaten with boiled rice as a dietary for treating scrofula, peritonitis, gonorrhea, etc. in the Hwanghae, North P'yongan, South P'yongan, North Hamgyong Provinces (Okamoto and Muramatsu 1922). **South Korea:** The powder of the roasted adults is eaten with boiled rice as a dietary for treating scrofula, peritonitis, gonorrhea, etc. in the North Chungcheong, South Gyeongsang, South Joella, South Joella, South Chungcheong, Gyeonggi, Kangwon Provinces (Okamoto and Muramatsu 1922).

Cicindela chinensis japonica The Japanese name is *nami-hanmyō*. In **Japan,** this species was used medicinally like *Meloe* spp. (Ono 1844).

Cicindela curvata **Mexico:** Native people belonging to the Chichimeca, Chinanteco, Mazahua, Mestizo, Mixe, Mixteco, Náhua, Otomi, Popolaca, Tarasco, Tlapaneca, Tzetzale, Tzottzile, Zapoteco, and Zoque tribes eat the larvae (Ramos-Elorduy and Pino 1989).

Cicindela roseiventris **Mexico:** Native people belonging to the Chichimeca, Chinanteco, Mazahua, Mestizo, Mixe, Mixteco, Náhua, Otomi, Popolaca, Tarasco, Tlapaneca, Tzetzale, Tzottzile, Zapoteco, and Zoque tribes eat the larvae (Ramos-Elorduy and Pino 1989).

Proagsternus sp. The larvae are fried in **Madagascar** (Decary 1937, Paulian 1943).

Curculionidae (Weevils) **(48)**

Anthonomus spp. **Colombia-Venezuela:** Native people of the Yukpa tribe eat the adults roasted (Ruddle 1973).

Aphiocephalus limbatus This species is fried with butter alone or together with garlic and parsley in **Madagascar** (Dufour 1987).

Aristobla approximator This species is eaten in **Thailand** (Leksawasdi 2008).

Arrhines hirtus People eat this species in **Thailand** (Hanboonsong et al. 2000).

Arrhines spp. People eat at least two species belonging to this genus in **Thailand** (Hanboonsong et al. 2000).

Astycus gestvoi People eat this species in **Thailand** (Hanboonsong et al. 2000).

Balaninus album (banana weevil) This species is eaten by the Ao-Naga people in Nagaland State, **India** (Meyer-Rochow 2005).

Balanius dentipes → *Curculio sikkimensis*

Behrensiellus glabratus **Indonesia** (West Papua State): Children living around the Pass Valley eat this species (Ramandey and van Mastrigt 2010).

Calandra → *Rhynchophorus*

Cnaphoscapus decorates This species is eaten in **Thailand** (Hanboonsong et al. 2000).

Cosmopolites sordida The adults are considered edible by the Ashuara and Shuara people in **Ecuador** (Onore 1997).

Crytotrachelus rufopectinipes birmanicus → *Cyrtotrachelus rufopectinipes birmanicus*

Curculio sikkimensis The Japanese name is *kurishigi-zōmushi*. The larvae are eaten roasted in the Yamaguchi Prefecture, **Japan** (Miyake 1919).

Cyrtopachelus → *Cyrtotrachelus*

Cyrtotrachelus buqueti borealis (giant bamboo weevil) **China:** The adults are roasted (Chen et al. 2009). **India:** The larvae are roasted by the Galo people in Arunachal Pradesh State (Kato and Gopi 2009). **Thailand:** This species is consumed (Hanboonsong 2000).

Cyrtotrachelus dochrous This species is eaten in **Thailand** (Hanboonsong et al. 2000).

Cyrtotrachelus longimanus **China:** The larvae and adults are roasted or fried (Ghesquière 1947). **Thailand:** This species is relished (Hanboonsong et al. 2000).

Cyrtotrachelus rufopectinipes birmanicus This species is considered as food item in **Thailand** (Hanboonsong 2010).

Dynamis borassi This species is considered edible by the Ashuara, Cofane, Huaorani, Quichua, Secoya and Shuara people in **Ecuador** (Onore 1997).

Dynamis nitidula The larvae and pupae are used as food article by the Ashuara, Cofane, Huaorani, Quichua, Secoya and Shuara people in **Ecuador** (Onore 1997).

Dynamis perryi → *Dynamis borassi*

Episomus aurivillius People eat this species in **Thailand** (Hanboonsong et al. 2000).

Episomus **sp.** A species belonging to this genus is consumed in **Thailand** (Hanboonsong 2010).

Eugnoristus monachus The larvae are eaten raw or fried in **Madagascar** (Decary 1937, Paulian 1943).

Hypodisa talaca The adults are used as delicacy in **Thailand** (Vara-asavapati et al. 1975).

Hypomeces squamosus (gold-dust weevil) **Sri Lanka:** This species is considered edible (Nandacena et al. 2010). **Thailand:** This species is eaten in the southern part of the country (Lumsa-ad 2001).

Larinus maculates → *Larinus onopordi*

Larinus mellificus **Iran** and **Iraq:** The cocoons are used as a sweetener or medicine. They are called *trehale* manna (Bodenheimer 1951).

Larinus nidificans → *Larinus mellificus*

Larinus onopordi **Iran** and **Iraq:** The cocoons are used as a sweetener or medicine. They are called *trehale* manna (Bodenheimer 1951).

Larinus rudicollis The cocoons are used as a sweetener or medicine in **Israel** (Bodenheimer 1951, Capiomont and Leprieur 1874, Pierce 1915), and in **Syria** (Capiomont and Leprieur 1874, Pierce 1915).

Larinus syriacus **Iran** and **Iraq:** The cocoons are boiled and the water is used for curing respiratory diseases (Bodenheimer 1951).

Macrochirus longipes **China:** The salted adults are treated as foodstuff. In Guangzhou, some restaurants serve this weevil (Umeya 1994).

Metamasius cinnamominus **Ecuador:** The adults are eaten by the Ashuara and Shuara people (Onore 1997). **Venezuela:** The larvae, pupae and the adults are consumed by the Piaroa and Yanomamo people (Paoletti and Dufour 2005).

Metamasius dimidiatipennis The adults are used as food item by the Ashuara and Shuara people in **Ecuador** (Onore 1997).

Metamasius hemipterus The adults are considered edible by the Ashuara and Shuara people in **Ecuador** (Onore 1997).

Metamasius sericeus The adults are eaten by the Ashuara and Shuara people in **Ecuador** (Onore 1997).

Metamasius spinolae The Mexican name is *gusano del nopal*. In **Mexico,** native people of the Chichimeca, Mazateco, Mestizo, Mixteco, Náhua, Otomi, Popolaca and Zapoteco tribes eat the larvae and pupae (Ramos-Elorduy de Conconi et al. 1984, Ramos-Elorduy and Pino 1989).

Metamasius **sp.** The larvae belonging to this genus are roasted by the Yanomamo people in Alto Orinoco, **Venezuela** (Araujo and Beserra 2007).

Onaphoscapus decorates People eat this species in **Thailand** (Rattanapan 2000).

Otidognathus davidis The adults are fried in **China** (Chen and Feng 1999).

Pachyrrhynchus moniliforis This species is used as a food item on the islands of Occidental Mindoro, the **Philippines** (DeFoliart 2002).

Pollendera atomaria People eat this species in **Thailand** (Hanboonsong et al. 2000).

Polyclaeis equestris People boil the adults after removing the wings until the water evaporates, and then roast them in **RSA** (Quin 1959).

Polyclaeis plumbeus People boil the adults after removing the wings until water evaporates, and then roast them in **RSA** (Quin 1959).

Rhyna **sp.** The larvae are eaten raw or fried in **Madagascar** (Decary 1937, Paulian 1943).

Rhinostomus barbirostris (bearded weevil) **Brazil:** The larvae are consumed by the Suri people (Setz 1991). **Ecuador:** The larvae and pupae are treated as foodstuff by the Ashuara, Awa, Cofane, Huaorani, Negr. Esmerald, Quichuas, Shuara, and Siona people (Onore 1997, Coinbra 1984). **Venezuela:** The larvae and pupae are eaten by the Guajibo, Hoti and Piaroa people (Paoletti and Dufour 2005).

Rhynchophorus bilineatus → *Rhychophorus ferrugineus papuanus*

Rhynchophorus chinensis **China:** The adults are roasted or fried (Ghesquièré 1947). **India:** This species was also eaten (Hope 1842). The **Philippines:** The larvae are eaten raw, roasted, simmered or prepared as a *saté* (Gourou 1948, Revel 2003, Robson and Yen 1976).

Rhynchophorus cruentatus (palmetto weevil) **Amazonia** and the **USA:** The larvae of this species are considered edible (Ghesquière 1947).

Rhnychophorus ferrugineus (Asian palm weevil) **India:** The larvae are eaten baked or as a chutney by the Karbi and Rengma Naga people (Meyer-Rochow 2005, Ronghang and Ahmed 2010). **Indonesia:** The Javanese name is *ulat sagu*. The larvae and adults are used as food item (van der Burg 1904, Nonaka 1999b, Tommaseo Ponzetta and Paoletti 1997). **Malaysia:** The larvae are eaten raw, cooked in porridge with thin slices of ginger or stir-fried with shallots and soy sauce. Sometimes they are skewered on a small stick and are then grilled or toasted lightly. The pupae and adults are also consumed but are not as popular as the larvae (Bragg 1990, Khen and Unchi 1998, Chung et al. 2002). **Myanmar:** The larvae are cooked (Ghosh 1924). **PNG:** The larvae are called *saksak binatang* in Pidgin. They are eaten raw or boiled (May 1984). People raise the larvae in *sagu* palm (*Metroxylum sagu*) logs cut down and left in forests (Mitsuhashi 1997). The **Philippines:** The larvae are eaten raw, roasted, boiled or used to prepare a *saté* (Revel 2003). **Sri Lanka:** This species is considered as edible (Nandacena et al. 2010). **Thailand:** People eat the larvae cooked (Mungkorndin 1981, Hanboonsong et al. 2000).

Rhynchophorus ferrugineus papuanus (black palm weevil) **PNG:** The Onabasulu people eat the larvae by wrapping them in banana leaves and cooking on heated stones (Meyer-Rochow 2005). **Indonesia** (West Papua State): This species is also consumed (Ramandey and van Mastrigt 2010).

Rhynchophorus palmarum **(PL.5-2) Amazonia:** The larvae are widely used as food (Ghesquière 1947). **Argentine:** The larvae are fried (Cowan 1865). **Barbados:** The larvae are roasted (Schomburgk 1847–1848). **Bolivia-Peru-Brazil-Colombia:** Native people of the Peban tribe eat the larvae (Cutright 1940, Posey 1987, Steward and Métraux 1963, Ealand 1915). **Bolivia:** Native people of the Siriono tribe eat the larvae fried (Cutright 1940). **Brazil:** The larvae are eaten by the Bororo (Albisetti and Venturelli 1962), Caingua (Métraux 1963), Cocama (Métraux 1963), Maku (Milton 1984) and Surui (Coimbra 1984) people. The Yanomamo people raise the larvae of this species in the palm trees cut down in forests (Chagnon 1968). **Colombia-Venezuela:** Native people of the Bari tribe collect the larvae by placing bait logs of palm trees (*Jessenia bataua*) in forests, and eat the larvae (Beckerman 1977). **Colombia:** Native people of the Tatuyo tribe eat the larvae. They raise the larvae in palm trees that are cut down (Dufour 1987). **Ecuador:** The larvae and pupae are eaten by the Ashuara, Awa, Cofane, Huaorani, Negr. Esmerald, Quichua, Secoya, Shuara, Siona Secoya, Siona and Tsachila people (Onore 1997). **Guyana-Brazil:** Native people of the Roucouyen tribe relish the larvae (Bancroft 1769). **Mexico:** Native people of the Chantale, Huasteco, Lacandone, Maya, Mestizo, Náhua, Tlapaneca, and Totonaca tribes eat the larvae and adults (Ramos-Elorduy de Conconi et al. 1984, Ramos-Elorduy and Pino 1989). **Paraguay:** The Ache people eat the larvae (Hawkes et al. 1982). The Guayaki people raise the larvae of this species in palm logs that are cut down (Clastres 1972, Hurtado et al. 1985). The larvae are eaten raw or roasted after removing the heads and intestines, or used to make a type of sauce (Chagnon 1868, Milton 1997, Steward and Métraux 1963). **Peru-Colombia:** Native people of the Bora tribe relish the larvae (Balick 1988). **Peru:** Native people of the Yameo tribe make a sauce from the larvae, adding red pepper and corn powder (Ealand 1915, Steward and Métraux 1963, Balic 1988). People in modern times also enjoy the barbequed larvae (Mitsuhashi 2010) **(PL V-2). Suriname:** The larvae are fried with very little butter and salt, or spit-roasted. People extract fat from the larvae, and make a type of butter, which can be used for cooking (Stedman 1796). **Venezuela:** The larvae are roasted by the people of the following tribes; Bari (Beckerman 1977), Baniwa, Curripaco, Guajibo, Hoti, Piaroa, Ynomamo (Paoletti and Dufour 2005) and Yekuana (Araujo and Beserra 2007). The larvae are roasted after removing their intestines (Spruce 1908). The adults are also eaten by the Piaroa (Paoletti and Dufour 2005). **West Indies:** The larvae are roasted adding spice, or boiled and served with orange or lemon juice (Labat 1722). The larvae are also consumed in **Martinique** (Hearn 1890), and in **Trinidad** and **Tobago** (Provancher 1890).

Rhynchophorus phoenicis (raphia weevil) **Angola:** The larvae are considered edible (Oliveira et al. 1976). **Benin:** Mainly in southern area, the Fon people eat the larvae (Tchibozo et al. 2005). **Cameroon:** The larvae are stewed, fried, or made into brochettes and grilled (Grimaldi and Bikia 1985). People cut down raphia palms, and leave them for two to three-weeks in order to get as many larvae of this species (Balinga et al. 2004). **DRC:** The larvae are consumed by the Ngandu people (Takeda 1990). **Ghana:** The larvae from this species are collected from oil palm, and are eaten as an important protein source instead of beef or fish (Dei 1989). **India:** The larvae are baked or as a chutney by the Karbi and Rengma Naga people (Ronghang and Ahmed 2010). **Mozambique:** The larvae have been eaten from the olden days as a traditional food (Ghesquiére 1947). **Nigeria:** The larvae are fried without adding oil (Fasoranti and Ajiboye 1993). **Sierra Leone:** People eat larvae taken from the oil palm. These are considered as the larvae of *Rhynchophorus phoenicis* (Hunter 1984). **PRC:** People eat the larvae raw, roasted or smoked (Bani 1995, Moussa 2004). In **Uganda,** people remove intestines from the larvae, and then frizzle with onion, salt and curry (Menzel and D'Aluiso 1998).

Rhynchophorus quadranglulus In **Africa,** this species is also consumed like *R. phoenicis* (Wattanapongsiri 1966).

Rhynchophorus schach **PNG:** The Gidra people eat the larvae (Ohtsuka et al. 1984). **Thailand:** The larvae are eaten (Bristowe 1932).

Rhynchophorus **spp. Mexico:** The beetle larvae of several species belonging to this genus are raised as foods (Ramos-Elorduy 1997). Several species belonging to this genus are used as food in **Madagascar** (Decary 1937, Paulian 1943), **PNG** (May 1984, Meyer-Rochow 1973), **Thailand** (Hamboonsong et al. 2001), **Uganda** (Menzel and D'Aluisio 1998), and in **Vietnam** (Bréhion 1913).

Scyphophorus acupunctatus (agave weevil) The Mexican name is *botija*. In **Mexico,** people of the Hñähñu tribe eat the larvae and pupae (Ramos-Elorduy de Conconi et al. 1984, Aldasoro Maya 2003).

Scyphophorus interstitialis → *Scyphophorus acupunctatus*

Scyphophorus **sp.** A species belonging to this genus is eaten in **Mexico** (Ramos-Elorduy et al. 1998).

Sepiomus → *Episomus*

Sipalinus aloysii-sabaudiae The larvae are cooked or roasted in **Tanzania** (Harris 1940).

Sitophilus granaries (granary weevil) **Greece:** This beetle was found in a sort of pea excavated from the Akrotiri Ruin, and was thought to be consumed by ancient people (Panagiotakopulu and Buchland 1991).

Tanymeces **sp.** A species belonging to this genus is used as food in **Thailand** (Hanboonsong et al. 2000).

Xanthochellus **sp.** A species belonging to this genus is considered edible in **Thailand** (Leksawasdi 2008).

Dermestidae (Larder beetles) **(1)**

Anthrenus **sp.** The **USA:** A species of larder beetles belonging to this genus seemed to be consumed by the aboriginal people in Nevada State, as it was found in the coprolites excavated from the Lovelock Cave of Nevada State (BC 1,200) (Heizer and Napton 1969).

Thylodrias contractus **Mexico:** In olden times (about BC 2,300), aboriginal people in the Tamaulipas State seemed to eat this species, as these beetles were found in the coprolites excavated from the caves in plateaus of the state (Callen 1969).

Dynastidae (32)

Allomyrina dichotoma (Japanese rhinoceros beetle) **China:** The larvae and adults are considered as foodstuff (Chen and Feng 1999). This species was used when delivery was difficult (Li 1596), and is also used as an anodyne (Zimian et al. 1997). **India:** The larvae and adults are roasted or boiled after removing appendages in Arunachal Pradesh State (Chakravorty et al. 2011). **Japan:** The Japanese name is *kabuto-mushi*. In the times of war, some people tried to eat the larvae. It was, however, concluded that the larvae are not suitable to eat because of the bad smell and hard skin (Esaki 1958).

Ancognatha castanea The larvae are eaten in **Ecuador** (Onore 1997).

Ancognatha jamesoni People eat the larvae in **Ecuador** (Onore 1997).

Ancognatha vulgaris The larvae are eaten in **Ecuador** (Onore 1997).

Ancognatha **sp. Colombia:** Native people of the Tatuyo tribe eat a species belonging to this genus (Dufour 1987).

Augosoma centaurus The larvae are consumed in **Cameroon** (Bodenheimer 1951), **DRC** (Takeda 1990) and **PRC** (Bani 1995).

Camenta **sp.** The larvae of a species belonging to this genus are used as food item in **Angola** (DeFoliart 2002).

Chalcosoma atlas **Indonesia:** The Javanese name is *kumbang tanduk*. In Java Island, people eat the larvae fried (Lukiwati 2010). **Malaysia** (Sabah state): This species is eaten (Chung et al. 2002). **Thailand:** This species is eaten (Leksawasdi 2010).

Chalcosoma moellenkampi (three-horned beetle). **Indonesia:** In the Kalimantan State, only the adults are consumed. The larvae are not eaten (Chung 2010). **Malaysia:** This species is consumed in the Sabah State. The adults are usually roasted after removing the hard and spiny parts (Chung et al. 2002).

Dynastes hercules **Brazil:** The larvae are used as a food item (Costa Neto and Ramos-Erolduy 2006). **Colombia:** The larvae are considered edible by the Tukanoan people (Paoletti and Dufour 2005). **Ecuador:** The larvae and adults are eaten by the Huaorani people (Onore 1997).

Dynastes hyllus The larvae are used as a food item in **Mexico** (Ramos-Elorduy and Pino 2002).

Dynastes **sp.** The larvae of a species belonging to this genus is a food item in **Mexico** (Ramos-Elorduy and Pino 2002).

Eupatorus gracilicornis **Thailand:** This species is considered edible (Hanboonsong et al. 2000).

Golofa (Golofa) imperialis The larvae, pupae, and the adults are used as foodstuff in **Mexico** (Ramos-Elorduy and Pino 2002). This species is raised by the Mazahua, Otomi and Nahua tribes as food for humans (Ramos-Elorduy 2009).

Golofa pusilla This species is raised by the Mazahua, Otomi and Nahua tribes in **Mexico** as food for humans (Ramos-Elorduy 2009).

Golofa (Golofa) tersander The larvae, pupae, and the adults are eaten in **Mexico** (Ramos-Elorduy and Pino 2002).

Golofa unicolor → *Proagolofa unicolor*

Golopha aeacus The larvae are consumed by the Otavalo and Quichua people in **Ecuador** (Onore 1997).

Golopha aegeon The larvae are considered edible by the Otavalo and Quichua people in **Ecuador** (Onore 1997).

Heterogomphus bourcieri The larvae are treated as a foodstuff by the Quichuas people in **Ecuador** (Onore 1997).

Megaceras crassum In **Colombia,** native people of the Tatujo tribe consumed the larvae and adults (Dufour 1987).

Megaceras **sp.** The larvae of a species belonging to this genus are considered a foodstuff in **Amazonia** (Ratcliffe 1990) and in **Brazil** (Costa Neto and Ramos-Elorduy 2006).

Megasoma actaeon **(PL V-3)** The larvae and adults are used as food by the Kayapo people in **Amazonia** and in **Brazil** (Posey 1987).

Megasoma anubis The larvae are eaten in **Brazil** (Netolizky 1920).

Megasoma elephas The larvae are fried or roasted by the Karbi and Rengma Naga people in Assam State, **India** (Ronghang and Ahmed 2010).

Megasoma hector → *Megasoma anubis*

Oryctes boas **Nigeria:** The larvae are fried (Fasoranti and Ajiboye 1993). **Sierra Leone:** This species was eaten by the natives (Hope 1842). The larvae and adults of this species are also food source in **DRC** (Chinn 1945), **PRC** (Bani 1995) and in **RSA** (Bodenheimer 1951).

Oryctes centaurus In **PNG:** The Onabasulu people eat this species (Meyer-Rochow 1973 and 2005).

Oryctes monoceros The larvae are consumed in **RSA** (Bergier 1941).

Oryctes nasicornis This species is considered edible in **Madagascar** (Bergier 1941).

Oryctes owariensis **Cape Coast:** This species was considered good for eating by the natives (Hope 1842). The larvae are also used as food in **DRC** (Chinn 1945), **Ghana** (Hope 1842), **PRC** (Bani 1995) and **RSA** (Bodenheimer 1951).

Oryctes rhinoceros **China:** This species is eaten (Chen and Feng 1999). **India:** This species is fit for eating (DeFoliart 2002). **Malaysia:** The larvae, pupae and the adults are considered as food source by the people of Sabah State. The adults are usually roasted after removing the hard and spiny parts (Chung et al. 2002). **Myanmar:** The Karen people are very fond of eating the larvae. They remove the intestines and fry them (Ghosh 1924). **PNG:** The larvae and adults are eaten (May 1984). **Thailand:** The Laotian name for the larvae is *mang bough*, for adults is *mang kwang*. The larvae, pupae and the adults are roasted (Bristowe 1932, Sangpradub 1982). This species is also eaten in the **Philippines** (Gibbs et al. 1912), **Solomon Islands** (Bernatzik 1936), **Sri Lanka** (Nandacena et al. 2010).

Oryctes **spp. India:** The adults of a species belonging to this genus are an article of food roasted by the Nishi people in Arunachal Pradesh State (Singh et al. 2007). The larvae of some species belonging to this genus are also fried in **Benin** (Tchibozo et al. 2005) and **Guinea** (Bodenheimer 1951). **PNG:** The Chuave people eat the larvae of some species belonging to this genus (Meyer-Rochow 1973).

Podischnus agenor In **Colombia-Venezuela**, native people of the Yukpa tribe eat the larvae and adults raw or spit-roasted (Ruddle 1973).

Praogolofa unicolor The adults are eaten fried by the Otavalo, Pilahuine, Quichua and Saraguro people in **Ecuador** (Onore 1997 and 2005, Paoletti and Dufour 2005).

Scapanes **sp.** A species belonging to this genus is considered edible in **PNG** (Meyer-Rochow 1973).

Strategus aloeus aloeus In **Mexico,** the adults are used as a food item in Tulancalco, Hidalgo (Ramos-Elorduy 2006).

Strategus julianus In **Mexico,** a medicinal drink is made from this insect for curing impotence (Ramos-Elorduy de Conconi and Pino 1988).

Strategus **sp.** The larvae and adults of a species belonging to this genus are considered as foodstuff by the Kayapo people in **Brazil** (Possey 1987).

Trypoxylus dichotomous → *Allomyrina dichotoma*

Xylotrupes dichotomus → *Allomyrina dichotoma*

Xylotrupes gideon **India:** The adults are roasted or boiled in Arunachal Pradesh State (Chakravorty et al. 2011). **Indonesia** (West Papua State):

The people living around the Arfak Mountains in Birdhead Peninsula and at Walmakin the central mountain range eat this species (Ramandey and van Mastrigt 2010). **Laos:** The Laotian is name: *meang kham*. This species is consumed (Boulidam 2008). **Malaysia:** This species is eaten in the Sabah State. The adults are usually roasted after removing the hard and spiny parts (Chung et al. 2002). **Myanmar:** The Karen people are very fond of eating the larvae. They remove the intestines and fry them (Ghosh 1924), **PNG:** The larvae are relished by the Chuave people (Meyer-Rochow 2005). **Sri Lanka:** This species is used as food (Nandacena et al. 2010). **Thailand:** The Laotian name is *mang kwang*. People of the Lahu tribe like eating the larvae roasted with a touch of chili sauce on them (Sirinthip and Black 1987). **Vietnam:** This species is treated as foodstuff (Yhoung-Aree and Viwatpanich 2005).

Xylotrupes gideon siamensis This species is used as a foodstuff in **Thailand** (Utsunomiya and Masumoto 1999).

Xylotryctes **spp. Mexico:** The larvae and pupae of several species belonging to this genus are eaten (Ramos-Elorduy de Conconi et al. 1984). **Thailand:** A species belonging to this genus is consumed (Leksawasdi 2010).

Dytiscidae (Predacious diving beetles) **(43)**

Acilius **sp.** A species belonging to this genus is eaten by the Han people in **China** (Ramos-Elorduy et al. 2009).

Agabus flavipennis → *Gaurodytes fulvipennis*

Copelatus **sp.** The adults of a species belonging to this genus are fried in **Thailand** (Hanboonsong et al. 2000).

Cybister bengalensis The adults of this species is considered edible in **China** (Hoffmann 1947).

Cybister binotatus This species is eaten by the Bantu people in **Africa** (Ramos-Elorduy et al. 2009).

Cybister brevis The Japanese name is *kuro-gengorō*. In **Japan,** the adults are roasted, with a touch of miso paste. Frying with oil, boiling with soy sauce or salt water were also common (Takagi 1929e). The larvae and adults are used for treating various diseases such as tuberculosis, gonorrhea, syphilis, etc. (Umemura 1943, Shiraki 1958).

Cybister distinctus This species is eaten in **Senegal, Sierra Leone** and **DRC** (Ramos-Elorduy et al. 2009).

Cybister ellipticus This species is used as food in the **USA** (Ramos-Elorduy et al. 2009).

Cybister explanatus **Mexico:** The larvae, pupae and the adults are used as foodstuff (Ramos-Elorduy and Pino 1989). The **USA:** In olden times this species seemed to be consumed by the native people in Nevada State, as this species was found in the coprolites excavated from the Hidden Cave and Lovelock Cave, Nevada (Ambro 1967). It is known that Native Americans eat the adults (Roust 1967).

Cybister flavocinctus **China:** This species is eaten by the Han people (Ramos-Elorduy et al. 2009). **Mexico:** The Mexican name is *cucaracha de agua*. The larvae and adults are used as foodstuff by the Matlazinca, Mazahua and Nahuatl people (Ramos-Elorduy et al. 2009).

Cybister frimbiolatus **China:** This species is considered delicacy by the Han people (Ramos-Elorduy et al. 2009). **Mexico:** This species is eaten by the Maya, Mazahua, Mixtec, Nahuatl, Otomi, Tlapaneco, Yutoaztec and Zapotec people (Ramos-Elorduy et al. 2009).

Cybister guerini **China:** The adults are an article of food (Hoffmann 1947). **Indonesia:** This species is considered good for eating by the Toraja people (Ramos-Elorduy et al. 2009).

Cybister hova The adults are fried by the Malagasy people in **Madagascar** (Decary 1937, Paulian 1943, Ramos-Elorduy et al. 2009).

Cybister insignis This species is used as a food item by the Galoas, Nkomis and Irungos people in **Gabon** (Ramos-Elorduy et al. 2009).

Cybister japonicus **China:** The adults are condisered as a food source by the Han people (Hoffmann 1947). This species is also used to promote blood circulation (Zimian et al. 1997). **Japan:** The Japanese name is *gengorō*. In the Iwate, Akita, Fukushima, Chiba, Yamanashi, Nagano, and Gifu Prefectures, the adults are skewered after removing the tails, and then roasted with soy sauce. They are also fried, or boiled with salt (Miyake 1919). The pupae were also eaten by frying with oil and soy sauce in the Nagano Prefecture (History of Nagano Publishing Body 1988). The charred adults were used for treating stomach diseases or convulsive fits (Miyake 1919). **North** and **South Korea:** The adults are roasted after removing the wings and legs (Okamoto and Muramatsu 1922).

Cybister lewisianus **China:** The adults are eaten (Hoffmann 1947). **Sri Lanka:** This species is eaten (Nandacena et al. 2010). **Thailand:** The Thai name is *maeng-tub-tao*. The adults are fried (Hanboonsong et al. 2000). **Laos,** **Myanmar** and **Vietnam:** This species is also consumed (Yhoung-Aree and Viwatpanich 2005).

Cybister occidentalis This species is treated as an article of food in **Cameroon, China** and **Mexico** (Ramos-Elorduy et al. 2009).

Cybister operosus This species is considered as foodstuff by the Malagasy people in **Madagascar** (Ramos-Elorduy et al. 2009).

Cybister owas → *Cybister hova*

Cybister rugosus People eat the adults fried in **Thailand** (Chen and Wongsiri 1995). And **Laos, Myanmar, Vietnam** (Yhoung-Aree and Viwatpanich 2005).

Cybister singulatus This species is eaten by the Han people in **China** (Ramos-Elorduy et al. 2009).

Cybister sticticus This species is eaten by the Han people in **China** (Ramos-Elorduy et al. 2009).

Cybister sugillatus The adults are used as food in **China** (Hoffmann 1947), and are used as a remedy for frequent urination (Inagaki 1984).

Cybister tripunctatus (small diving beetle) **China:** The adults are considered edible (Hu and Zha 2009). This species is medicinally used to promote blood circulation (Zimian et al. 2005). **Indonesia:** This species is a food source (Ramos-Elorduy et al. 2009). **Sri Lanka:** This species is consumed as food (Nandacena et al. 2010). **Thailand:** The larvae and adults are used as a food item. People make "*mang-eed*" from the collected beetles, and eat them after removing the adult wings (Vara-asavapati et al. 1975).

Cybister tripunctatus asiaticus **China:** This species is used as a remedy for frequent urination (Inagaki 1984). **Thailand:** The adults are fried (Hanboonsong et al. 2000).

Cybister tripunctatus orientalis **China:** The adults are cooked with salt, and then dried. They taste like dried squids (Esaki 1942, Yasumatsu 1965, Mao 1997). This species is also used to improve blood circulation as well as frequent urination (Inagaki 1984, Zimian et al. 1997). **Japan:** The Japanese name is *kogatano-gengorō*. The adults are roasted with soy sauce in the Nagano Prefecture (Takagi 1929e). The larvae and adults are used for treating various diseases such as tuberculosis, gonorrhea, syphilis, etc. (Umemura 1943, Shiraki 1958). **North** and **South Korea:** This species is used as a traditional medicine (Pemberton 2005). **Thailand:** The adults are eaten fried (Masumoto and Utsunomiya 1997).

Cybister ventralis **China:** This species is used as a remedy for frequent urination (Inagaki 1984).

Cybister **sp. India:** The adults of a species belonging to this genus are eaten boiled or roasted by the Nishi people in Arunachal Pradesh (Singh et al. 2007). **Malaysia:** A species belonging to this genus is consumed in the Sabah State (Chung et al. 2002). Some species belonging to this genus are also eaten in following countries: **Cambodia, Laos, Myanmar** and **Vietnam** (Ramos-Elorduy et al. 2009), **Thailand** (Lumsa-ad 2001).

Cypris **sp.** People eat the adults of a species belonging to this genus in southern **Thailand** (Lumsa-ad 2001, Leksawasdi 2010).

Dytiscus (Macrodytes) circumflex This species is used as food by the Arabic, Berber and Sefardi people in **Africa** (Ramos-Elorduy et al. 2009).

Dytiscus habilis This species is used as a food item by the Han people in **China** (Ramos-Elorduy et al. 2009).

Dytiscus marginalis (great diving beetle) **China:** The adults are kept in salt, and then dried (Bodenheimer 1951). **Turkey:** The adults are considered edible (İncekara and Türkez 2009).

Dytiscus marginicollis This species is eaten by the Chol, Maya, Tzeltal, Zapotec and Zoque people in **Mexico** (Ramos-Elorduy et al. 2009).

Dytiscus validus **China:** This species is fit for eating (Ramos-Elorduy et al. 2009). **Japan:** The Japanese name is *ko-gengorōmodoki*. This species is used as foodstuff (Ramos-Elorduy et al. 2009).

Dytiscus **sp. India:** The adults of some species belonging to this genus are boiled or roasted by the Nishi people in Arunachal Pradesh State (Singh et al. 2007). Some species belonging to this genus are also eaten in following countries: **Cameroon, China, Japan** and **Mexico** (Ramos-Elorduy et al. 2009).

Eretes sticticus (twinpoes diving beetle) **China:** This species is treated as a food item by the Han People (Ramos-Elorduy et al. 2009). **India:** The larvae and adults are used as foodstuff (Essig 1947). **Myanmar:** The adults are sun-dried for special cuisines (DeFoliart 2002). **Sri Lanka:** This species is eaten (Nandacena et al. 2010). **Thailand:** The adults are fried (Hanboonsong et al. 2000). **Laos, Vietnam:** This species is also eaten (Yhoung-Aree and Viwatpanich 2005).

Eretes sticus → *Eretes sticticus*

Gaurodytes fulvipennis This species is consumed by the Han people in **China** (Ramos-Elorduy et al. 2009).

Hololepta (Hololepta) guidnis This species is used as a food item by the Maya, Nahuatl, Otopame and Zapotec people in **Mexico** (Ramos-Elorduy et al. 2009).

Hololepta **sp.** A species belonging to this genus is eaten by the Mixtec people in **Mexico** (Ramos-Elorduy et al. 2009).

Hydaticus rhantoides People eat the adults fried in **Thailand** (Hanboonsong et al. 2000).

Laccophilus apicalis This species is consumed by the Nahuatl and Otomi people in **Mexico** (Ramos-Elorduy et al. 2009).

Laccophilus pulicarius People eat the adults fried in **Thailand** (Hanboonsong et al. 2000).

Laccophilus **sp. India:** The adults of a species belonging to this genus are fried in oil by the Meeteis people in Nagaland State (Meyer-Rochow 2005). **Mexico:** A species belonging to this genus is used as foodstuff by the Nahuatl and Yutoaztec people (Ramos-Elorduy et al. 2009).

Megadytes gigantean This species is considered edible by the Chol, Mixe, Tzeltal, Tzotzil, Zapotec and Zoque people in **Mexico** (Ramos-Elorduy et al. 2009).

Megadytes giganteus The adults are eaten by the Zoque people in **Mexico** (Ramos-Elorduy et al. 2009).

Megadytes **sp.** In **Mexico,** the larvae of a species belonging to this genus, whose Mexican name is *haba de agua,* are eaten by the Maya, Mazahua, Nahuatl, Otomi, Otopame and Zapotec people (Ramos-Elorduy et al. 2009). The adults are also consumed (Ramos-Elorduy et al. 1998).

Platynectes guttula This species is eaten by the Han people in **China** (Ramos-Elorduy et al. 2009).

Rhantaticus congestus The adults are fried in **Thailand** (Hanboonsong et al. 2000).

Rhantus atricolor The Mexican name is *cucarachita.* In **Mexico,** the larvae and adults are food article (Ramos-Elorduy et al. 1998).

Rhantus consimilis This species is treated as food item by the Malagasy people in **Madagascar** (Ramos-Elorduy et al. 2009). This species is also considered edible by the Otomi and Nahuatl people in **Mexico** (Ramos-Elorduy et al. 2009).

Rhantus pulverosus The Japanese name is *hime-gengorō.* In **Japan,** the larvae and adults were used for treating various diseases such as tuberculosis, gonorrhea, syphilis, etc. (Umemura 1943, Shiraki 1958).

Rhantus **sp.** In **Mexico,** the adults of a species belonging to this genus are eaten (Ramos-Elorduy and Pino 1990).

Thermonectes bsilaria This species is considered as a food source by the Huatesco, Mixtec, Nahuatl, Otomi, Popolaca, Totonaco, Yutoaztec, and Zapotec people in **Mexico** (Ramos-Elorduy et al. 2009).

Thermonectes marmoratus This species is treated as foodstuff by the Huasteco, Nahuatl, Otomi, Totonaco, Yutoaztec and Zapotec people in **Mexico** (Ramos-Elorduy et al. 2009).

Thermonectes **sp.** A species belonging to this genus is considered edible by the Nahua, Otomi, Popolaca, Totonaco and Zapotec people in **Mexico** (Ramos-Elorduy et al. 2009).

Elateridae (Click beetles) **(6)**

Cardiophorus aequabilis This species is used as food item by people in Arunachal Pradesh State, **India** (Singh et al. 2007).

Chalcolepidius laforgei The larvae and adults are edible in **Mexico** (Ramos-Elorduy and Pino 2002).

Chalcolepidius mexicanus → *Pyrophorus mexicanus*

Chalcolepidius pellucens → *Pyrophorus pellucens*

Chalcolepidius rugatus The larvae and adults are eaten in **Mexico** (Ramos-Elorduy and Pino 2002).

Pyrophorus mexicanus The adults are eaten in Chiapas, **Mexico** (Ramos-Elorduy and Pino 2002).

Pyrophorus pellucens The adults are eaten in Chiapas, **Mexico** (Ramos-Elorduy and Pino 2002).

Pyrophorus **sp.** A species belonging to this genus is used as foodstuff in **Dominica** and **Haiti (Hispaniola Island)** (Gilmore 1963).

Tetralobus flabellicornis The larvae are consumed in **CAR** (Berensburg 1907).

Gyrinidae (Whirligig beetles) **(6)**

Dineutus marginatus The Japanese name is *tao-mizusumashi*. The adults were used for treating various diseases such as colds, fever, stomach diseases, etc. in **Japan** (Umemura 1943).

Dineutus orientalis The Japanese name is *ō-mizusumashi*. In **Japan,** the adults are said to be good for curing swelling (Shiraki 1958).

Gyrinus curtus The Japanese name is *ko-mizusumashi*. In **Japan,** eating the adults raw, pulverized or roasted is said to be good for treating colds, fever, stomach disorders, etc. (Miyake 1919).

Gyrinus japonicus The Japanese name is *hime-mizusumashi*. In **Japan,** this species was used for treating various diseases such as, colds, stomach disorders, fever, etc. (Kuwana 1930, Watanabe 1982).

Gyrinus parcus This species is eaten by the Huasteco, Nahuatl, Otomi, Popolaca, Tarascan, Totonaco, Yutoaztec and Zapotec people in **Mexico** (Ramos-Elorduy et al. 2009).

Gyrinus (Oreogyrinus) plicatus This species is considered edible by the Huasteco, Nahuatl, Otomi, Popolaca, Totonaco, Yutoaztec and Zapotec people in **Mexico** (Ramos-Elorduy et al. 2009).

Helmidae (Riffle beetles) **(2)**

Austrelmis chilensis The adults are used to make chichi soup in **Chile** (Netolitzky 1920), and in **Peru** (Brygoo 1946).

Austrelmis condimentarius The adults are used to make chichi soup in **Chile** (Netolitzky 1919), and in **Peru** (Brygoo 1946). **Mexico:** This species is consumed by the Maya, Tojolabal, Tzeltal, Zapotec and Zoque people (Ramos-Elorduy et al. 2009).

Elmis chilensis → *Austrelmis chilensis*

Elmis condimentarius → *Austrelmis condimentarius*

Haliplidae (Crawling water beetles) **(3)**

Haliplus punctatus This species is eaten by the Huasteco, Nahuatl, Otomi, Popolaca Totonaco, Yutoaztec and Zapotec people in **Mexico** (Ramos-Elorduy et al. 2009).

Haliplus **sp.** A species belonging to this genus is considered as food item by the Amuzgo, Mixtec, Nahuatl and Trapaneco people in **Mexico** (Ramos-Elorduy et al. 2009).

Peltodytes mexicanus This species is used as foodstuff by the Mixtec, Nahuatl, Otomi, Yutoaztec and Zapotec people in **Mexico** (Ramos-Elorduy et al. 2009).

Peltodytes ovalis This species is treated as a food source by the Mixtec, Nahuatl, Otomi, Yutoaztec and Zapotec people in **Mexico** (Ramos-Elorduy et al. 2009).

Histeridae (Hister beetles) **(0)**

Homolepta **sp.** The larvae of a species belonging to this genus are considered as foodstuff in Oaxaca State, **Mexico** (Ramos-Elorduy de Conconi et al. 1984).

Hydrophilidae (Water scavenger beetles) **(16)**

Berosus **sp.** A species belonging to this genus is eaten by the Nahuatl and Yutoaztec people in **Mexico** (Ramos-Elorduy et al. 2009).

Dilobodeus **sp.** A species belonging to this genus is used as an article of food by the Maya, Tojolabal, Tzeltal, Zapotec and Zoque people in **Mexico** (Ramos-Elorduy et al. 2009).

Dibolocelus **sp.** A species belonging to this genus is eaten by the Cora, Huichol and Tepehua people in Mexico (Ramos-Elorduy et al. 2009).

Hydrobiomorpha spinicollis People eat the adults fried in **Thailand** (Hanboonsong et al. 2000).

Hydrochara **sp.** The adults of a species belonging to this genus are considered good for eating in **Thailand** (Chen and Wongsiri 1995).

Hydrophilus acuminatus **China:** The adults are cooked with salt (Hoffmann 1947, Chen and Feng 1999). **India:** This species is eaten by the Parsis and Sijs people (Ramos-Elorduy et al. 2009). **Japan:** The Japanese name is *gamushi*. In the Yamagata, Iwate and Nagano Prefectures, the adults are skewered and then roasted with soy sauce, or cooked with soy sauce and sugar, or fried, or immersed in vinegar (Miyake 1919). This species is used as a remedy for frequent urination (Inagaki 1984). **North** and **South Korea:** The adults are roasted (Okamoto and Muramatsu 1922). **Tibet:** This species is treated as foodstuff (Hoffmann 1947).

Hydrophilus affinis **Japan:** The Japanese name is *ko-gamushi*. The larvae are used for treating convulsive fits in children, whooping cough, etc. (Koizumi 1935). **Laos:** The Laotian name is *meang ee tao*, and the species is eaten (Boulidam 2008).

Hydrophilus bilineatus **China:** The adults are cooked with salt (Hoffmann 1947). **Thailand:** The adults are fried (Hanboonsong et al. 2000). **Vietnam:** The adults are cooked with salt (Hoffmann 1947). People of the Bana, Cham, Co-ho, Ede, Hoa, Kumer, Mong, Nung, San Chay, Tay, and Thai tribes eat this species (Ramos-Elorduy et al. 2009).

Hydrophilus bilineatus cashimirensis **China:** This species is used as a remedy for frequent urination (Inagaki 1984). **Thailand:** The adults are eaten fried (Masumoto and Utsunomiya 1997). **Laos, Myanmar** and **Vietnam:** This species is consumed (Yhoung-Aree and Viwatpanich 2005).

Hydrophilus (Stethoxus) cavisternus **China:** The adults are cooked with salt (Hoffmann 1947). **Thailand:** The adults are fried (Hanboonsong et al. 2000). **Vietnam:** People of the Bana, Cham, Co-ho, Ede, Hoa, Kumer, Mong, Nung, San Chay, Tay, and Thai tribes eat this species (Ramos-Elorduy et al. 2009).

Hydrophilus (Dytiscus) hastatus **China:** The adults are roasted or fried (Hoffmann 1947). **Vietnam:** This species is eaten by people of the Bana, Cham, Co-ho, Ede, Hoa, Kumer, Mong, Nung, San Chay, Tay, and Thai tribes (Ramos-Elorduy et al. 2009). This species is also used as food item in **Cambodia, Laos, Myanmar** and **Thailand** (Ramos-Elorduy et al. 2009).

Hydrophilus marginatus This species is consumed by the Soninke and Wolof people in **Senegal** (Ramos-Elorduy et al. 2009).

Hydrophilus olivaceus **India:** This species is eaten by the Parsis and Sijs people (DeFoliart 2002, Ramos-Elorduy et al. 2009). **Sri Lanka:** This species is considered as a food item (Nandacena et al. 2010). **Thailand:** The adults are fried (Hanboonsong et al. 2000).

Hydrophilus pallidipalpis → Hydrophilus acuminatus

Hydrophilus piceus The adults are eaten in **Turkey** (İncekara and Türkez 2009).

Hydrophilus picicornis In the **Philippines,** people of the Aeta, Iloko, Austronesian, Visayas, Tagalog, Manobo and Negrito tribes eat this species (Ramos-Elorduy et al. 2009).

The adults are boiled after removing the wings and legs (Gibbs et al. 1912).

Hydrophilus senegalensis This species is relished by the Soninke and Wolof people in **Senegal** (Ramos-Elorduy et al. 2009).

Hydrophilus sp. India: The adults of a species belonging to this genus are roasted by the Nishi people in Arunachal Pradesh State (Singh et al. 2007). Thailand: The adults of a species belonging to this genus are fried (Leksawasdi and Jirada 1983, Leksawasdi 2008).

Hydrous → *Hydrophilus*

Sternolophus rufipes Japan: The Japanese name is *hime-gamushi*. The larvae were used for treating convulsive fits in children, whooping cough, etc. (Koizumi 1935). Thailand: The adults are fried (Hanboonsong et al. 2000).

Stethoxus acuminatus → *Hydrophilus acuminatus*

Tropisternus mexicanus Mexico: This species is consumed by the Nahuatl, Yutoaztec and Zapotec people (Ramos-Elorduy et al. 2009). Panama: This species is eaten by the Ngobe, Kunas, Wounan and Brbris people (Ramos-Elorduy et al. 2009).

Tropisternus sublaevis This species is treated as a food item by the Mixtec, Nahuatl, Otomi, Yutoaztec and Zapotec people in Mexico (Ramos-Elorduy et al. 2009).

Tropisternus tinctus The larvae, pupae, and the adults are used as foodstuff in Mexico (Ramos-Elorduy et al. 1985). People of the Maya, Mixtec, Nahuatl, Otomi, Yutoaztec and Zapotec tribes eat this species (Ramos-Elorduy et al. 2009).

Tropisternus sp. A species belonging to this genus is eaten by the Mixtec, Nahuatl, Otomi, Yutoaztec and Zapotec people in Mexico (Ramos-Elorduy et al. 2009).

Ipidae (Bark beetles) **(3)**

Sphaerotrypes yunnanensis The larvae are frizzled with spice in China (Chen and Feng 1999).

Lampyridae (Fireflies) **(2)**

Luciola cruciata (Genji-firefly) The Japanese name is *genji-botaru*. In Japan, the adults were used for treating various diseases such as colds, stomach aches, difficulty in urination, etc. (Koizumi 1935, Umemura 1943, Shiraki 1958).

Luciola lateralis The Japanese name is *heike-botaru*. In Japan, the adults were used for treating whooping cough, stomach disorders, fever, etc. (Shiraki 1958).

Lariidae (3)

Algarobius spp. In **Mexico**, the larvae and pupae are used as food (Bell and Castetter 1937). In the **USA**, the larvae, pupae and the adults of a species called mesquite beetle are desiccated for eating (Sutton 1988).

Caryobruchus scheelaea **Colombia-Venezuela:** Native people of the Yukpa tribe eat the larvae spit-roasted, or smothered (Ruddle 1973).

Caryobruchus **spp.** People eat the larvae raw, fried, or roasted in **Brazil** (Coimbra 1984) and in **Venezuela** (Coimbra 1984, Ruddle 1973).

Neltumis **spp.** In the **USA,** the larvae, pupae and the adults of several species belonging to this genus, are desiccated for preserving insect food (Sutton 1988).

Pachymerus cardo The larvae are eaten raw or fried in **Brazil** (Coimbra 1984).

Pachymerus nucleorum **Brazil:** The larvae are eaten raw or fried (Costa Neto 2000).

Pachymerus **sp.** The larvae of a species belonging to this genus are used as a food item in **Brazil** (Costa-Neto 2003b).

Lucanidae (Stag beetles) (20)

Calodes cuvera This species is consumed in Arunachal Pradesh State, **India** (Singh et al. 2007).

Calodes siva This species is consumed in Arunachal Pradesh State, **India** (Singh et al. 2007).

Cladognathus serricornis The larvae are fried in **Madagascar** (Decary 1937, Paulian 1943).

Cyclommatus albersi This species is considered as a food item by people of Arunachal Pradesh State, **India** (Singh et al. 2007).

Cyclommatus pehengenesis chiangmainesis The larvae and adults are roasted by the Nishi people in Arunachal Pradesh State in **India** (Singh et al. 2007).

Dorcus **spp. India:** A species belonging to this genus is consumed by the Nishi and Galo people. They eat the larvae and adults by roasting or boiling. They also make a type of chutney from the beetle. When the adults are eaten, people remove the antennae and other appendages (Chakravorty et al. 2011). **Malaysia:** Some species belonging to this genus are used as food in the Sabah State (Chung et al. 2002).

Eurytrachelus titanus The adults are considered edible by the Karbi and Rengma Naga people in the Assam State, **India** as a chutney or roasted (Ronghang and Ahmed 2010).

Hexathrius forsteri The adults is roasted in Arunachal Pradesh State, **India** (Singh et al. 2007).

Lucanus cantor This species is eaten in **India** (Singh et al. 2007).

Lucanus cervus (European stag beetle) **Italy:** The larvae known as *cossus* by the Roman nations, and were consumed (Hope 1842).

Lucanus elaphus The adults are eaten by the Karbi and Rengma Naga people in Assam State, **India** as a chutney or roasted (Ronghang and Ahmed 2010).

Lucanus laminifer This species is edible in Arunachal Pradesh State, **India** (Singh et al. 2007).

Lucanus maculifemoratus The Japanese name is *miyama-kuwagata*. The larvae and adults were used medicinally in **Japan** for treating rheumatism, swelling, difficult delivery, etc. (Umemura 1943, Shiraki 1958).

Lucanus **sp.** The larvae of a species belonging to this genus are treated as foodstuff in **Mexico** (Ramos-Elorduy and Pino 2002).

Odontolabris cuvera The adults are eaten by the Karbi and Rengma Naga people in Assam State, **India** as a chutney or roasted (Ronghang and Ahmed 2010).

Odontolabis gazella The larvae and adults are considered edible by people of the Nishi and Galo tribes in **India**. The larvae are fried slightly in oil and are added to boiled vegetables. When the adults are eaten, the antennae and other appendages are removed before eating (Chakravorty et al. 2011).

Odontolabis **spp. Indonesia:** Some species belonging to this genus are used as food. Only the adults are consumed in the Kalimantan State. The larvae are not eaten (Chung 2010). **Malaysia:** Some species belonging to this genus are considered edible in the Sabah State (Chung et al. 2002).

Prismognathus angularis angularis The Japanese name is *oni-kuwagata*. **Japan:** The larvae and adults were medicinally used for treating rheumatism, swelling, difficult delivery, etc. (Umemura 1943, Shiraki 1958).

Prosopocoilus inclinatus inclinatus The Japanese name is *nokogiri-kuwagata*. The larvae and adults were medicinally used in **Japan** for treating rheumatism, swelling, difficult delivery, etc. (Umemura 1943, Shiraki 1958).

Prosopocoilus serricornis This species is eaten in **Madagascar** (Decary 1937).

Prosopocoilus **sp.** The larvae and adults are consumed by people of the Nishi and Galo tribes in **India**. The adults are usually roasted after removing the antennae and other appendages (Chakravorty et al. 2011).

Psalidoremus inclinatus → *Prosopocoilus inclinatus inclinatus*

Sphaenognathus feisthamelii The larvae are used as food item by the Quichua and Saraguro people in **Ecuador** (Onore 1997).

Sphaenognathus lindenii The larvae are relished by the Quichua people in **Ecuador** (Onore 1997).

Sphaenognathus metallifer The larvae are considered good for eating by the Canari people in **Ecuador** (Onore 1997).

Lyctidae (Powder post beetles) **(1)**

Meligethes aeneus The larvae are fried or as a chutney by the Karbi and Rengma Naga people in Assam State, **India** (Ronghang and Ahmed 2010).

Meloidae (Oil beetles) **(12)**

Alosimus tenuicornis **Morocco:** This species is used as an ingredient for sauce (Benhalima et al. 2003).

Epicauta gorhami (bean blister beetle) **China:** This species was used to induce abortion (Li 1596). **Japan:** The Japanese name is *mame-hanmyō*. This species is said effective for dropsy (Inagaki 1984).

Epicauta megalocephala **Japan:** The Japanese name is *chōsen-mamehanmyō*. The adults were considered good for treating gonorrhea, swelling, etc. (Shiraki 1958). **North Korea:** The powder of the dried adults is used for treating scrofula, gonorrhea, difficulty in urination, etc. in Hwanghae, North Hamgyong, South Hamgyong and South P'yongan Provinces (Okamoto and Muramatsu 1922). **South Korea:** The powder of the dried adults is good for treating scrofula, gonorrhea, difficulty in urination, etc. in North Chungcheong and North Gyeongsang Provinces (Okamoto and Muramatsu 1922).

Lytta caraganae **China:** This species was used to induce abortion (Li 1596).

Lytta vesicatoria **Europe:** This species was a source of cantharidine (Clausen 1954). **Roma Antiqua:** Around AD 0, people used this species medicinally, although they knew of the toxicity of this beetle (Plinius AD 77–79).

Meloe dugesi The adults are used medicinally in **Mexico** (Ramos-Elorduy and Pino 2002).

Meloe laevis The adults are one of folk medicine in **Mexico** (Ramos-Elorduy and Pino 2002).

Meloe nebulosus The adults are used medicinally in **Mexico** (Ramos-Elorduy and Pino 2002).

Meloe violaceus **North Korea:** The powder of the dried adults is effective for treating syphilis in the Hwanghae Province (Okamoto and Muramatsu 1922).

Myrabris cichorii (lesser blister beetle) **China:** This species was used medicinally more than 2000 years ago (Shennung Ben Ts'ao King; Read 1941). It has been considered to promote blood circulation, and is said to be effective against lung cancer (Zimian et al. 1997). Recently, it was found that the extract from this beetle contains anti-cancer substances (Verma and Prasad 2012). **India:** This species is eaten in Arunachal Pradesh State (Singh et al. 2007).

Mylabris himalayaensis This species is consumed by people of Arunachal Pradesh State in **India** (Singh et al. 2007).

Mylabris phalerata (large blister beetle) **China:** This species is used to promote blood circulation, and is said to be effective against lung cancer (Zimian et al. 1997).

Mordellidae (1)

Metoeus stanus **Japan:** This species is a parasite of a wasp, *Vespula flavipes lewisii* (Hymenopter, Vespidae). The larvae and adults of this beetle are sometimes found in tinned wasp larvae (cooked with soy sauce and sugar). Most people eat this beetle with wasp larvae without knowing the contamination of the parasites (Shida 1959).

Nitidulidae (Sap beetles) (0)

Epuraea sp. The adults are roasted by the Nishi people in Arunachal Pradesh State, **India** (Singh et al. 2007).

Noteridae (Burrowing water beetles) (0)

Suphisellus sp. A species belonging to this genus is eaten by the Huasteco, Nahuatl, Otomi, Tarascan and Yutoaztec people in **Mexico** (Ramos-Elorduy et al. 2009).

Passalidae (Bess beetles) (9)

Aceraius helferi This species is appreciated as food by people of Arunachal Pradesh State in **India** (Singh et al. 2007).

Aceraius **spp. Malaysia:** This species is a source of food in Sabah State (Chung et al. 2002).

Aulacocyclus bicuspi This species is eaten by the people of Arunachal Pradesh State in **India** (Singh et al. 2007).

Odontotaenius sp. The larvae and adults of a species belonging to this genus are consumed by the Nishi and Galo tribes in **India**. The adults are roasted or fried after removing the wings (Chakravorty et al. 2011).

Oileus reinator The larvae and pupae are used as food in Oaxaca State, **Mexico** (Ramos-Elorduy de Conconi et al. 1984).

Oileus sp. Mexico: Native people of the Chontale, Huasteco, Huave, Lacandone, Maya, Mestizo, and Náhua tribes eat this species (Ramos-Elorduy and Pino 1989).

Passalus interruptus This species is eaten in India (Brygoo 1946), Sri Lanka (Nandacena et al. 2010) and Suriname (Brygoo 1946). The larvae are also consumed in Colombia and Paraguay (DeFoliart 2002).

Passalus sp. The larvae of a species, whose Mexican name is *gusano del palo*, belonging to this genus are eaten in Mexico (Ramos-Elorduy et al. 1998).

Passalus (Passalus) interstitialis The larvae and pupae are regarded as food in Mexico (Ramos-Elorduy and Pino 2002).

Passalus (Passalus) punctiger The Mexican name is *bechano*. The larvae and pupae are consumed in Oaxaca State, Mexico (Ramos-Elorduy de Conconi et al. 1984).

Passalus (Passalus) sp. Mexico: Native People of the Chontale, Huasteco, Huave, Lacandone, Maya, Mestizo, Náhua, Trapaneca, and Totonaca tribes eat the larvae and pupae of a species belonging to this genus (Ramos-Elorduy and Pino 1989 and 2002).

Passalus (Pertinax) punctarostriatus The larvae and pupae are eaten in Mexico (Ramos-Elorduy and Pino 2002).

Paxillus leachi The larvae and pupae are used as foodstuff in Mexico (Ramos-Elorduy and Pino 1990).

Veturius sinuosus The larvae are relished by the Tukanoan people in Colombia (Dufour 1987).

Psephenidae (0)

Psephemus sp. Japan: The Japanese name is *hirata-doromushi*. The larvae are a member of *zazamushi* (see Tricoptera-Leptoceridae-*Stenopsyche sauteri*). They are cooked with soy sauce and sugar in the Nagano Prefecture (Torii 1957).

Ptinidae (Spider beetles) (0)

Ptinus sp. USA: A species of spider beetles belonging to this genus seemed to have been consumed by the aboriginal people in Nevada State, as it was found in the coprolites excavated from the Lovelock Cave of Nevada State (BC 1,200) (Heizer and Napton 1969).

Scarabaeidae (202)

Adoretus compressus Sri Lanka: This species is eaten (Nandacena et al. 2010). Thailand: The Laotian name is *mang ee noon*. The people of the Lao tribe eat this species (Bristowe 1932).

Adoretus convexus The Laotian name is *mang ee noon*. The people of the Lao tribe eat this species in **Thailand** (Bristowe 1932).

Adoretus cribatus People eat this species in **Thailand** (Leksawasdi 2010).

Adoretus pachysomatus This species is used as food source in **Thailand** (Masumoto and Utsunomiya 1997).

Adoretus spp. Several species belonging to this genus are edible in **Thailand** (Hanboonsong et al. 2000).

Agestrata orichalca People eat this species in **Thailand** (Hanboonsong et al. 2000).

Agestrata sp. The adults are roasted by the Nishi people in Arunachal Pradesh State, **India** (Singh et al. 2007).

Amphimallon assimile → *Rhizotrogus assimile*

Amphimallon pini → *Rhizotrogus pini*

Ancognatha castanea The larvae are regarded as a food item by the Canari, Otavalo, Pilahuine, Quichua, Salazaca and Saraguro people in **Ecuador** (Onore 2005).

Ancognatha jamesoni The larvae are eaten by the Quichua people in **Ecuador** (Onore 2005).

Ancognatha vulgaris The larvae are appreciated as food by the Canari, Otovalos and Quichua people in **Ecuador** (Onore 1997 and 2005).

Ancognatha sp. The fried larvae are eaten in **Colombia** (DeFoliart 2002).

Anomala anguliceps People eat this species in **Thailand** (Hanboonsong et al. 2000).

Anomala antiqua People eat this species in **Thailand** (Hanboonsong et al. 2000).

Anomala bilunulata People eat this species in **Thailand** (Utsunomiya and Masumoto 1999).

Anomala blaisei This species is considered as foodstuff in **Thailand** (Utsunomiya and Masumoto 2000).

Anomala cantori This species is recognized as a food item in **Thailand** (Utsunomiya and Masumoto 2000).

Anomala chalcites People eat this species in **Thailand** (Hanboonsong et al. 2000).

Anomala chlrochelys This species is eaten in **Thailand** (Leksawasdi 2008).

Anomala concha This species is accepted as a food source in Sabah State, **Malaysia** (Chung et al. 2002).

Anomala corpulenta (metallic-green beetle) The larvae are eaten in **China** (Chen and Feng 1999). This species is used as an anodyne (Zimian et al. 1997).

Anomala coxalis This species is consumed in Sabah State, **Malaysia** (Chung et al. 2002).

Anomala cuprea This species is used as a diuretic or to cure internal bleeding in **China** (Inagaki 1984).

Anomala cupripes **Thailand:** People eat this species (Hanboonsong et al. 2000). **Laos, Myanmar** and **Vietnam:** This species is eaten (Yhoung-Aree and Viwatpanich 2005).

Anomala exoleta (red-brown beetle) **China:** This species is used as an anodyne (Zimian et al. 1997).

Anomala fusikibia This species is consumed in **Thailand** (Utsunomiya and Masumoto 2000).

Anomala laotica People eat this species in **Thailand** (Hanboonsong et al. 2000).

Anomala lasiocnemis This species is considered as an article of food in Sabah State, **Malaysia** (Chung et al. 2002).

Anomala latefemorata People eat this species in Sabah State, **Malaysia** (Chung et al. 2002).

Anomala lignea People appreciate this species as foodstuff in **Thailand** (Hanboonsong et al. 2000).

Anomala pallida People eat this species in **Thailand** (Hanboonsong 2010).

Anomala parallera This species is regarded as foodstuff in **Thailand** (Utsunomiya and Masumoto 2000).

Anomala puctulicollis This species is eaten in **Thailand** (Utsunomiya and Masumoto 2000).

Anomala rugosa This species is considered as food item in **Thailand** (Utsunomiya and Masumoto 2000).

Anomala scherei This species is recognized as a food article in **Thailand** (Hanboonsong et al. 2000).

Anomala shanica This species is eaten in **Thailand** (Hanboonsong et al. 2000).

Anomala vuilletae This species is a food resource in **Thailand** (Hanboonsong et al. 2000).

Anomala **spp. India:** The adults of a species belonging to this genus are roasted or boiled by people of Arunachal Pradesh State (Chakravorty et al. 2011). **Thailand:** Some species belonging to this genus are consumed (Leksawasdi 2008).

Anoplognathus viridiaeneus The larvae are eaten in **Australia**. They were said to be palatable and wholesome (Hope 1842).

Aphodius **sp.** In **North Korea,** the powder of the dried adults is used for treating epilepsy, difficult delivery, etc. in the Hwanghae and North Hamgyong Provinces (Okamoto and Muramatsu 1922).

Aphodius (Pharaphodius) crenatus People eat this species in **Thailand** (Hanboonsong et al. 2000).

Aphodius (Pharaphodius) marginellus People relish this species in **Thailand** (Hanboonsong et al. 2000).

Aphodius (Pharaphodius) putearius People eat this species in **Thailand** (Hanboonsong et al. 2000).

Aphodius (Pharaphodius) **sp.** A species belonging to this genus is consumed in **Thailand** (Hanboonsong 2010).

Apogonia **sp.** People eat a species belonging to this genus in southern **Thailand** (Lumsa-ad 2001, Yhoung-Aree and Viwatpanich 2005).

Ateuches sacer The adults are eaten in **Egypt** (Bodenheimer 1951).

Brachylepis bennigseri This species is consumed in **Cameroon** (Seignobos et al. 1996).

Brahmina mikado This species is used as food item in **Thailand** (Hanboonsong et al. 2000).

Brahmina parvula This species is eaten in **Thailand** (Hanboonsong et al. 2000).

Canthon humectus → *Chanton humectus hidalguensis*

Cassolus humeralis This species is regarded as food in Arunachal Pradesh State, **India** (Singh et al. 2007).

Cathacius birmanicus → *Catharsius birmanicus*

Cathacius molossus → *Catharsius molossus*

Catharsius birmanicus People eat this species in **Thailand** (Hanboonsong et al. 2000).

Catharsius molossus **China:** People eat the larvae and adults (Chen and Feng 1999), and is also used as medicine for treating various diseases (Nanba 1980). **Thailand:** This species is a food article (Hanboonsong et al. 2000).

Catharsius **sp.** The adults of a species belonging to this genus are recognized as a food item by the Galo people in **India**. Wet paste made from this insect is used as a remedy for diarrhea (Chakravorty et al. 2011).

Chaetadoretus cribratus People eat this species in **Thailand** (Hanboonsong et al. 2000).

Chanton humectus hidalguensis This species is treated as a foodstuff in Hidalgo State, **Mexico** (Ramos-Elorduy 2006). The Hñähñu people eat this species for treating whooping cough (Aldasoro Maya 2003).

Clavipalpus antisanae The larvae are considered good for eating the Quichua people in **Ecuador** (Onore 1997).

Coelosis biloba The larvae are regarded as food the Awa and Tsachila people in **Ecuador** (Onore 1997).

Copris **(s.str.)** *carinicus* People eat this species in **Thailand** (Hanboonsong 2010).

Copris corpulentus This species is eaten in **India** (Singh et al. 2007) and **Thailand** (Utsunomiya and Masumoto 2000).

Copris furciceps This species is used as foodstuff in **India** (Singh et al. 2007) and **Thailand** (Utsunomiya and Masumoto 1999).

Copris **(s.str.)** *nevinsoni* This species is considered as an article of food in **Thailand** (Hanboonsong 2010).

Copris (Paracopris) puctulatus People eat this species in **Thailand** (Hanboonsong 2010).

Copris punctatus This species is relished in Arunachal Pradesh State, **India** (Singh et al. 2007).

Copris (Microcopris) reflexus People eat this species in **Thailand** (Hanboonsong 2010).

Copris sinicus This species is appreciated as a food item in **Thailand** (Utsunomiya and Masumoto 1999).

Copris (Microcopris) vittalisi This species is accepted as food in Arunachal Pradesh State, **India** (Singh et al. 2007).

Copris (Paracopris) **sp.** A species belonging to this genus is eaten in **Thailand** (Hanboonsong et al. 2000).

Cotilis **sp.** A species belonging to this genus is roasted in **Indonesia** (West Papua State) (Ramandey and van Mastrigt 2010).

Cotinis mutabilis **var.** *oblicua* This subspecies is eaten in **Mexico** (Ramos-Elorduy 2006).

Cyclocephala borealis (northern masked chafer) The adults are good for eating in the **USA** (Sutton 1988).

Cyclocephala dimidiata The adults are roasted in the **USA** (Sutton 1988).

Cyclocephala fasciolata **Thailand:** The larvae, pupae, and the adults are considered edible (Leksawasdi 2008). **Mexico:** The larvae and adults are used as food item (Ramos-Elorduy and Pino 2002).

Cyclocephala melanocephala → *Cyclocephala dimidiata*

Cyclocephala villosa → *Cyclocephala borealis*

Democrates burmeisteri The larvae are eaten by the Quichuas and Salazacas people in **Ecuador** (Onore 1997).

Dicranocephalus wallichi The larvae and adults are recognized as food in **China** (Hu and Zha 2009).

Diplognatha gagates **Cameroon:** The Mofu people eat this species throughout the year (Seignobos et al. 1996).

Empectida tonkinensis This species is eaten in **Thailand** (Utsunomiya and Masumoto 2000).

Euchloropus laetus This species is eaten in **Thailand** (Leksawasdi 2008).

Eulepida anatina The larvae are appreciated as food in **Zimbabwe** (Chavanduka 1975).

Eulepida mashona The larvae are consumed in **Zimbabwe** (Chavanduka 1975).

Eulepida nitidicollis The larvae are used as a food source in **Zimbabwe** (Chavanduka 1975).

Exolontha castanea This species is consumed in **Thailand** (Hanboonsong et al. 2000).

Exopholis hypoleuca This species is treated as foodstuff in Sabah State, **Malaysia** (Chung et al. 2002).

Exopotus **sp.** A species belonging to this genus is used as a food item in **Thailand** (Sangpradub 1982).

Geniatosoma nignum The larvae are relished in **Brazil** (Carrera 1992, Milton 1984).

Gnathocera **sp.** The larvae of a species belonging to this genus are regarded as food item in **DRC** (DeFoliart 2002).

Goliathus cacicus This species is consumed as food in **CAR** (Bergier 1941).

Goliathus cameronensis This species is appreciated as food source in **CAR** (Bergier 1941).

Goliathus goliathus (Goliath beetles) **(PL V-4)** This species is eaten in **CAR** (Bergier 1941, Roodt 1993).

Goliathus regius This species is used as a food article in **CAR** (Bergier 1941).

Goliathus **sp.** A species belonging to this genus is consumed in **DRC** (DeFoliart 2002).

Gymnopleurus aethiops This species is regarded as a food item in **Thailand** (Yhoung-Aree and Viwatpanich 2005).

Gymnopleurus melanarius People eat this species in **Thailand** (Hanboonsong et al. 2000).

Gymnopleurus mopsus **China:** This species is used for treating various diseases (Nanba 1980).

Gymnopleurus sinuatus **North Korea:** The pulverized larvae and adults are used for treating convulsive fits in the Hwanghae Province (Okamoto and Muramatsu 1922). **South Korea:** The pulverized larvae and adults are used for treating convulsive fits in North Gyeongsang and North Joella Province (Okamoto and Muramatsu 1922).

Heliocopris bucephalus **China:** This species is used for treating various diseases (Nanba 1980). **India:** This species is eaten by people of Arunachal Pradesh State (Singh et al. 2007). **Laos:** The Laotian name is *boa* or *duang chud chii*. This species is eaten (Boulidam 2008). **Myanmar:** People of the Shan tribe dig dung-ball from below the earth, and eat the pupae (Ghosh 1924). **Thailand:** This species is cooked with vegetables, used for soup, or added to a curry (Mungkorndin 1981).

Heliocopris dominus **(PL V-5a and 5b)** This species is eaten in **Laos, Myanmar, Thailand** and **Vietnam** (Yhoung-Aree and Viwatpanich 2005).

Heliocopris **sp.** The adults of a species belonging to this genus are consumed by people living in the northern part of **Thailand**. This beetle is expensive (DeFoliart 2002).

Heteronychus lioderes People eat this species in **Thailand** (Hanboonsong et al. 2000).

Holotrichia cephalotes People eat this species in **Thailand** (Utsunomiya and Masumoto 1999).

Holotrichia diomphalia **China:** This species was used medicinally for more than 2000 years (Shennung Ben Ts'ao King; Read 1941).

Holotrichia diomphilia **China:** This species is used as a diuretic as well as an anodyne (Nanba 1980, Zimian et al. 1997).

Holotrichia hainanensis People eat this species in **Thailand** (Hanboonsong et al. 2000).

Holotrichia lata The larvae and adults are used as foodstuff in **China** (Hu and Zha 2009).

Holotrichia nigricollis People eat this species in **Thailand** (Hanboonsong et al. 2000).

Holotrichia nigricollis rubricollis People consume this species in **Thailand** (Utsunomiya and Masumoto 1999).

Holotrichia morosa **China:** This species is used as an anodyne (Zimian et al. 1997).

Holotrichia oblita The larvae are eaten in **China** (Chen and Feng 1999).

Holotrichia ovata The larvae and adults are treated as foodstuff in **China** (Hu and Zha 2009).

Holotrichia parallela The larvae are used as a food item in **China** (Chen and Feng 1999).

Holotrichia pruinosella This species is a food article in **Thailand** (Leksawasdi 2008).

Holotrichia siamensis This species is a food item in **Thailand** (Leksawasdi 2010).

Holotrichia sinensis The larvae and adults are a food source in **China** (Hu and Zha 2009).

Holotrichia srobiculata The larvae and adults are used as food item in **China** (Hu and Zha 2009).

Holotrichia szechuanensis The larvae and adults are treated as foodstuff in **China** (Hu and Zha 2009).

Holotrichia titanis **China:** This species is used as an anodyne (Zimian et al. 1997).

Holotrichia **spp. Thailand:** At least two species belonging to this genus are fried (Leksawasdi and Jirada 1983). **Laos, Myanmar** and **Vietnam:** Some species belonging to this genus are eaten (Yhoung-Aree and Viwatpanich 2005).

Hoplosternus malaccensis → Melolontha malaccensis

Lachnosterna **sp.** This species is fried in the **USA** (Howard 1915 and 1916).

Lepidiota anatina → Eulepida anatina

Lepidiota bimaculata The adults are steamed or roasted with salt, in **Thailand** (Mungkorndin 1981, Utsunomiya and Masumoto 1999).

Lepidiota discidens This species is a food source in **Thailand** (Leksawasdi 2008).

Lepidiota hauseri This species is considered as food item in **Thailand** (Masumoto and Utsunomiya 1997).

Lepidiota hypoleuca The adults are roasted in **Indonesia** (Java) (Bodenheimer 1951).

Lepidiota mashona → Eulepida mashona

Lepidiota nitidicollis → Eulepida nitidicollis

Lepidiota punctum The adults are roasted after removing the wings and legs, or boiled in the **Philippines** (Gibbs et al. 1912).

Lepidiota stigma **India:** The adults are eaten roasted by the Nishi people in Arunachal Pradesh State (Singh et al. 2007). **Indonesia:** In Kalimantan State, only the adults are consumed. The larvae are not eaten (Chung 2010). **Sri Lanka:** This species is considered as food item (Nandacena et al. 2010). The adults are widely relished in northern and eastern **Thailand** (Jonjuapsong 1996).

Lepidiota stigma **var. *alba* Thailand:** The Laotian name: *mang ee noon*. This species is eaten by frying (Bristowe 1932, Jonjuapsong 1996). People around Songkhla in the southern area eat the adults by cooking after removing the wings, legs and intestines (Nakao 1964).

Lepidiota vogeli The adults are fried in **PNG** (Jolivet 1971).

Lepidiota **sp.** Some species belonging to this genus are eaten in **India** (Chakravorty et al. 2011) and **Thailand** (DeFoliart 2002).

Leucopelaea albescens The larvae and adults are consumed by the Otavalo, Quichuas and Salazacas people in **Ecuador** (Onore 1997).

Leucopholis irrorata The adults are boiled or roasted after removing the wings and legs in the **Philippines** (Gibbs et al. 1912).

Leucopholis pulverulenta The adults are boiled or roasted after removing the wings and legs in the **Philippines** (Gibbs et al. 1912).

Leucopholis rorida The adults are roasted in **Indonesia** (Java) (Bodenheimer 1951).

Leucopholis staudingeri This species is eaten in Sabah State, **Malaysia** (Chen et al. 2002).

Leucopholis **sp.** The adults are widely consumed in northern and eastern **Thailand** (Jonjuapsong 1996).

Leucophosis → Leucopholis

Liatongus affinis People eat this species in **Thailand** (Hanboonsong et al. 2000).

Liatongus (Paraliatongus) rhadamistus People eat this species in **Thailand** (Hanboonsong et al. 2000).

Liatongus tridentatus This species is eaten in **Thailand** (Hanboonsong et al. 2000).

Liatongus venator People eat this species in **Thailand** (Hanboonsong et al. 2000).

Liatongus (Paraliatongus) rhadamistus This species is eaten in **Thailand** (Hanboonsong et al. 2000).

Liocola brevitarsis **China:** This species is used as an anodyne (Zimian et al. 1997).

Macrodactylus lineaticollis This species is used as food in **Mexico** (Ramos-Elorduy et al. 2006).

Maladera **sp.** A species belonging to this genus is recognized as food source in **Thailand** (Leksawasdi 2008).

Megistophylla andrewesi The adults are considered as good food item in **Thailand** (Masumoto and Utsunomiya 1997).

Melolontha aprilina **Italy:** Farmers in Lombardia ate the abdomens of the adult beetles (Ealand 1915).

Melolontha hypoleuca The adults are eaten in **Indonesia** (Java) (Hope 1842).

Melolontha malaccensis The adults are consumed in **Thailand** (Masumoto and Utsunomiya 1997).

Melolontha melolontha (cockchafer or maybug) **Czech:** People around Pilsen make a soup from the adult beetles (Bejsak 1992).

Melolontha **sp.** In **Mexico**, the larvae of a species called *gallina ciega* in Mexican and belonging to this genus are eaten (Ramos-Elorduy and Pino 1990).

Microtrichia **sp.** A species belonging to this genus is considered edible in **Thailand** (Jamjanya et al. 2001).

Mimela ferreroi The adults are food item in **Thailand** (Hanboonsong et al. 2000).

Mimela ignistriata The adults are good for eating in **Thailand** (Utsunomiya and Masumoto 2000).

Mimela linping The adults are fit for eating in **Thailand** (Utsunomiya and Masumoto 1999).

Mimela luciaula **North Korea:** The infusion of the larvae and adults or the liquid obtained by squeezing the adults are used for treating malaria, paralysis, stomach disorders, etc. in Hwanghae, North P'yongan, South P'yongan and North Hamgyong Provinces (Okamoto and Muramatsu 1922). **South Korea:** The infusion of the larvae and adults or the liquid obtained by squeezing the adults are used for treating malaria, paralysis, stomach disorders, etc. in Gyeonggi, North Gyeongsang, South Gyeongsang and North Joella Provinces (Okamoto and Muramatsu 1922).

Mimela schneideri This species is used as foodstuff in **Thailand** (Masumoto and Utsunomiya 1997).

Mimela schulzei The adults are food item in **Thailand** (Hanboonsong et al. 2000).

Mimela splendens The Japanese name is *koganemushi*. In **Japan,** the larvae were used for treating swelling, malaria, stomach disorders, etc. (Shiraki 1958).

Mimela sp. A species belonging to this genus is considered good for eating in **Thailand** (Leksawasdi 2008).

Miridiba tuberculipennis obscura The adults are appreciated as a food in **Thailand** (Hanboonsong et al. 2000).

Oniticellus cinctus The adults are recognized as food in **Thailand** (Hanboonsong et al. 2000).

Onitis castaneous This species is considered as food item in Arunachal Pradesh State in **India** (Singh et al. 2007).

Onitis falcatus The adults are regarded as foodstuff in **Thailand** (Masumoto and Utsunomiya 1997).

Onitis feae This species is treated as foodstuff in Arunachal Pradesh State in **India** (Singh et al. 2007).

Onitis kiuchii → *Onitis falcatus*

Onitis niger People eat this species in **Thailand** (Hanboonsong et al. 2000).

Onitis subopacus **India:** This species is eaten by people of Arunachal Pradesh State (Singh et al. 2007b). **Thailand:** People eat this species (Hanboonsong et al. 2000).

Onitis virens The Laotian name is *mang chew*. In **Thailand,** Laotian people cook the pupae with vegetables, make soup from the pupae, or put the pupae into a curry (Bristowe 1932, Mungkorndin 1981).

Onitis sp. A species belonging to this genus is eaten in **Thailand** (DeFoliart 2002).

Onthophagus avocetta People eat this species in **Thailand** (Hanboonsong et al. 2000).

Onthophagus bonasus People consume this species in **Thailand** (Hanboonsong et al. 2000).

Onthophagus khonmiinitnoi People use this species for meal in **Thailand** (Hanboonsong et al. 2000).

Onthophagus luridipennis This species is a food source in **Thailand** (Hanboonsong et al. 2000).

Onthophagus mouhoti This species is considered as foodstuff in **Thailand** (Jamjanya et al. 2001).

Onthophagus orientalis People eat this species in **Thailand** (Hanboonsong et al. 2000).

Onthophagus papulatus People consumed this species in **Thailand** (Hanboonsong et al. 2000).

Onthophagus proletarius This species is recognized good for eating in **Thailand** (Hanboonsong et al. 2000).

Onthophagus ragoides People eat this species in **Thailand** (Hanboonsong 2010).

Onthophagus rectecornutus This species is used as foodstuff in **Thailand** (Hanboonsong et al. 2000).

Onthophagus sagittarius People consume this species in **Thailand** (Hanboonsong et al. 2000).

Onthophagus seniculus People eat this species in **Thailand** (Utsunomiya and Masumoto 2000).

Onthophagus taurinus This species is accepted as a food source in **Thailand** (Hanboonsong et al. 2000).

Onthophagus tragoides This species is eaten in **Thailand** (Hanboonsong et al. 2000).

Onthophagus tragus People eat this species in **Thailand** (Hanboonsong et al. 2000).

Onthophagus tricornis People use this species as a food item in **Thailand** (Hanboonsong et al. 2000).

Onthophagus trituber People consume this species in **Thailand** (Hanboonsong et al. 2000).

Onthophagus **sp.** A species belonging to this genus is eaten in **Thailand** (Hanboonsong 2010).

Oxycetonia jucunda The larvae are fried in **China** (Chen and Feng 1999).

Pachnessa **sp.** People eat a species belonging to this genus in **Thailand** (Hanboonsong et al. 2000).

Pachnoda marginata **Cameroon:** The Mofu people consume this species throughout the year (Seignobos et al. 1996).

Pachylomera fermoralis The larvae are fried in **Zambia** (Mbata 1995).

Paragymnopleurus aethiops → *Gymnopleurus aethiops*

Paraphytus hindu This species is eaten by the people of Arunachal Pradesh State in **India** (Singh et al. 2007).

Pelidnota nigricauda The larvae and adults are fried by the Quichua people in **Ecuador** and **Venezuela** (Onore 1997 and 2005).

Pelidnota (*Chalcoplethis*) **sp.** The larvae of a species belonging to this genus are eaten by the Yanomamo people in **Venezuela** (Paoletti and Dufour 2005).

Phyllophaga crinite This species is consumed in the **USA** (Essig 1934).

Phyllophaga fusca This species is used as foodstuff in the **USA** (Essig 1934).

Phyllophaga mexicana The larvae are reared by people of the Mazahua, Otomi and Nahua tribes in **Mexico** as a food for humans (Ramos-Elorduy 2009).

Phyllophaga rubella The Mexican name is *gusano de la tierra*. The larvae are consumed in the temperate zones of **Mexico** (Ramos-Elorduy and Pino 1989).

Phyllophaga rugipennis The larva, pupae and the adults are considered as foodstuff in **Mexico** (Ramos-Elorduy and Pino 2002).

Phyllophaga **spp.** The Mexican name is *gusano de la tierra*. In **Mexico,** native people of the Náhua and Otomi tribes eat the larva, pupae and the adults of several species belonging to this genus (Ramos-Elorduy de Conconi et al. 1984, Ramos-Elorduy and Pino 1989 and 2002). **Thailand:** Some species belonging to this genus are treated as foodstuff (Jamjanya et al. 2001).

Platycoelia forcipalis The larvae are eaten fried in **Ecuador** (Onore 1997).

Platycoelia lutescens The larvae and the adults are fried in **Ecuador** (Onore 2005).

Platycoelia parva The larvae are eaten in **Ecuador** (Onore 1997).

Platycoelia rufosignata The larvae are eaten in **Ecuador** (Onore 1997).

Platygenia barbata This species is considered edible in **DRC** (Adriaens 1951).

Platygenia **spp.** The larvae of several species belonging to this genus are used as food item in **Africa** (tropical area) (Ghesquiére 1947).

Polyphylla crinite The larvae are recognized edible in the **USA** (Essig 1934 and 1965).

Polyphylla laticollis The larvae are used as food item in **China** (Chen and Feng 1999).

Polyphylla tonkinensis This species is eaten in **Thailand** (Hanboonsong et al. 2000).

Polyphylla **sp.** The larvae and adults of a species probably belonging to this genus are roasted after removing the antennae and appendages in Arunachal Pradesh State, **India** (Chakravorty et al. 2011).

Popillia femoralis This species is an important food insect in **Cameroon** (DeLisle 1944, DeFoliart 1989).

Proagolofa unicolor The larvae and adults are eaten in **Ecuador** (Onore 2005).

Proagosternus **sp.** A species belonging to this genus is eaten in **Madagascar** (Decary 1937).

Propomacrus **sp.** The adults are used as foodstuff by the Nishi people of Arunachal Pradesh State in **India**. The beetles are smoked, roasted or boiled after removing the wings (Chakravorty et al. 2011).

Protaetia aerate The larvae are fried in **China** (Chen and Feng 1999).

Protaetia fusca This species is used as a food article in **Thailand** (Leksawasdi 2008).

Protaetia **sp.** A species belonging to this genus is considered edible in **Thailand** (Hanboonsong et al. 2000).

Psilophosis **sp.** In **Thailand,** a species belonging to this genus is fried after removing the wings. People sometimes grind the beetles to make a spicy sauce, or put them into bamboo shoot soup or red ant egg soup (Vara-asavapati et al. 1975).

Pyronota festiva The Maori name is *manuka*. In **New Zealand**, this species is said to have found occasional acceptance as human food (Meyer-Rochow and Changkija 1997).

Rhizotrogus assimilis **Italy:** Farmers in Lombardia ate the abdomens of the adult beetles (Ealand 1915).

Rhizotrogus pini **Romania:** In previous Moldavia and Walachia, people ate the adults as a common food (Hope 1842).

Scaptodera rhadamistus → *Liatongus (Paraliatongus) rhadamistus*

Scarabaeus molossus The larvae are pulverized after roasting in **China** (Donovan 1842).

Scarabaeus perigrinus **China:** This species is used for treating various diseases (Terada 1933).

Scarabaeus sacer The adults were eaten in **Egypt** mainly by women who wanted a degree of plumpness. As being plump was considered a charming factor (Hope 1842, Ealand 1915). In the community of the Bedouin people, a ceremony initiating a boy to manhood takes place when a boy reaches the age of 11 or 12. At the ceremony people let the boys eat the adult beetles of this species (Bristowe 1932).

Sophrops abscessus People eat this species in **Thailand** (Hanboonsong et al. 2000).

Sophrops bituberculatus People consumed this species in **Thailand** (Hanboonsong et al. 2000).

Sophrops excises This species is eaten in **Thailand** (Hanboonsong et al. 2000).

Sophrops foveatus This species is used as a food article in **Thailand** (Hanboonsong et al. 2000).

Sophrops opacidorsalis This species is a food source in **Thailand** (Hanboonsong et al. 2000).

Sophrops paucisetosa This species is accepted as food in **Thailand** (Hanboonsong et al. 2000).

Sophrops rotundicollis People eat this species in **Thailand** (Hanboonsong et al. 2000).

Sophrops simplex This species is used as food in **Thailand** (Hanboonsong et al. 2000).

Sophrops spancisetosa This species is used as a food item in **Thailand** (Masumoto and Utsunomiya 1997).

Sophrops **sp.** At least two species belonging to this genus are considered edible in **Thailand** (Hanboonsong 2010).

Strategus aloeus **Mexico:** Native people of the Hñähñu tribe eat this species (Ramos-Elorduy de Conconi et al. 1984, Aldasoro Maya 2003).

Strategus **sp. Mexico:** The larvae are eaten in the Nayarit and Chiapas State (Ramos-Elorduy 2006).

Trematodes tenebriodes **China:** This species is used as a diuretic or as a remedy for internal bleeding (Inagaki 1984).

Tricholespis **sp.** The larvae of a species belonging to this genus are fried in **Angola** (Bergier 1941) and **Madagascar** (Decary 1937, Paulian 1943).

Xyloryctes teuthras The larvae, pupae and the adults are consumed in **Mexico** (Ramos-Elorduy and Pino 2002).

Xyloryctes thestalus The larvae, pupae and the adults are used as foodstuff in **Mexico** (Ramos-Elorduy and Pino 2002).

Xyloryctes ensifer The larvae, pupae and the adults are eaten in **Mexico** (Ramos-Elorduy and Pino 2002).

Xyloryctes **spp.** The Mexican name is *escarabajo rinoceronte*. In **Mexico,** native people of the Náhua, and Otomi tribes eat some species belonging to this genus (Ramos-Elorduy and Pino 1989).

Scolitidae (Bark beetles) **(3)**

Spaerotrypes yunnanensis The larvae are considered as edible in **China** (Chen and Feng 1999).

Tomicus piniperda The larvae are consumed as food in **China** (Chen and Feng 1999).

Xyleborus emarginatus The larvae are food source in **China** (Chen and Feng 1999).

Silvaniidae (Flat-grain beetles) **(0)**

Oryzaephilus sp. **Greece:** This beetle was found in a type of pea excavated from the Akrotiri Ruin, and was thought to be consumed as foodstuff by ancient people (Panagiotakopulu and Buchland 1991).

Tenebrionidae (Darkling beetles) **(11)**

Asida rughosissima **Mexico:** This species is used as a remedy for insomnia (Aldasoro Maya 2003).

Blaps rhynchoptera **China:** The alcoholic extract of this species is said to be used as an antifebrile (Umeya 2004).

Blaps sulcata The adults are eaten raw in **Egypt** (Bequaert 1921), **Saudi Arabia** (DeFoliart 2002), **Tunisia** (DeFoliart 2002), and in **Turkey** (Ealand 1915).

Blaps sp. Women in **Turkey** ate the adults of some beetles belonging to this genus when they wished to put on weight. As plump women were considered the ideal beauty (Shaw 1738).

Eleodes sp. **Mexico:** The larvae are used as a food article (Ramos-Elorduy and Pino 2002). The Hñähñu people in Hidalgo State eat this species as a remedy for hyperemia of the trachea (Aldasoro Maya 2003).

Martianus dermestoides → *Palembus dermestoides*

Palembus dermestoides **China:** The adults are eaten raw as a medicine (Mao 1997). This species is raised, and is used as a cordial (Zimian et al. 1997). **Japan:** The Japanese name is *kyūryū-gomimushidamashi*. While the adults were eaten raw expecting a medicinal effect in many prefectures, the effect however was doubtful (Mainichi News Paper 1979). The **Philippines:** This species was a popular food item in the early 1970s (Adalla and Cervancia 2010).

Pimelia sp. Women in **Turkey** ate the adults of some beetles belonging to this genus when they wished to put on weight. As plump women were considered the ideal beauty (Shaw 1738).

Stenomorpha sp. Mexico: This species is used as a remedy for insomnia (Aldasoro Maya 2003).

Tenebrio molitor (meal worms) (PL VI-1) China: It is recommended that this species be raised on a large scale as food for domestic animals and fowls. The larvae and pupae are considered as food item by people (Chen and Feng 1999, Chin and Chin 2006, Liú 2006). Mexico: The larvae and pupae are consumed in the Jalisco, Michoacan, Hidalgo and DF States (Ramos-Elorduy 2009), and are raised as food (Ramos-Elorduy 1997). Thailand: This species is used as food item (Jamjanya et al. 2001). The USA: The larvae are fried or used to make candy (Huyghe 1992) (PL VI-2).

Tenebrio obscurus The larvae and pupae are consumed as food item in China (Hu and Zha 2009).

Tenebrio sp. Saudi Arabia: The adults of a species belonging to this genus are eaten raw or fried (Hope 1842). Turkey: Women ate the adults of some beetles belonging to this genus when they wished to put on weight. As plump women were considered ideally beautiful (Shaw 1738).

Tribolium castaneum (red flour beetle) This species is eaten in the Jalisco, Michoacan, Hidalgo and DF States in Mexico (Ramos-Elorduy 2009), and is raised as food (Ramos-Elorduy 1997).

Tribolium confusum China: The larvae and pupae are treated as food item (Hu and Zha 2009). Mexico: This species is eaten in Jalisco, Michoacan, Hidalgo and DF States (Ramos-Elorduy 2009), and is raised as food (Ramos-Elorduy 1997).

Ulomoides dermestoides The larvae and adults are considered as edible in Saudi Arabia (Costa-Neto 1999).

Zopherus jourdani The adults are consumed in Mexico (Ramos-Elorduy and Pino 2002).

Zophobas morio This species is raised as food in Mexico (Ramos-Elorduy 1997).

Zophobas spp. Venezuela: The Yekuana people living around Alto Orinoco, eat the larvae roasted (Araujo and Beserra 2007). Mexico: Some species belonging to this genus are consumed in Jalisco, Michoacan, Hidalgo and DF States (Ramos-Elorduy 2009).

Trictenotomidae (0)

Trictenotoma sp. The adults are smoked, roasted or boiled after removing the wings by the people of Arunachal Pradesh State in India (Chakravorty et al. 2011).

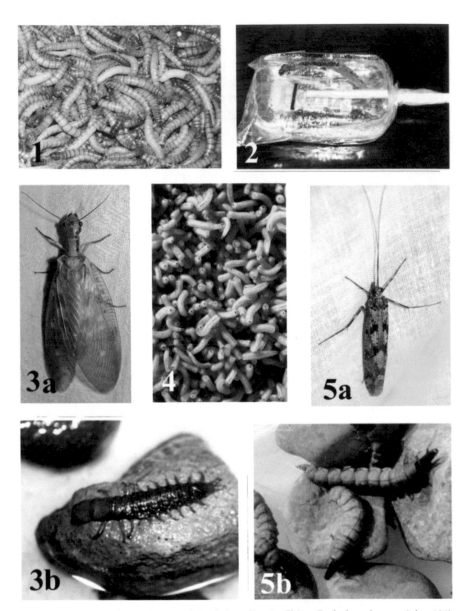

Plate VI. 1. Mass cultured larvae of *Tenebrio molitor* in China. Body length: mm (cf. p.131); 2. An USA made candy containing a larva of *T. molitor* (cf. p.131); 3. *Protohermes grandis*. a: Adult. Body length: 40 mm, wing span: 100 mm, b: Larvae of *P. grandis*. Body length: 60 mm (cf. p.133); 4. Mass cultured larvae of *Musca domestica* in China. Body length: 10 mm (cf. p. 135); 5. *Stenopsyche marmorata*. a: an adult. Body length: 18 mm, Wing span: 50 mm. b: A larva of *S. marmorata*. Body length: 45 mm (cf. p.140).

13. NEUROPTERA [7]

Corydalidae (Dobsonflies) **(6)**

Acanthacorydalis orientalis In **China**, the minorities of Yunnan Province eat the larvae and adults fried in oil or frizzled after removing the head, tails and the intestines. Often people add a bit of salt on them before eating. Otherwise, they boil the larvae, remove the head and the intestines, and fry in oil or frizzle (Chen and Feng 1999, Chen et al. 2009).

Corydalus armatus This species is eaten in **Peru** (Paoletti and Dufour 2005).

Corydalus cornutus The larvae are used as food item by the Tzeltzales Tzoltziles people in **Mexico** (Ramos-Elorduy and Pino 2002).

Corydalus peruvianus This species is consumed in **Peru** (Paoletti and Dufour 2005).

Corydalus **spp. Colombia-Venezuela:** Native people of the Yukpa and Yekuana tribes eat the larvae of several species belonging to this genus (Ruddle 1973, Araujo and Beserra 2007).

Parachauliodes japonicas The Japanese name is *kurosuji-hebitonbo*. In **Japan,** this species live in the same area as *Protohermes grandis*. The larvae are occasionally caught together with *P. grandis*. They are eaten in the same way as *P. grandis* (Komuro 1968).

Protohermes grandis **(PL VI-3a and 3b)** The Japanese name is *hebitonbo*. The larvae are eaten cooked in the Nagano Prefecture, **Japan** (Katagiri and Awatsuhara 1996). The larvae are well known as an insect medicine. They have been used for a long time to reduce hyperactivity in children. The chemical analysis, however, did not show specific active substance(s) for children's hyperactivity (Umemura 1943).

Myrmeleontidae (Antlions) **(1)**

Haenomyia micans The Japanese name is *usuba-kagerō*. The larvae were used for treating whooping cough, beriberi, etc. in **Japan** (Koizumi 1935, Umemura 1943).

14. SIPHONAPTERA [1]

Pulicidae (1)

Pulex irritans In **Canada,** this species is highly esteemed as food by the Inuits (Brygoo 1946).

15. DIPTERA (Two-winged flies) **(52)**

Calliphoridae (Blow-flies) **(2)**

Calliphora megacephala (oriental latrine fly) **China:** This species is used as an antidote or as an aid for digestion (Zimian et al. 1997). **India:** The larvae are eaten boiled by the Nishi people in Arunachal Pradesh State (Singh et al. 2007). **Thailand:** The Laotian people rear this fly, and eat the larvae together with vegetable salad (Kuwabara 1997a,b).

Calliphora lata The Japanese name is *ō-kurobae*. In **Japan,** the larvae were used for treating stomach disorders, jaundice, fever, etc. (Koizumi 1935, Umemura 1943).

Chrysomyia megacephala → *Calliphora megacephala*

Cecidomyiidae (Gall midges) **(1)**

Pseudasphondylia matatabi The Japanese name is *matatabi-mitamabae*. In **Japan,** the gall, which is produced on the fruits of silvervine by this midge, is used as an anodyne, a tonic, and as a remedy for lumbago, stomach disorders, etc. (Watanabe 1982).

Chaoboridae (5)

Chaoborus (*=Neochaoborus*) *anomalus* In **Uganda,** when the outbreak of this flies occur, people capture large amounts of the adults, and make a block called *kungu* for preserving food (Bergeron et al. 1988, MacDonald 1956).

Chaoborus edulis In **Uganda,** the collected adults are pressed to make a block as food (Platt 1980).

Chaoborus pallidipes This species is eaten in **Uganda** (Bergier 1941).

Procladius umbrosus In **Uganda,** when the outbreak of this flies occur, people capture large amounts of the adults, and make a block called *kungu* for preserving food (Bergeron et al. 1988).

Sayomyia pallidipes → *Chaoborus pallidipes*

Tanypus guttatipennis In **Uganda,** when the outbreak of this flies occur, people captured large amounts of the adults, and make a block called *kungu* as a preserving food (Bergeron et al. 1988).

Drosophilidae (Vinegar flies) **(1)**

Drosophila melanogaster (vinegar fly) This species is eaten by the Mazahua, Otomi and Tepehua people in DF, **Mexico** (Ramos-Elorduy 2009). The flies are raised as food (Ramos-Elorduy 1997).

Dryomyzidae (Dryomyzid flies) **(1)**

Dryomyza formosa The Japanese name is *bekkō-bae*. **Japan:** The larvae were used for treating stomach disorders, jaundice, fever, etc. (Koizumi 1935, Umemura 1943).

Ephydridae (4)

Ephydra californica → *Hydropyrus hians*

Ephydra cinerea In the **USA**, this species coexists with *Hydropyrus hians* in the Great Salt Lake and other areas, but are much smaller than *H. hians*. Native American eat the pupae (Ebeling 1986).

Ephydra gracilis → *Ephydra cinerea*

Ephydra hians → *Hydropyrus hians*

Ephydra macellaria In the **USA,** native Americans living around the Great Salt Lake, dry the pupae and eat them in winter (Ebeling 1986).

Ephydra subopaca → *Ephydra macellaria*

Gymnopa tibialis The larvae are eaten in **Mexico** (FD) (Ramos-Elorduy de Conconi et al. 1984).

Hydropyrus hians **Mexico:** The Mexican name is *gusano verde*. Native people of the Mestizo tribe eat the larvae (Ramos-Elorduy and Pino 1989). This species is raised as food (Ramos-Elorduy 1997). **USA:** Native Americans belonging to the following tribes consume this species; Washoe, Northern Paiute, Owens Valley Paiute, Panamint (Fowler 1986).

Mossilus tibialis → *Gymnopa tibialis*

Hypodermatidae (2)

Hypoderma bovis (ox warble-flies) **Canada:** Dog Rib Indians living around Lake Athabasca like to eat the raw larvae parasitized under the caribou's skin (Fladung 1924).

Hypoderma lineate → *Hypoderma bovis*

Oedemagena tarandi In **Canada,** the Inuits living in the Arctic Circle eat the larvae of this species parasitizing in the caribou (Overstreet 2003).

Hyppoboscidae (1)

Hyppobosca longipennis **China:** This species was used to cure small-pox, etc. (Li 1596).

Muscidae (3)

Fannia canicularis The Japanese name is *hime-iebae*. In **Japan,** the larvae were used for treating stomach disorders, jaundice, fever, etc. (Koizumi 1935, Umemura 1943).

Musca domestica (common house-flies) **China:** This species is widely mass-produced as a feed for domestic animals **(PL VI-4)**. For humans, it is reported that a nutrition-rich substance was extracted from the larvae (DeFoliart 2002). **Japan:** The Japanese name is *ie-bae.* The larvae were used for treating stomach disorders, jaundice, fever, etc. (Koizumi 1935, Umemura 1943). **Mexico:** The Mexican name is *gusano del queso*. This species is eaten by the Mazahua, Otomi and Tepehua people in DF, **Mexico** (Ramos-Elorduy 2009). The flies are raised as food for humans (Ramos-Elorduy 1997).

Musca stabulans The Japanese name is *ō-iebae*. In **Japan,** the larvae were used for treating stomach disorders, jaundice, fever, etc. (Koizumi 1935, Umemura 1943).

Oestridae (2)

Cephenemyia phobifer In **Canada,** this species is eaten raw (Overstreet 2003).

Cephenemyia trompe In **Canada** and **Alaska,** the **USA,** the Nunamiut people eat this species raw (Overstreet 2003).

Piophilidae (1)

Piophila (=Tyrophagus) casei (cheese-skippers) **England, France** and **Italy:** It is said that cheese lovers are particularly partial to cheese attacked by this species (Hope 1842). At present, this custom is seen in Sardegna Island, Aosta Valley, Piemonte and Carabria. A famous maggot cheese is the product of Sardegna, "Casu Marzu" (M.G. Paoletti, pers. comm.). **Mexico:** This species is raised as food (Ramos-Elorduy 1997).

Rhagionidae (snipe flies) (0)

Atherix **sp.** This species was considered as flies, which were eaten by the Pit River Indians in the **USA** (Aldrich 1912), but later Essig (1947 and 1965) corrected the species to Californian salmonfly, *Pteronarcys californica* by the results of his survey.

Simuliidae (Black flies) (2)

Simulium aureohirtum **Thailand:** This species is considered as food item (Leksawasdi 2010).

Simulium rubithorax **Venezuela-Brazil:** Native people of the Yanomamo tribe eat the larvae by wrapping them with banana leaves and roasting (DeFoliart 2002).

Simulium **sp. Thailand:** The Karen people consume a species belonging to this genus (Leksawasdi 2010).

Stratiomyidae (Soldier flies) (3)

Helmetia illucens **Malaysia:** The larvae are eaten raw by some Kadazandusuns while drinking their locally-brewed beer, known as *tapai* in Sabah State (Chung et al. 2002). **Mexico:** In olden times (about BC 200–300), aboriginal people in the Tamaulipas State seemed to eat the larvae of this species, as the larvae were found in the coprolites excavated from the cave in the plateau of the state (McFadden 1966). This species is raised as food (Ramos-Elorduy 1997).

Campylostoma sp. The Mexican name is *gusano blanco*. In **Mexico,** the larvae are eaten (Ramos-Elorduy et al. 1998).

Chryschlorina spp. In **Venezuela** and **Colombia,** native people of the Yukpa tribe consumed the larvae of several species belonging to this genus (Ruddle 1973).

Copestylum anna The Mexican name is *gusano plano*. In **Mexico,** the larvae are used as a food item (Ramos-Elorduy et al. 1998).

Copestylum haggi The Mexican name is *gusano plano*. In **Mexico,** the larvae are used as an article of food (Ramos-Elorduy et al. 1998).

Syrphidae (flower flies, hover flies) **(3)**

Copestylum haaggii The Mexican name is *gusanos planos de maguey*. In **Mexico,** the larvae are eaten (Ramos-Elorduy de Conconi et al. 1984).

Eristalis tenx (drone fly) The Japanese name is *hana-abu*. In **Japan,** the larvae were used for remedy of irregular menstruation (Koizumi 1935, Umemura 1943).

Eristalis sp. The Mexican name is *cola de ratón*. In **Mexico,** the larvae are eaten (Ramos-Elorduy et al. 1998).

Ornidia obesa **(=*obescens*)** (green hover fly) In **Mexico,** this species is raised as food (Ramos-Elorduy 1997).

Tabanidae (Horse flies) **(12)**

Atylotus horvathi **China:** This species is used to promote blood circulation (Nanba 1980).

Tabanus amaenus **China:** This species is used to promote blood circulation (Nanba 1980).

Tabanus budda **China:** This species is used as a suppressor of swelling (Zimian et al. 1997).

Tabanus chrysurus The Japanese name is *akaushi-abu*. In **Japan,** the adults were used for remedy of irregular menstruation, abortion, epidemic conjunctivitis, etc. (Koizumi 1935, Umemura 1943).

Tabanus kiansuensis (Jiangsu horse fly) **China:** This species is used as an antidote, an antifebrile, or as a suppressor of swelling. It is also said to be effective against hepatic cancer (Zimian et al. 1997).

Tabanus kinoshitai **China:** This species is said to promote blood circulation (Nanba 1980).

Tabanus mandarinus (Chinese horse fly) **China:** This species is used as a suppressor of swelling, and is said to be effective against hepatic cancer (Zimian et al. 1997). **Japan:** The Japanese name is *shirofu-abu*. The adults were used for remedy of irregular menstruation, abortion, epidemic conjunctivitis, etc. (Koizumi 1935, Umemura 1943).

Tabanus rubidus **China:** This species is intaken to promote blood circulation (Nanba 1980).

Tabanus rufidens The Japanese name is *yamato-abu*. In **Japan,** the adults were used for remedy of irregular menstruation, abortion, epidemic conjunctivitis, etc. (Koizumi 1935, Umemura 1943).

Tabanus trigonus The Japanese name is *ushi-abu*. In **Japan,** the adults were used for remedy of irregular menstruation, abortion, epidemic conjunctivitis, etc. (Koizumi 1935, Umemura 1943).

Tabanus tropicus **North Korea:** The dried or roasted adults, or the infusion of the adults are considered effective for unhealthy blood (Okamoto and Muramatsu 1922).

Tabanus yao **China:** This species is said good to promote blood circulation (Nanba 1980).

Tabanus **sp.** This species is used as a traditional medicine in **North** and **South Korea** (Pemberton 2005).

Tachinidae (1)

Blepharipa zebina (silkworm tachina fly) The Japanese name is *kaikonouji-bae*. The larvae are eaten in the Nagano Prefecture, **Japan** (Kurumi 1918).

Sturmia sericariae → *Blepharipa zebina*

Tephritidae (Fruit flies) (2)

Anastrepha ludens (Mexican fruit fly) This species is considered as a food item by the Mazahua, Otomi and Tepehua people in DF, **Mexico** (Ramos-Elorduy 2009). The flies are raised as food (Ramos-Elorduy 1997).

Anastrepha **spp. Ecuador:** The larvae of several species belonging to this genus are often found in the fruits of guava, and people obliviously eat the infested fruits with the larvae inside (Onore 2005).

Ceratitis capitata (Mediterranean fruit fly) **Ecuador:** The larvae are often found in citrus fruits, and people eat the infested fruits oblivious of the larvae inside (Onore 2005).

Tipulidae (Crane flies) (6)

Antocha (Proantocha) spinifera **Japan:** This species is a member of *zazamushi* (see Tricoptera-Leptoceridae-*Parastenopsyche sauteri*). In the

Nagano Prefecture, this species is eaten cooked (Nagano Fishery Experiment Station, Suwa 1985).

Holorusia hespera → *Holorusia rubiginosa*

Holorusia rubiginosa In the **USA**, this species is abundant in late winter and early spring in California, and is consumed when most other foods are scarce (Essig 1965).

Tipula derbyi In the **USA**, this species is abundant in late winter and early spring in California, and is used as foodstuff when most other foods are scarce (Essig 1965).

Tipula paludosa The pupae infected by *Cordyceps* fungus are medicinally used in **China** (Chen and Feng 1999).

Tipula quaylii In the **USA**, this species is abundant in late winter and early spring in California, and is used as foodstuff when most other foods are scarce (Essig 1965).

Tipula simplex In the **USA**, this species is abundant in late winter and early spring in California, and is used as foodstuff when most other foods are scarce (Essig 1965).

16. TRICHOPTERA (Caddisflies) (10)

Calamoceratidae (Comblipped casemake caddisflies) **(0)**

Phylloicus **sp.** The Yekuana people in Alto Orinoco area, **Venezuela** eat the larvae of a species belonging to this genus (Araujo and Beserra 2007).

Hydropsychidae (Net-spinning caddisflies) **(4)**

Cheumatopsyche brevilineatus The Japanese name is *kogata-shimatobikera*. This species is a member of *zazamushi* (see *Stenopsyche griseipennis*). The larvae are cooked with soy sauce in the Nagano Prefecture, **Japan** (Nagano Fishery Experiment Station 1985).

Hydropsyche orientalis → *Hydropsyche ulmeri*

Hydropsyche ulmeri The Japanese name is *urumā-shimatobikera*. This species is a member of *zazamushi* (see *Stenopsyche griseipennis*). The larvae are cooked with soy sauce in the Nagano Prefecture, **Japan** (Nagano Fishery Experiment Station 1985).

Hydropsyche **sp. Japan:** A species belonging to this genus is a member of *zazamushi* (see *Stenopsyche griseipennis*). The larvae are cooked with soy sauce in the Nagano Prefecture (Torii 1957).

Leptonema **spp. Colombia-Venezuela:** Native people of the Yukpa and Yekuana tribes eat the larvae of several species belonging to this genus

(Ruddle 1973, Araujo and Beserra 2007). **Mexico:** The larvae are eaten in Veracruz State (Ramos-Elorduy de Conconi et al. 1984).

Macronema radiatum The Japanese name is *ō-shimatobikera*. In **Japan,** this species is a member of *zazamushi* (see *Stenopsyche griseipennis*). The larvae are cooked with soy sauce in the Nagano Prefecture (Nakai 1988).

Oecetis disjunta The Mexican name is *gusano de casa.* The larvae are eaten by the Tzeltzales Tzoltziles people in **Mexico** (Ramos-Elorduy and Pino 2002).

Leptoceridae (Long-horned caddisflies) **(2)**

Parastenopsyche sauteri → *Stenopsyche sauteri*

Stenopsyche griseipennis → *Stenopsyche marmorata*

Stenopsyche marmorata **(PL VI-5a and 5b)** The Japanese name is *higenaga-kawatobikera*. In **Japan**, this species comprises most of *zazamushi* (mixture of aqueous insects which live in the shallow part of a river). It consists of larvae or adults of Ephemeroptera, Odonata, Plecoptera, Neuroptera, Tricoptera and Hemiptera. The collection of *zazamushi* is carried out in winter, December to February (Mitsuhashi 1997, 2003 and 2005). The larvae of this species are cooked with soy sauce and sugar in the Nagano Prefecture (Katagiri and Awazuhara 1996).

Stenopsyche sauteri The Japanese name is *chabane-higenagakawatobikera*. This species is a member of *zazamushi* (see *Stenopsyche griseipennis*). In **Japan,** the larvae are cooked with soy sauce in the Nagano Prefecture (Torii 1957).

Triplectides **sp.** The Yekuana people in Alto Orinoco area, **Venezuela** eat the larvae of a species belonging to this genus (Araujo and Beserra 2007).

Limnephilidae (Northern caddisflies) **(0)**

Allophylax **sp. Japan:** This species is a member of *zazamushi* (see *Stenopsyche griseipennis*). The larvae are cooked with soy sauce in the Nagano Prefecture (Torii 1957).

Odontoceridae (Mortarjoint casemaker caddislies) **(0)**

Marilia **sp.** The Yekuana people in Alto Orinoco area, **Venezuela,** eat the larvae of a species belonging to this genus (Araujo and Beserra 2007).

Phryganeidae (Large caddisflies) **(3)**

Neuronia melaleuca The Japanese name is *gomafu-tobikera*. In **Japan,** this species is a member of *zazamushi* (see *Stenopsyche griseipennis*). The larvae are cooked with soy sauce in the Fukushima and Nagano Prefectures (Takagi 1929f).

Neuronia regina The Japanese name is *murasaki-tobikera*. This species is a member of *zazamushi* (see *Stenopsyche griseipennis*). In **Japan,** the larvae are cooked with soy sauce in the Fukushima and Nagano Prefectures (Takagi 1929f).

Phryganea japonica The Japanese name is *tsumaguro-tobikera*. This species is a member of *zazamushi* (see *Stenopsyche griseipennis*). In **Japan,** the larvae are cooked with soy sauce in the Fukushima and Nagano Prefectures (Takagi 1929f).

Rhyacophilidae (Free-living caddis flies) **(1)**

Rhyacophila nigrocephala The Japanese name is *munaguro-nagaretobikera*. This species is a member of *zazamushi* (see *Stenopsyche griseipennis*). In **Japan,** the larvae are cooked with soy sauce in the Nagano Prefecture (Mitsuhashi, unpublished data).

Rhyacophila **sp.** This species is a member of *zazamushi* (see *Stenopsyche griseipennis*). In **Japan,** the larvae are cooked with soy sauce in the Nagano Prefecture (Torii 1957).

17. LEPIDOPTERA [386]

Arctiidae (Tiger moths) **(6)**

Amastus ochraceator The Mexican name is *gusano de los palos*. In **Mexico,** the larvae are eaten by the Náhuatl, Yutoazteca and Otomi people in Zongolica, Veracruz States (Ramos-Elorduy et al. 2011).

Arctia caja americana The larvae are consumed in the **USA** (Arnett 1985).

Diacrisia obliqua **India:** This species is used as a food item by the Ao-Naga people in Nagaland State (Meyer-Rochow and Changkija 1997). **Sri Lanka:** This species is eaten (Nandasena et al. 2010).

Elysius superba The Mexican name is *gusano del palo mulato*. In **Mexico,** the larvae are treated as food by the Náhuatl, Yutoazteca and Otomi people in Zongolica, Veracruz States (Ramos-Elorduy et al. 2011).

Estigmene acrea (salt-marsh caterpillar) The adults are considered good for eating by the Náhuatl, Yutoazteca and Otomi people in Zongolica, Veracruz States, **Mexico** (Ramos-Elorduy et al. 2011).

Pelochyta cervina The adults are eaten by the Náhuatl, Yutoazteca and Otomi people in Zongolica and Veracruz States, **Mexico** (Ramos-Elorduy et al. 2011).

Bombycidae (Silkworm moths) **(4)**

Andraca bipunctata The pupae are consumed in **China** (Hu and Zha 2009).

Bombyx mandarina **China:** The pupae are considered as a food item (Hu and Zha 2009). **Japan:** The Japanese name is *kuwako*. The charred pupae and cocoon were used for treating whooping cough, convulsive fits, etc. (Miyake 1919).

Bombyx mori (commercial silkworm) **China:** The silkworm pupae can be obtained from silk-reeling factories in the Jiangsu and Zhejiang regions. People eat them by deep-frying or stir-frying with chives (Zhi-Yi 2005). The silkworm is also raised as food, and is used also for treating rheumatism and other diseases. The larvae infected by a parasitic fungus, *Beauveria bassiana*, is said to be effective against lung cancer (Zimian et al. 1997). **India:** People consume the larvae and pupae (Pathak and Rao 2000). **Japan:** The Japanese name is *kaikoga*. The silkworm is one of the most popular insect foods in Japan. The larvae are roasted with a bit of salt and in the Nagano Prefecture (Miyake 1919). The pupae are eaten in many prefectures (Akita, Niigata, Yamagata, Saitama, Shizuoka, Nagano, Gifu, Aichi, Mie, Wakayama, Nara, Kyoto, Fukui, Ishikawa, Shimane, Tottori, Okayama, Hiroshima, Yamaguchi, Ehime, Kochi, Oita, and Miyazaki). People eat them raw, or by roasting, dipping into soy sauce, cooking with soy sauce, as well as in sugar-soy sauce, or by adding a bit of salt (Miyake 1919). The adults are roasted and seasoned with salt or soy sauce, or cooked with soy sauce in the Saitama and Nagano Prefectures (Miyake 1919). The silkworm is nutrient rich insect, and was consumed often when there was a scarcity of food at the end of the World War II. It is said that the nutritional value of three pupae is equivalent to that of a fowl egg. At that time people extracted oil from the pupae for cooking (Umemura 1943). The silkworm pupae have high riboflavin (Vitamin B_2) content, and it was extracted for officinal use during the war time (Ishimori 1944). At present, cans of cooked silkworm flavored with soy sauce are commercially available (Mitsuhashi 1997, 2003 and 2005). **North** and **South Korea:** The larvae are commonly eaten boiled with salt water. Canned cooked pupae are commercially available (Nonaka 1991). The larvae infected by a parasitic fungus, *Beauveria bassiana*, are dried, pulverized and used to cure various diseases (Okamoto and Muramatsu 1922). **Laos:** The Laotian name is *dak dir*. The pupae are eaten (Nonaka 1999a). **Mexico:** This species is consumed in Hidalgo, Oaxaca and Tlaxcala States. This species is raised as food (Ramos-Elorduy 1997 and 2009). **Myanmar:** The pupae are eaten (Yhoung-Aree and Viwatpanich 2005). **Sri Lanka:** This species is eaten (Nandasena et al. 2010). **Thailand:** The Thai name is *dak-dae-mai*. The pupae are a common food in the northern or north eastern area of this country (Hanboonsong 2010). **Vietnam:** The pupae are relished (Yhoung-Aree and Viwatpanich 2005) **(PL VII-1).**

Rondotia menciana **China:** From the very old remains located at Shanxi Province (BC 2,500–2,000), the cocoons of this species were found. A tip

Plate VII. 1. Fried *Bombyx mori* pupae with vegetables in Da Lat, Viet Nam (cf. p.142); 2. *Aegiale hesperiaris*. Upper: An adult. Body length: 30 mm, wing span: 75 mm. Lower: a larva fried. Body length: 65 mm (cf. p.151); 3. *Erionota torus*. Upper: an adult. Wing span 70 mm (Photo: A. Miyagi, Okinawa, Japan) (cf. p.151). Lower: a larva. Body length: 40 mm (cf. p.152); 4. Fried larvae of *Xyleutes redtembacheri*. Body length: 35 mm (cf. p.144); 5. *Dendrolimus spectabilis* larva on a pine branch. Body length: 6 mm (Photo: A. Shimazu, Tokyo, Japan) (cf. p.153); 6. Boiled and dried larvae of *Goninbrasia belina*, sold in South Africa. Body length: 50 (cf. p.167).

of the cocoons was cut off, and the pupae were taken out through the cut opening and were probably eaten by people (Nunome 1979).

Theophia mandarina → *Bombyx mandarina*

Brahmaeidae (Brahmin moths) **(2)**

Brahmaea japonica (ligustrum moths) The Japanese name is *ibotaga*. **Japan:** The larvae are roasted with a bit of sugar-soy sauce, or skewer-roasted in the Chiba Prefecture (Miyake 1919). The roasted larvae were highly prized as a medicine for tuberculosis (Kuwana 1930).

Dactylocerus lucina The larvae are eaten in **Africa** (Malaisse 2005).

Carposinidae (Fruitworm moths) **(1)**

Carposina niponensis The pupae are consumed in **China** (Hu and Zha 2009).

Castniidae (Sun moths) **(6)**

Castnia chelone In **Mexico,** native people of the Hñähñu tribe living in Hidalgo State eat this species (Aldasoro Maya 2003).

Castnia daedalus The larvae are considered edible by the Huaorani and Quichuas people in **Ecuador** (Onore 2005, Paoletti and Dufour 2005).

Castnia licoides The larvae are treated as a foodstuff by the Huaorani and Quichuas people in **Ecuador** (Onore 2005, Paoletti and Dufour 2005).

Castnia licus The larvae are used as food item by the Huaorani and Quichuas people in **Ecuador** (Onore 2005, Paoletti and Dufour 2005).

Castnia synparamides chelone The Mexican name is *gusano del junquillo*. In **Mexico,** the larvae are eaten by the Náhuatl and Otomi people in Hidalgo State (Ramos-Elorduy et al. 2011).

Eupalamides cyparissias The larvae are eaten in **Ecuador** (Onore 2005, Paoletti and Dufour 2005).

Cossidae (Carpenter moths) **(20)**

Catoxophylla cyanauges The adults are relished in **Australia** (DeFoliart 2002).

Comadia redtenbacheri **(PL VII-4)** The Mexican name is *gusano rojo de maguey*. In **Mexico,** a reddish purple larva was put into a bottle of mescal, a kind of liqueur, in order to indicate the content of sufficient concentration of alcohol. Native people of the Hñähñu tribe living in Hidalgo State eat the larvae (Aldasoro Maya 2003). People of different tribes in many states used to fatten the larvae collected from agave plants with tortilla (Ramos-Elorduy 2009).

Cossus chinensis The larvae are considered edible in **China** (Hu and Zha 2009).

Cossus cossus The larvae are considered good for eating in **China** (Hu and Zha 2009).

Cossus hunanensis The larvae are recognized as a food item in **China** (Hu and Zha 2009).

Cossus insularis The Japanese name is *hime-bokutō*. In **Japan,** the larvae were used for treating convulsive fits, fever, malnutrition, etc. (Shiraki 1958).

Cossus jezoensis (oriental carpenter moth) The Japanese name is *bokutōga*. In **Japan,** the larvae are spread with soy sauce and roasted in the Ehime Prefecture (Takagi 1929h). The larvae were used to cure convulsive fits in children, fever, etc. (Watanabe 1982). **South Korea:** The dried larvae are pulverized, and used for treating epilepsy in the North Joella Province (Okamoto and Muramatsu 1922).

Cossus ligniperda **Roma Antiqua:** This species may be the insect called *Cossus* and was relished by the Roman epicures in Pliny's time (Hope 1842).

Cossus redtembacheri → *Comadia redtembacheri*

Cossus vicari(o)us → *Cossus jezoensis*

Cossus **sp.** A species belonging to this genus is eaten in **Australia** (DeFoliart 2002).

Duomitus leuconotus → *Xyleutes leuconotus*

Endoxyla biarpiti → *Xyleutes biarpiti*

Endoxyla eucalypti → *Xyleutes eucalypti*

Endoxyla leuchomochla → *Xyleutes leuchomochla*

Endoxyla **n. sp.** A species belonging to this genus is consumed in **Australia** (Smyth 1878).

Holcocerus vicarius This species was used in **China**, to promote blood circulation (Li 1596).

Xyleutes amphiplecta **Australia:** The larvae are called *witjuti* grub by the aboriginal people. They eat the larvae raw or boiled (Tindale 1962).

Xyleutes biarpiti The larvae are consumed in **Australia** (DeFoliart 2002).

Xyleutes boisduvali The larvae are considered good for eating in **Australia** (DeFoliart 2002).

Xyleutes capensis (castor bean borer) **Tanzania, Kenya:** The larvae are eaten raw (Schabel 2010).

Xyleutes eucalypti The larvae are esteemed as food item in **Australia** (DeFoliart 2002).

Xyleutes leuchomochla In **Australia,** the larvae are called *witjuti* grub by the aboriginal people. They eat the larvae raw or roasted. They also eat the adults raw, or by roasting on hot ashes (Campbell 1926, Tindale 1932).

Xyleutes (=Duomitus) leuconotus This species is used as foodstuff in **Thailand** (Kerr 1931).

Xyleutes redtembachi → *Comadia redtembacheri*

Xyleutes redtembacheri → *Comadia redtembacheri*

Zeuzera citurata The larvae bore into *Acacia* stems. In **Australia,** aborigines like to eat these larvae (Smyth 1878).

Zeuzera coffeae (red coffee borer) People in Ayutthaya district, **Thailand,** liked to eat the fried larvae, which bore into stems of *Sesbania roxburghii* or kapok from olden times (Kerr 1931).

Zeuzera eucalypti The larvae bore into *Acacia* stems. In **Australia,** people living on the Tasmania Island before the present aborigines ate the larvae from very ancient times (Noetling 1910).

Zeuzera multistrigata (oriental leopard moth) **Japan:** The Japanese name is *kimadara-kōmoriga*. The larvae are spread with soy sauce and roasted in the Ehime Prefecture (Takagi 1929h). The larvac were used to cure convulsive fits in children, fever, etc. (Watanabe 1982).

Zeuzera pyrina → *Zeuzera multistrigata*

Zeuzera **sp.** A species belonging to this genus is eaten in Sabah State, **Malaysia** (Chung et al. 2002).

Ctenuchidae (Wasp moths) **(1)**

Amata phegea → *Syntomis phegea*

Syntomis phegea **Italy:** Around Tramonti de Sotto in the Friurli district, people eat the abdomens of the adult moths raw in summer (Paoletti and Dreon 2005).

Danaidae (Monarchs) **(3)**

Danaus gilippus thersippus The Mexican name is *mariposa del tizmo.* The larvae are consumed by the Náhuatl and Otomi people in Hidalgo State, **Mexico** (Ramos-Elorduy et al. 2011).

Danaus plexippus (monarch butterflies, milkweed butterflies) **Mexico:** The adults are food item in the Chiapas, Hidalgo and Michoacán States (Ramos-Elorduy et al. 2011). **USA:** Earlier the larvae was used as foodstuff by the Miwok people (Ikeda et al. 1993).

Euploea hamata In **Australia,** aborigines greatly relish the larvae (Ealand 1915).

Hamadryas **sp.** The larvae of a species belonging to this genus are used as food by the Maya people in Yucatán district, **Mexico** (Ramos-Elorduy et al. 2011).

Eupterotidae (Giant lappet moths) **(2)**

Ochrogaster lunifer The larvae spin a large silken bag as a communal shelter. In **Australia**, the larvae and/or pupae are eaten in times of extreme hardship (Latz 1995, Isaacs 1987).

Panacela **sp.** The larvae are eaten in **Australia** (DeFoliart 2002).

Striphnopteryx edulis (edible monkey) This species is eaten in southern **Africa** (Bergier 1941).

Gelechiidae (Nebs and grounding moths) **(1)**

Pectinophora gossypiella (pink bollworm) In **China,** people extracted oil for cooking from the hibernating larvae, around the Shandong and Jiangsu Provinces (Chen and Feng 1999).

Platyedra gossypiella → *Pectinophora gossypiella*

Geometridae (Geometrid moths) **(7)**

Acronyctodes mexicanaria The larvae and pupae are consumed by the Yotoazteca, Náhuatl and Otomi people in **Mexico** (Ramos-Elorduy et al. 2011).

Biston marginata In **China**, the pupae infected by a fungus, *Cordyceps militaris,* are officinally used (Chen and Feng 1999).

Evis mexicanaria → *Acronyctodes mexicanaria*

Hemirophila atrilinata → *Phthonandria atrilineata*

Menophora atrilineata → *Phthonandria atrilineata*

Panthera pardalaria The larvae are eaten by the Náhuatl, Otomi and Yutoazteca people in Tlaxcala State, **Mexico** (Ramos-Elorduy et al. 2011).

Pantherodes pardalaria In ancient times, the larvae of this species were traded in **Mexico** (Ramos-Elorduy et al. 2011).

Phthonandria atrilineata **Japan:** The Japanese name is *kuwa-edashaku*. The larvae were used for treating meningitis, involuntary emission of semen, etc. (Shiraki 1958). **South Korea:** The roasted larvae are pulverized, and mixed with lard as a remedy for involuntary emission of semen, meningitis among children, etc. in the North Gyeongsang Province (Okamoto and Muramatsu 1922).

Synopsia mexicanaria The Mexican name is *pescaditos*. **Mexico:** The larvae are used as a food article in Mexico FD (Ramos-Elorduy de Conconi et al. 1984).

Gracillaridae (1)

Stomphosistis thraustica (Jatropha leaf miner) In southern Chhattisgarh district, **India**, the larvae are collected just before pupation, dried and pulverized. The powder is given with lukewarm water to lactating women in order to increase secretion of milk. The larvae are also used for treatment of common fever (Srivastava et al. 2009).

Hepialidae (Swift moths) (47)

Abantiades marcidus In **Australia**, the larvae, pupae and the adults are used as food by aborigines (Tindale 1932).

Aenetus virescens The Maori name is *puriri* moth. In **New Zealand**, this moth is said to be occasionally accepted as human food (Meyer-Rochow and Changkija 1997).

Endoclytaex crescens (common swift moth) The Japanese name is *Kōmoriga*. In the Nagano Prefecture, **Japan,** the larvae are skewer-roasted with salt or soy sauce. People also cook the larvae with soy sauce and sugar, or frizzle with oil, or pickle with vinegar (Miyake 1919). This species was used medicinally for treating various diseases including convulsive fits in children (Kuwana 1930).

Endoclyta sinensis (grape tree borer) **China:** This species was used to promote health (Li 1596). **Japan:** The Japanese name is *kimadara-kōmoriga*. People in the Saitama, Gunma, and Okayama Prefectures roast the larvae and add soy sauce or miso paste to them. This worm is also used as an insect medicine for treating various diseases. As it is a rich nutrient, the worms might ameliorate delicate children (Miyake 1919, Takagi 1929g).

Hepialus albipictus In **China**, the larvae infected by a fungus, *Cordyceps sinensis*, are used as a medicine (Chen and Feng 1999).

Hepialus altaicola In **China**, the larvae infected by a fungus, *Cordyceps sinensis*, are one of folk medicine (Chen and Feng 1999).

Hepialus armoricanus **China:** This species is used as a medicine for a cough, and is said to be effective against lung cancer. The larvae killed by a fungus *Cordyceps sinensis* are sold expensively as a officinal material, therefore the development of mass rearing methods for this species is needed (Zimian et al. 1997).

Hepialus baimaensis In **China**, the larvae infected by a fungus, *Cordyceps sinensis*, are treated as a medicine (Chen and Feng 1999).

Hepialus cingulatus In **China**, the larvae infected by a fungus, *Cordyceps sinensis*, are given to invalids (Chen and Feng 1999).

Hepialus deqinensis In **China**, the larvae infected by a fungus, *Cordyceps sinensis*, are used as a medicine (Chen and Feng 1999).

Hepialus deudi In China, the larvae infected by a fungus, *Cordyceps sinensis*, are administered to invalids (Chen and Feng 1999).

Hepialus dongyuensis In **China**, the larvae infected by a fungus, *Cordyceps sinensis*, are used as a medicine (Chen and Feng 1999).

Hepialus ferrugineus In **China**, the larvae infected by a fungus, *Cordyceps sinensis*, are treated as a medicine (Chen and Feng 1999).

Hepialus ganna In **China**, this species is used as a host of a *Cordyceps* fungus (Chen and Feng 1999).

Hepialus gonggaensis In **China**, the larvae infected by a fungus, *Cordyceps sinensis*, are given to invalids (Chen and Feng 1999).

Hepialus jinshaensis In **China**, the larvae infected by a fungus, *Cordyceps sinensis*, are appreciated as a medicine (Chen and Feng 1999).

Hepialus kangdingensis In **China**, the larvae infected by a fungus, *Cordyceps sinensis*, are used as a medicine (Chen and Feng 1999).

Hepialus kangdingroides In **China**, the larvae infected by a fungus, *Cordyceps sinensis*, are regarded as a medicine (Chen and Feng 1999).

Hepialus lijiangensis In **China**, the larvae infected by a fungus, *Cordyceps sinensis*, are esteemed as a medicine (Chen and Feng 1999).

Hepialus litangensis In **China**, the larvae infected by a fungus, *Cordyceps sinensis*, are used as a medicine (Chen and Feng 1999).

Hepialus luquensis In **China**, the larvae infected by a fungus, *Cordyceps sinensis*, are appreciated as a medicine (Chen and Feng 1999).

Hepialus macilentus In **China**, this species is recognized as a host of a *Cordyceps* fungus (Chen and Feng 1999).

Hepialus markamensis In **China**, the larvae infected by a fungus, *Cordyceps sinensis*, are treated as a medicine (Chen and Fen 1999).

Hepialus meiliensis In **China**, the larvae infected by a fungus, *Cordyceps sinensis*, are used as a medicine (Chen and Feng 1999).

Hepialus menyuanicus In **China**, the larvae infected by a fungus, *Cordyceps sinensis*, are considered as a medicine (Chen and Feng 1999).

Hepialus nebulosus In **China**, the larvae infected by a fungus, *Cordyceps sinensis*, are used as a medicine (Chen and Feng 1999).

Hepialus oblifurcus **China:** The larvae infected by a fungus, *Cordyceps sinensis*, are regarded as a medicine (Chen and Feng 1999). **Korea:** The larvae infected by a fungus, *Cordyceps sinensis*, are used as a medicine (Pemberton 2005).

Hepialus pratensis In **China**, the larvae infected by a fungus, *Cordyceps sinensis*, are appreciated as a medicine (Chen and Feng 1999).

Hepialus renzhiensis In **China**, the larvae infected by a fungus, *Cordyceps sinensis*, are given to invalids as a medicine (Chen and Feng 1999).

Hepialus sichuanus (Sichuan swing moth) In **China,** this species is said to be effective against lung cancer (Zimian et al. 1997).

Hepialus varians In **China**, the larvae infected by a fungus, *Cordyceps sinensis*, are used as a medicine (Chen and Feng 1999).

Hepialus xunhuaensis In **China**, the larvae infected by a fungus, *Cordyceps sinensis*, are treated as a medicine (Chen and Feng 1999).

Hepialus yeriensis In **China**, the larvae infected by a fungus, *Cordyceps sinensis*, are considered as a medicine (Chen and Feng 1999).

Hepialus yuloangensis In **China**, the larvae infected by a fungus, *Cordyceps sinensis*, are accepted as a medicine (Chen and Feng 1999).

Hepialus yunlongensis (Yunlong swing moth) This species is said to be effective against lung cancer in **China** (Zimian et al. 1997).

Hepialus yunnanensis In **China**, the larvae infected by a fungus, *Cordyceps sinensis*, are appreciated as a medicine (Chen and Feng 1999).

Hepialus yushuensis (Yushun swing moth) This species is said to be effective against lung cancer in **China** (Zimian et al. 1997).

Hepialus zhangmoensis In **China**, the larvae infected by a fungus, *Cordyceps sinensis*, are used as a medicine (Chen and Feng 1999).

Hepialus zhayuensis In **China**, the larvae infected by a fungus, *Cordyceps sinensis*, are regarded as a medicine (Chen and Feng 1999).

Hepialus zhongzhiensis In **China**, the larvae infected by a fungus, *Cordyceps sinensis*, are used as a medicine (Chen and Feng 1999).

Hepialus **sp. Ecuador** The larvae of a species belonging to this genus are eaten by the Negr. Esmerald and Quichuas people (Onore 2005, Paoletti and Dufour 2005).

Napialus hunanensis The larvae and pupae are consumed in **China** (Hu and Zha 2009).

Oxycanus **sp.** In **Australia**, the larvae are called *bardi* grubs, and the larvae, pupae and the adults are used as food item (DeFoliart 2002).

Pahssus excrescens → *Endoclyta excrescens*

Phassus signifier → *Endoclyta sinensis*

Phassus trajesa The Mexican name is *zatam*. The larvae and pupae are consumed by the people of Tzelzales Tzoltziles, Lacandones, Tojolabales, and Zoques tribes in **Mexico** (Ramos-Elorduy and Pino 2002).

Phassus triangularis The Mexican name is *gusano del tepozan*. The larvae are used as foodstuff by people of many tribes in **Mexico** (Ramos-Elorduy et al. 2011). The 6th instar larvae are collected and raised in several states (Ramos-Elorduy 2009).

Phassus **spp.** In **Mexico,** the larvae of a species called *gusanode la jarilla* in Mexican are eaten in Oaxaca and Puebla States (Ramos-Elorduy et al. 1998). The larvae of a species called *gusano del aile* in Mexican are also eaten widely (Ramos-Elorduy et al. 2011).

Schausiana trajesa → *Phassus trajesa*

Trictena argentata In **Australia,** aborigines eat large adults (the wing span is about 120 mm), as well as the larvae and pupae, by roasting on hot ashes (Tindale 1962, 1966).

Trictena argyrosticha In **Australia,** the larvae are called *bardi* grubs. Aborigines eat large adults (the wing span is about 150 mm), as well as the larvae and pupae, by roasting on hot ashes (Tindale 1962 and 1966).

Trictena atripalpis In **Australia,** this species is known as a *bardi* grub and is eaten by the aborigines (Naumann 1993).

Zelotypia stacyi (bent wing swift moth) In **Australia,** the larvae are one of the witchety grub, and are eaten by the aborigines (Tindale 1962).

Hesperiidae (Skippers) **(7)**

Achlyodes pallida The Mexican name is *saltadora*. The larvae are considered as food item by people of many tribes in **Mexico** (Ramos-Elorduy et al. 2011).

Acentrocneme hesperiaris → *Aegiale hesperiaris*

Aegiale hesperiaris The Mexican name is *gusano blanco de maguey*. In **Mexico,** the larvae are highly relished. They are regarded as food by people of many tribes (Ramos-Elorduy 2009, Aldasoro Maya 2003). The fried larvae are served in some restaurants in Mexico City (Ramos-Elorduy et al. 2011) **(PL VII-2)**. This species is raised as food (Ramos-Elorduy 1997). This species is said to be good for stomach disorders and rheumatism (Ramos-Elorduy de conconi and Pino 1988).

Aegiale kollari → *Aegiale heperiaris*

Teria agavis → *Aegiale hesperiaris*

Anchistroides nigrita **Malaysia:** In Sabah State, the pupae are eaten (Chung 2010).

Coeliades libeon The larvae of this species are consumed in **DRC** and **PRC** (Bani 1995).

Erionata thrax thrax → *Erionota torus*

Erionata torus (banana skipper, banana leaf-roller) **(PL VII-3). China:** The larvae are relished by the people of minorities in the Yunnan Province (Chen and Feng 1999). **Laos:** This species is treated as food (Nonaka 1999a). The larvae fried with oil are delicious (Iwano 2000). **Malaysia:** In Sabah State, the pupae are called *bingogo* by the Kadazandusun people, and are eaten raw or boiled until dry (Chung 2010). **Myanmar** and **Vietnam:** This species is also eaten (Yhoung-Aree and Viwatpanich 2005). **Thailand:** This species is considered good for eating (Hanboonsong 2010).

Megathymus yuccae (yucca giant skipper) In the **USA**, the larvae, about 2 inches long, are often roasted and eaten as a delicacy (Ebeling 1986).

Parnara guttata guttata (rice skipper) **China:** The pupae are food item (Hu and Zha 2009). **Japan:** The Japanese name is *ichimonji-seseri*. The larvae are cooked in seasoned water in the Nagano Prefecture (Mukaiyama 1987).

Teria agavis → *Aegiale hesperialis*

Hyblaeidae (1)

Hyblaea puera (teak caterpillar) The Javanese name is *enthung jati*. On the Java Island, **Indonesia,** some people eat the cocoons fried in coconut oil or African palm oil. Although a few people break out into an allergic rash after eating the cocoons, consuming the cocoons is believed to enhance vitality (Lukiwati 2010).

Lasiocampidae (Tent caterpillar moths) (17)

Bombycomorpha pallida The larvae are consumed in **RSA, Zambia** (Quin 1959) and **Zimbabwe** (Dube et al. 2013).

Borocera cajani → *Livethra cajani*

Borocera madagascariensis The larvae and pupae are considered edible in **Madagascar**. People raise this species on the natural host plant (Gade 1985).

Brachiostegia **sp.** A species belonging to this genus is regarded as a food item in **Zimbabwe** (Bodenheimer 1951).

Catalebeda jamesoni The pupae are treated as food in **Zambia** (Silow 1976).

Chatra grisea In Meghalaya State, **India,** sometimes an outbreak of this species occurs. People gather the cocoons and sell them at markets (Meyer-Rochow and Changkija 1997, Meyer-Rochow 2005).

Cnethocampa diegoi This species is eaten in **Madagascar** (Decary 1937).

Dendrolimus houi Minorities in the Yunnan Province, **China** eat this species, or use it as medicine. People gather the pupae, and fry them in oil or roast them. They are also used as material from which oil for cooking is extracted (Chen and Feng 1999).

Dendrolimus kikuchii Minorities in the Yunnan Province, **China** eat this species, or use it as medicine. People gather the pupae, and fry them in oil or roast them. They are also used as material from which oil for cooking is extracted (Chen and Feng 1999).

Dendrolimus punctatus Minorities in the Yunnan Province, **China** eat this species, or use it as medicine. People gather the pupae, and fry them in oil or roast them. They are also used as material from which oil for cooking is extracted (Chen and Feng 1999).

Dendrolimus punctatus wenshanensis Minorities in the Yunnan Province, **China** eat this species, or use it as medicine. People gather the pupae, and fry them in oil or roast them. They are also used as material from which oil for cooking is extracted (Chen and Feng 1999).

Dendrolimus spectabilis (pine moth, pine caterpillar) **(PL VII-5) China:** The larvae are used as a medicine for longevity (Umemura 1943). **Japan:** The Japanese name is *matsu-kareha*. The larvae are used as a medicine for the elixir of life (Umemura 1943). **South Korea:** A way to cook them has been devised which make the larvae a nutritious and delicious food (Kuwana 1930).

Dendrolimus superans **China:** The pupae and adults are consumed (Hu and Zha 2009). **Mexico:** The larvae are recognized as a foodstuff by the Náhuatl and Otomi people in Hidalgo (Ramos-Elorduy et al. 2011).

Gonometa postica The larvae and pupae are foodstuff in **RSA** (Quin 1959), **Zambia** (Silow 1976) and **Zimbabwe** (Dube et al. 2013).

Gonometa robusta The larvae are eaten in **Zimbabwe** (Bodenheimer 1951).

Gonometa **sp.** In Sud-Kivu, **DRC,** the larvae of a species belonging to this genus are boiled or grilled (Mushambanyi 2000).

Hypsoïdes diego → *Cnethocampa diegoi*

Hypsoïdes radama → *Rombyx radama*

Livethra cajani The pupae are killed in boiling water, and fried in **Madagascar** (DeFoliart 2002).

Malacosoma **spp.** (tent caterpillar moth) **India:** A species, whose vernacular name is *mesang-long*, is eaten by the Ao-Naga people in Nagaland State (Meyer-Rochow and Changkija 1997). The **USA:** The larvae of several species belonging to this genus were singed to remove the hairs and roasted by Native Americans (Essig 1949).

Mimopacha aff. *Knoblauchi* This species is consumed in **CAR** and **Zambia** (Silow 1976).

Mimopacha aff. *Knoblauchi* (twin line lappet) In **CAR** and **Zambia,** the Cokwe people call this species *ingongolila*. The larvae are not eaten because they have poisonous hairs. The pupae, however, are traditionally edible for the people of the Cokwe, Lucazi, Luvale, Mbunda, Nkangla, and Nkangala tribes (Silow 1976).

Pachymeta robusta → Gonometa robusta

Rombyx radama The pupae are considered edible in **Madagascar** (Decary 1937).

Limacodidae (Slug caterpillar moths) **(4)**

Cania bilineata The pupae are relished by people of the minorities in Yunnan Province, **China** (Chen and Feng 1999).

Cnidocampa flavescens (oriental moths) **China:** This species was used medicinally more than 2000 years (Shennung Ben Ts'ao King; Read 1941), and is still used as an antidote (Zimian et al. 1997). **Japan:** The Japanese name is *Iraga*. In the Gunma Prefecture, people roast the larvae and pupae, and add a touch of salt to them (Miyake 1919). The larvae, prepupae and the pupae were used for treating canker sores, etc. (Umemura 1943, Shiraki 1958).

Hadraphe ethiopica The larvae are eaten in **Zambia** (Malaisse 2005).

Monema flavescens → Cnidocampa flavescens

Thosea sinensis The pupae infected by a fungus, *Cordyceps militaris* are used as a medicine in **China** (Chen and Feng 1999).

Lycaenidae (Blues and coppers) **(1)**

Lyphyra brassolis The larvae of this butterfly eat the larvae of weaver ants, *Oecophylla smaragdina*. In southern **Thailand**, people eat the larvae of *L. brassolis* together with weaver ant larvae (Eastwood 2010).

Lymantriidae (Tussock moths) **(1)**

Rhypopteryx poecilanthes In **DRC,** this species is called *nsongi* in the Kicongo language, and the larvae are consumed (Latham 2003).

Megalopygidae (Crinkled flannel moths) **(1)**

Trosia **sp.** A species belonging to this genus is used medicinally in Bahia, **Brazil** (Cost-Neto and Pacheco 2005).

Noctuidae (Owlet moths and Underwings) **(27)**

Agrotis infusa (*Bogong* moth or *Bugong* moth) **Australia:** In olden times, the outbreak of this species was common in New South Wales State. The adults were collected in large amounts during aestivation at high mountains, and were roasted, or processed to make a kind of paste (Bennett 1834, Common 1954, Flood 1996).

Agrotis ypsilon **China:** This species is a host of a fungus, *Cordyceps* sp., and the infected larvae and pupae are medicinally used (Chen and Feng 1999). **Japan:** The Japanese name is *tamana-yaga*. The larvae were used for treating eye diseases or as an anodyne (Shiraki 1958).

Anomis flava The pupae are used as a food item in **China** (Hu and Zha 2009).

Ascalapha agarista → *Ascalapha odorata*

Ascalapha odorata The Mexican name is *mariposa del muerto*. The larvae and pupae are consumed by the Náhuatl, Tlapaneco and Amuzgo people in Oaxaca and Guerrero States, **Mexico** (Ramos-Elorduy de Conconi et al. 1984, Ramos-Elorduy et al. 2011).

Busseola fusca (African maize stalk) **Zambia:** This species is a traditional food for the Nkangala people (Silow 1976).

Elacodesprasinodes → *Nyodes parasinodes*

Erebus odoratus → *Ascalapha odorata*

Euxoa segetis **Zambia:** This species is a traditional food for the Nkangala people (Silow 1976).

Gerra sevorsa The Mexican name is *gusano del mai*. The larvae are eaten in Hidalgo, Veracruz States and Mexico DF, **Mexico** (Ramos-Elorduy et al. 2011).

Helicoverpa armigera (cotton bollworm) The larvae are considered as foodstuff in **Zambia** (DeFoliart 2002).

Helicoverpa zea (corn earworms) **Mexico:** The Mexican name is *gusano del elote*. Native people of the Hñähñu tribe living in Hidalgo State eat the larvae (Ramos-Elorduy et al. 1998, Aldasoro Maya 2003). This species is raised as food for humans (Ramos-Elorduy 1997). The **USA:** Indians such as the Northern Pomo tribe eat the larvae. They killed the larvae by dipping them in cold water, and roasted them on hot ashes or boiled (Barrett 1936).

Heliothis armigera → *Helicoverpa armigera*

Heliothis obsoleta **Zambia:** This species is a traditional food for the Nkangala people Silow (1976).

Heliothis zea → *Helicoverpa zea*

Homoncocnemis fortis In the **USA**, Indians relish the hairless larvae which are dried and roasted (Swezey 1978).

Hyblea puera In **Indonesia**, the larvae are eaten on the Java Island (van der Burg 1904).

Hydrillodes lentalis → Hydrillodes morosa

Hydrillodes morosa In **China**, the feces excreted by the larvae, which eat leaves of *Platgcary strobilacea*, are drunk by people as a type of tea. It has medicinal effects on various diseases (Thémis 1997, Cheng and Feng 1999).

Laphygma exigua → Spodoptera exigua

Laphygma frugiperda → Spodoptera frugiperda

Latebraria amphipyrioides The larvae are consumed in Chiapas, Hidalgo, Oaxaca, Puebla, Veracruz States and Mexico DF, **Mexico.** In the humid-tropical area, people pickle the larvae to give them a flavor similar to herrings (Ramos-Elorduy et al. 2011).

Mamestra brassicae **Japan:** The Japanese name is *yotōga.* The larvae were used for treating lumbago (Shiraki 1958). **North Korea:** The larvae are eaten raw, or the powder of the larvae is drunk with a liquer as a remedy for lumbago in the South Hamgyong Province (Okamoto and Muramatsu 1922).

Mocis punctularis → Mocis repanda

Mocis repanda **Colombia:** The larvae are wrapped in leaves and roasted by the Yukpa people (Ruddle 1973). **Venezuela:** The larvae are used as food by the Yukpa people (Ruddle 1973).

Naranga aenescens The pupae are consumed in **China** (Hu and Zha 2009).

Nyodes prasinodes The larvae are foodstuff in **DRC** (Malaisse and Parent 1980).

Plusia gamma **France:** When an outbreak of this species occurred, people ate the larvae. Some people were reported to get food poisoning. A famous natural scientist, Reaumur, however, denied the toxicity of this worm, and recommended eating vegetables while eating the worms (Reaumur 1734–42).

Plodenia litura → Spodoptera litura

Prodenia sp. The larvae of a species belonging to this genus, is eaten in **DRC** (Malaisse 1997).

Sesamia inferens (Asiatic pink stem borer) The larvae and pupae are used as foodstuff in **China** (Hu and Zha 2009).

Sphingomorpha chlorea (sundowner moth) The Cokwe name is *mbundasese* or *vandasese*. People of the Cokwe, Kangali, Lucazi, Luvale, Mbunda, Nkangala, and Yauma tribes in **Zambia** eat the larvae (Silow 1976).

Spodoptera exempta (nutgrass armyworms) The larvae are consumed in **Zambia** (Mbata 1995).

Spodoptera exigua (beet armyworms) **China:** The pupae are used as food item (Hu and Zha 2009). **Mexico:** The Mexican name is *gusano soldado*. The larvae are regarded as foodstuff by the Yutoazteca, Náhuatl and Otomi people in Mexico DF (Ramos-Elorduy et al. 2011). **Zambia:** The larvae are consumed (Mbata 1995).

Spodoptera litura The pupae are used as food item in **China** (Hu and Zha 2009).

Spodoptera frugiperda (fall armyworms) **Mexico:** The Mexican name is *gusano elotero*. The larvae are regarded as food by the Yutoazteca, Náhuatl, Otomi, Otopame, Mazahua and Matlazinca people in Hidalgo, Tlaxcala States and Mexico FD (Ramos-Elorduy de Conconi et al. 1984, Ramos-Elorduy et al. 2011). This species is raised as food (Ramos-Elorduy 1997). **Colombia:** The larvae are wrapped in leaves and roasted by the Yukpa people (Ruddle 1973). The **USA:** The larvae are used as food by Indians of the arid areas of the west (Ebeling 1986). **Venezuela:** The larvae are eaten by the Yukpa people (Ruddle 1973).

Spodoptera **sp.** The larvae of a species called *gusano soldado* in Mexican are considered as common food by the Yutoazteca, Náhuatl and Otomi people in **Mexico DF, Mexico** (Ramos-Elorduy et al. 2011).

Strigops grandis In **Australia,** the large and fat larvae are esteemed by aborigines (Simmonds 1885).

Thysania agrippina The Mexican name is *mariposa águila*. In **Mexico,** the larvae are eaten by the Maya, Tzotzil, Tzeltal, Chol, Lacandón and Tojolabal people in Chiapas State (Ramos-Elorduy et al. 2011).

Notodontidae (Prominents, Puss moths) **(19)**

Acherontia atropus In **DRC,** this species is called *munsona sona* in the Kikongo language and is consumed (Toirambe Bamoninga 2007).

Anaphe ambrizia → *Anaphe reticulata*

Anaphe gribodoi This species is regarded as traditional food in **DRC** (Takeda 1990).

Anaphe imbrasia → *Anaphe panda*

Anaphe infracta → *Anaphe panda*

Anaphe panda This species is widely consumed in the eastern, southern and central areas of Africa. **DRC:** The larvae are used as food item by the Ngandu people (Takeda 1990). **Nigeria:** The Yoruba people eat the larvae (Fasoranti and Ajiboye 1993, Dalziel 1937, Ene 1963). **Zambia:** The Mbunda name is *liungu luanda*. People of the Ayisenga, Cokwe, Kangali, Lucazi, Lukolwe, Lunda, Luvale, Mashasha, Mbunda, Ndembu, Nkoya, and Yauma tribes eat the larvae (Silow 1976). This species is also eaten in **Tanzania** (Harris 1940, Bodenheimer 1951) and **Zimbabwe** (DeFoliart 2002).

Anaphe panda nathalia → *Anaphe panda*

Anaphe reticulata The larvae are recognized as food in the southern areas of Africa, especially in **Nigeria** (Banjo et al. 2006).

Anaphe venata (African silkworm) The larvae are used as foodstuff in western **Nigeria** (Ashiru 1988), **CAR** (Hoare 2007) and **Zambia** (Silow 1976).

Anaphe **spp. DRC:** A species called *nkankiti* by the Kikongo people is considered edible (Toirambe Bamoninga 2007). Some species belonging to this genus are used as foodstuff in **Cameroon, PRC** (Takeda 1990) and **Equatorial Africa** (Bergier 1941).

Antheua insignata In **DRC,** this species called *tukoto* by the Kibemba people, and the larvae are consumed (Malaisse 1997).

Antheua sp. A species, whose Kikongo name is *malomba loka,* is considered as food item in **DRC** and the central area of **Africa** (Toirambe Bamoninga 2007).

Busseola fusca The adults are recognized as food in **Africa** (DeFoliart 2002).

Cerurina marshalli The larvae are consumed in **CAR** (Malaisse 2005).

Craniophora ligustri The Japanese name is *ibota-kenmon*. In **Japan,** the larvae were eaten raw, or roasted and pulverized were used for treating tuberculosis (Miyake 1919).

Desmeocraea **sp.** The larvae of a species belonging to this genus, is considered edible in **Zambia** (Silow 1976).

Drapetides uniformis The larvae are called *tulongwe* in Kikongo, and are appreciated as food item in **DRC** and the southern central areas of Africa (Malaisse 1997).

Elaphrodes lacteal The larvae are called *tunkubiu* in Kikopngo, and are used as foodstuff in **DRC**, the southern central areas of Africa, and the northern central areas of Africa and **Zambia** (Malaisse 1997, Malaisse and Parent 1980).

Homonococnemis fortis The **USA:** Native Americans of the western Pomo tribe eat this species (Bean and Theodoratus 1978, Swezey 1978).

Ipanaphe carteri The larvae are eaten in **Africa** (Malaisse and Lognay 2003).

Leucodonta bicoloria This species is a host of a fungus *Cordyceps militaris*, and the infected pupae are used as a medicine in **China** (Chen and Feng 1999).

Loptoperyx uniformis → *Drapetides uniformis*

Notodonta dembowskii This species is a host of a fungus *Cordyceps militaris*, and the infected pupae are used as a medicine in **China** (Chen and Feng 1999).

Onophalera lacteal → *Elaphrodes lacteal*

Phalera assimilis This species is a host of a fungus *Cordyceps militaris*, and the infected pupae are folk medicine in **China** (Chen and Feng 1999).

Phalera bucephala This species is a host of a fungus *Cordyceps militaris*, and the infected pupae are given to invalids in **China** (Chen and Feng 1999).

Pheosigna insignata → *Antheua insignata*

Rhenea mediata This species is boiled, fried or desiccated in Shaba State, **DRC** (Malaisse and Parent 1980, Mapunzu 2002 and 2004).

Semidonta biloba This species is a host of a fungus *Cordyceps militaris*, and the infected pupae are used as a medicine in **China** (Chen and Feng 1999).

Nymphalidae (Brush-footed butterflies) **(13)**

Anartia fatima The larvae are considered as food in **Mexico** (Ramos-Elorduy and Pino 2002).

Brassolis astyra The larvae are used as foodstuff by the Huaorani and Quichuas people in **Ecuador** (Onore 2005).

Brassolis sophorae **Brazil:** The larvae are consumed by the Nhambiquara people in Moto Grosso State (Costa Neto and Ramos-Elorduy 2006). **Ecuador:** The larvae are treated as food by the Haorani and Quichuas people (Onore 2005).

Caligo memtion The larvae are considered as a food source by the Lacandones people in **Mexico** (Ramos-Elorduy and Pino 2002).

Charaxes jasius The Mexican name is *mariposa del madrono. In* **Mexico,** the larvae make a silk bag (a communal cocoon), and about 400 larvae pupate in a bag. People eat the larvae, but also preserve of this species (Ramos-Elorduy de Conconi 1987).

Chlosyne lacinia lacinia The Mexican name is *gusanito.* The larvae are consumed by the Maya, Tzotzil, Tzeital, Chol, Lacandon and Tojolabal people in Chiapas State, **Mexico** (Ramos-Elorduy et al. 2011).

Cymothe → *Cymothoe*

Cymothoe aranus → *Cymothoe aramis*

Cymothoe caenis (migratory glider) In central Africa, especially **DRC,** the larvae of this butterfly are used as foodstuff (Toirambe Bamoninga 2007).

Junonia lavinia The larvae are treated as food by the Mames people in **Mexico** (Ramos-Elorduy and Pino 2002).

Morpho sp. The larvae of a species called *pepen* in Mexican are considered edible by the Mames people in Hidalgo State, **Mexico** (Ramos-Elorduy and Pino 2002).

Nymphalis antiopia antiopia (mourning cloak) The larvae are regarded as a food item by the Otopame, Mazahua, Matlazinca, Náhuatl, and Totonaco people in Puebla State, **Mexico** (Ramos-Elorduy et al. 2011).

Panacea prola In **Ecuador**, native people of the Quichua tribe eat the larvae. The people boil the larvae in salt water, and then fry them (Onore 1997 and 2005).

Pareuptychia metaleuca The Mexican name is *gusano gordo*. The larvae are relished by many people in Chiapas, Oaxaca, Puebla and Veracruz States in **Mexico** (Ramos-Elorduy et al. 2011).

Vanessa annabella The larvae and pupae are recognized edible by the Náhuatl and Otomi people in Hidalgo State, **Mexico** (Ramos-Elorduy et al. 2011).

Vanessa virginiensis The Mexican name is *gusano del llano*. The larvae and pupae are eaten by the Náhuatl and Otomi people in Hidalgo State, **Mexico** (Ramos-Elorduy et al. 2011).

Papilionidae (Swallowtails, Apollos) **(7)**

Papilio lagleizei The larvae of this species are gregarious. In **PNG,** this species is a popular food item in many places such as Garaina in the Central Province, Karimui in the Simbu Province and Koinambe in the Western Highlands (Parsons 1999).

Papilio machaon (swallowtail, old world swallowtail) This species was used to mitigate pain of the small intestines in **China** (Li 1596). It is also used for curing stomach disorders (Zimian et al. 1997).

Papilio multicaudata multicaudata → *Pterourus multicaudata multicaudata*

Papilio polytes The pupae are eaten in **China** (Hu and Zha 2009).

Papilio polyxenes (eastern black swallowtail, American black swallowtail) The larvae are consumed by the Tzeltzales Tzoltziles people in **Mexico** (Ramos-Elorduy and Pino 2002).

Papilio xuthus The pupae are regarded as food item in **China** (Hu and Zha 2009).

Papilio **sp.** The larvae of a species belonging to this genus are considered edible in **CAR** (Malaisse 2005).

Protographium philolaus philolaus The larvae are eaten by the Maya people living in the Yucatán district, **Mexico** (Ramos-Elorduy et al. 2011).

Pterourus multicaudata multicaudata (two tailed swallowtail) The Mexican name is *mariposa de colores*. The adults are used as foodstuff by the Náhuatl and Otomi people in Hidalgo State, **Mexico** (Ramos-Elorduy et al. 2011).

Pieridae (Whites and yellows) (16)

Catasticta flisa flisa The Mexican name is *mariposa del tejocote*. The larvae are consumed by the Yutoazteca, Náhuatl and Otomi people in Mexico D.F. area, **Mexico** (Ramos-Elorduy et al. 2011).

Catasticta nimbice nimbice The larvae are eaten by the Náhuatl and Otomi people in Hidalgo State, **Mexico** (Ramos-Elorduy et al. 2011).

Catasticta teutila teutila The Mexican name is *gusano del caplín*. The larvae are appreciated as food by people of many tribes in Oaxaca State, **Mexico** (Ramos-Elorduy et al. 1998).

Catopsilia pomona (cassia butterfly, lemon migrant) The pupae are considered a delicacy in **Thailand** (Khun 2008).

Eucheira socialis socialis The Mexican name is *gusano del madroño*. In **Mexico**, native people belonging to the Chinanteco, Mazahua, Mazateco, Mestizo, Mixe, Mixteco, Náhua, Popolaca, Tarahumara, Tzetzale, Tzottzile, and Zapoteco tribes eat this species (Ramos-Elorduy and Pino 1989). This species is collected as delicious food and many colonies have become extinguished in some areas. The rearing of this butterfly is being developed, and the butterfly is raised as food (Ramos-Elorduy and Pino 1989, Ramos-Elorduy 1997).

Eucheira socialis westwoodi The Mexican name is *gusano del madroño*. The larvae are eaten by the Tepehuano and Tarahumara people in Durango State, **Mexico** (Ramos-Elorduy et al. 2011).

Eurema lisa (little sulfur) The larvae are consumed by the Lacandones people in **Mexico** (Ramos-Elorduy and Pino 2002).

Eurema salome jamapa The larvae are used as food item by the Totonaco and Huasteco people in Veracruz State, **Mexico** (Ramos-Elorduy et al. 2011).

Leptophobia aripa elodia The larvae are used as food item by the Otopame, Mazahua and Maltazinca people around Valle de Mexico, **Mexico** (Ramos-Elorduy et al. 2011).

Phoebis agarithe agarithe The Mexican name is *gusano pinto*. The larvae are considered good as food by the Maya people in Yaucatán area, **Mexico** (Ramos-Elorduy et al. 2011).

Phoebis philea philea The larvae are consumed by the Otomi and Tarasco people in **Mexico** (Ramos-Elorduy et al. 2011).

Phoebis sennae macellina The larvae are considered as foodstuff by Otopame, Mazahua and Matlazinca people in **Mexico** (Ramos-Elorduy et al. 2011).

Pieris brassicae In **Mexico,** this species is raised as food (Ramos-Elorduy 1997).

Pieris rapae (small white, cabbage white) This species is a host of a fungus *Cordyceps* sp., and the infected larvae are used as a medicine in **China** (Chen and Feng 1999). The pupae are one of food item (Hu and Zha 2009).

Pieris **sp.** The larvae of a species belonging to this genus are used as foodstuff by the Tzeltzales Tzoltziles people in **Mexico** (Ramos-Elorduy and Pino 2002).

Pontia protodice The larvae are consumed by the Otopame, Mazahua, and Matlazinca people in Valle de Mexico area, **Mexico** (Ramos-Elorduy et al. 2011).

Synchloe callidice The larvae are recognized as food by the Tzeltzales Tzoltziles people in **Mexico** (Ramos-Elorduy and Pino 2002).

Psychidae (Bagworm moths) **(4)**

Clania minuscula → *Eumeta minuscula*

Clania moddermanni → *Eumeta cervina*

Deborrea malagassa The larvae and pupae of this species are eaten in **Madagascar** (Decary 1937).

Eumeta cervina **DRC:** The Medje people living in *Ituri* forests relish the larvae (Bequaert 1921). **Equatrial Africa:** This species is considered edible (Bergier 1941). **RSA:** The Thonga people eat the larvae (Junod 1927).

Eumeta minuscula (tea bagworm) The Japanese name is *cha-minoga*. In **Japan**, the infusion of the larvae was used as a medicine for heart diseases. The roasted larvae and adults were used for treating lung diseases (Kuwana 1930).

Eumeta rougeoti The larvae of this species are considered as good food in **DRC** (DeFoliart 2002).

Eumeta **sp.** The larvae of a species belonging to this genus are considered as food item in **Africa** (Pagezy 1975).

Oiketicus **spp.** The bags of some species belonging to this genus are used to cure asthma in Bahia State, **Brazil** (Costa-Neto and Pacheco 2005).

Pyralidae (Pyralid moths) **(20)**

Aglossa dimidiata In **China,** people of the Tong tribe living in the Yunnan Province make tea from the larval excrete (Thémis 1997, Chen and Feng 1999).

Brihaspa astrostigmella The larvae and pupae are eaten raw or cooked in **Vietnam** (Tiêu 1928).

Chilo fuscidentalis → *Omphisa fuscidentalis*

Chilo luteellus **China:** This species was used to prevent vomiting of milk by babies (Li 1596).

Chilo simplex → *Chilo suppressalis*

Chilo suppressalis (rice stem borer) **China:** The larvae and pupae are consumed (Hu and Zha 2009). **Japan:** The Japanese name is *nikameiga*. This species was once a notorious pest of rice plants, but its population density decreased markedly after the World War II by excess application of insecticides. The larvae are cooked in seasoned water, roasted with a bit of salt on them, or fried without coating in the Yamaguchi and Nagano Prefectures (Takagi 1929h, Kuwana 1930, Mukaiyama 1987).

Chilo **sp.** The larvae of a species resembling *Omphisa fuscidentalis* are fried in **China** (Chen and Feng 1999).

Cnaphalocrocis medinalis The larvae and pupae are considered edible in **China** (Hu and Zha 2009).

Conogethes punctiferalis (peach moth) The Japanese name is *momonogomadara-nomeiga*. The larvae are used as medicine for convulsive fits in children or for treating indigestion in **Japan** (Kuwana 1930).

Dichocrocis punctiferalis The pupae are eaten in **China** (Hu and Zha 2009).

Galleria mellonella (greater wax moth) In the **USA**, the larvae are fried. It is said that this species is one of the tastiest insects (Huyghe 1992).

Glyphodes pyloalis **Japan:** The Japanese name is *kuwa-nomeiga*. The larvae were used for treating meningitis, acute peritonitis, etc. (Shiraki 1958). **North Korea:** The roasted larvae are pulverized, and mixed with lard as a remedy for colds, convulsive fits, acute peritonitis, etc. in the Hwanghae and North P'yongan Provinces (Okamoto and Muramatsu 1922).

Laniifera cyclades The Mexican name is *gusano del nopal*. In **Mexico**, native people of the Hñähñu, Chichimeca, Mestizo, Mixteco, Náhuatl, Otomi, Popolaca, and Zapoteco tribes eat the larvae (Ramos-Elorduy and Pino 1989, Aldasoro Maya 2003).

Myelobia **(Morpheis)** *smenintha* The larvae are considered as food item in **Brazil** (Schorr and Schmitz 1975).

Numonia pyrivorella The Japanese name is *nashimadara-meiga*. In **Japan**, the larvae and pupae were used for curing babies crying during the night (Shiraki 1958).

Omphisa fuscidentalis **China:** The larvae are well known insect food in the Yunnan Province. The most popular way to eat them is to fry them. The larvae are sold at markets, and the dish is served at some restaurants in cities (Chen and Feng 1999). **Laos:** The Laotian name is *daung-nor-mai*. The larvae are eaten (Boulidam 2008). **Thailand:** The Thai name is *rod-duan*. The fried larvae are greatly relished (Yhoung-Aree et al. 1997).

Ostrinia fumalis The larvae and pupae are esteemed as food in **China** (Hu and Zha 2009).

Ostrinia furnacalis (oriental corn borers) The larvae are fried or roasted in **China** (Chen and Feng 1999).

Plodia interpunctlla (Indian meal moth) The pupae are used as a food item in **China** (Hu and Zha 2009).

Proceras venosatum (striped stem borer) **China:** This species is used as an antidote (Zimian et al. 1997).

Pyrausta polygoni The Japanese name is *ai-nomeiga*. In **Japan,** the larvae are roasted in the Iwate and Tochigi Prefectures (Miyake 1919).

Schoenobius incertulas → *Scirpophaga incertulas*

Scirpophaga incertulas (yellow rice borers) **China:** The larvae and pupae are consumed (Hu and Zha 2009). **Japan:** The Japanese name is *sanka-meiga*. The larvae are roasted or cooked with soy sauce in the Yamaguchi Prefecture (Takagi 1929h).

Sylepta derogate The pupae are treated as food in **China** (Hu and Zha 2009).

Tryporyza incertulas → *Scirpophaga incertulas*

Saturniidae (Giant silkworm moths) **(104)**

Actias luna The Mexican name is *gusano manchado.* In **Mexico,** the larvae are appreciated as a food item by the Tarahumara people (Ramos-Elorduy et al. 2011).

Actias truncatipennis The Mexican name is *gusano gordo.* In **Mexico,** the larvae are relished by the Tarahumara people (Ramos-Elorduy et al. 2011).

Antheraea assamensis (muga silkworm moth) **India:** The pupae are consumed baked or as a chutney by the Karbi and Rengma Naga people in Assam State (Ronghang and Ahmed 2010). **Sri Lanka:** This species is eaten (Nandasena et al. 2010).

Antheraea assamica → *Antheraea assamensis*

Antheraea mylitta (tussah silk moth) **India:** In southern **India**, this species is relished (Ealand 1915).

Antheraea paphia **India:** Breeders of wild silk worm such as this species are known to regard the pupae in the cocoon as a delicacy, and eat it when the silk has been reeled off (Maxwell-Lefroy 1906). **Sri Lanka:** This species is consumed (Nandasena et al. 2010).

Antheraea pernyi (Chinese oak silk moth) **China:** Usually, the pupae are boiled or fried. Sometimes, oil is extracted from the pupae for cooking purpose (Chen and Feng 1999). This species is used as a sedative (Zimian et al. 1997). **Japan:** The Japanese name is *sakusan*. The pulverized cocoon was used to cure meningitis (Kuwana 1930). **Mexico:** This species is raised as food (Ramos-Elorduy 1997).

Antheraea polyphemus The larvae and pupae are esteemed as food item by the Tzeltzales Tzoltziles people in **Mexico** (Ramos-Elorduy and Pino 2002).

Antheraea polyphemus mexicana The larvae are appreciated as good food item by the Náhuatl, Yutoazteca and Otomi people in Veracruz State, **Mexico** (Ramos-Elorduy et al. 2011).

Antheraea roylei → Antheraea paphia

Antheraea yamamai yamamai (Japanese silk moth) The Japanese name is *yamamayu*. In **Japan,** the roasted larvae, pupae and the adults, or the infusion of the larvae are used for treating various diseases (Kuwana 1930, Watanabe 1982).

Antherina suraka The larvae are eaten in **Madagascar** (DeForiart 2002).

Anthocera monippe → Melanocera monippe

Anthocera teffraria The larvae are used as a food item in **Gabon** (Bergier 1941).

Anthocera zambezina This species is considered as a food source in **Zambia** (Ghaly 2009, Dube et al. 2013) and **Zimbabwe** (Dube et al. 2013).

Anthocera **spp.** The larvae are considered as food item in **Gabon** (Bergier 1941).

Argema mimisae (African moon-moth) The larvae are esteemed as foodstuff in **RSA** and in other countries in South Africa (Krop 1899).

Argemia **sp.** A species belonging to this genus is boiled or roasted in South Kivu State, **DRC** (Mushambanyi 2000).

Arsenura armida The Mexican name is *gusano del jonote*. In **Mexico**, native people of the Chontale, Huasteco, Huave, Lacandone, Maya, Mestizo, Mixteco, Náhua, Tlapaneca, Totonaca, and Zapoteco tribes eat the larvae. In the humid-tropical area, people pickle the larvae to give them a flavor similar to herrings (Ramos-Elorduy and Pino 1989).

Araesura polyodonta The larvae are eaten by the Náhuatl and Totonaco people in Puebla State, **Mexico** (Ramos-Elorduy et al. 2011).

Arsenura richardsonii The larvae are consumed by people of some tribes in Chiapas and Hidalgo States, **Mexico** (Ramos-Elorduy et al. 2011).

Athletes gigas The larvae are considered as food item in **DRC** and **Zimbabwe** (Dube et al. 2013) and also in **South Central Africa** (Malaisse 1997).

Athletes semialba The larvae are eaten in **DRC** and **South Central Africa** (Malaisse 1997), and **Zimbabwe** (Dube et al. 2013).

Attacuscynthia → *Samia cynthia*

Attacus ricini → *Samia ricini*

Automeris **sp.** This species is used as food item by the Tukanoan people in **Colombia** (Paoletti and Dufour 2005).

Bunaea alcinoe (common emperor) **DRC:** The Kibemba name is *mubambagoma*. The larvae are eaten (Malaisse 1997). **RSA:** The Ronga and Thonga people eat the larvae (Junod 1927). **Zambia:** People of the Cokwe, Lucazi, Luvale, Mbunda, and Nkangala tribes eat the larvae (Silow 1976). This species is also consumed in **Angola, Cameroon, Gabon, Mozambique, Tanzania** and **Zimbabwe** (Silow 1976, DeFoliart 2002, Hoare 2007).

Bunaea aslauga The larvae are regarded as food item in **Tanzania** (Harris 1940).

Bunaea **sp.** The larvae of a species belonging to this genus are considered edible in **Zimbabwe** (DeFoliart 2002).

Bunaeopsis aurantiaca The consumption of this species is common in southern central **Africa**. **DRC:** The Kibemba name is *kawanatengo*. The larvae are eaten (Malaisse 1997). **Zambia:** The Pende name is *mbula*. The Pende and Sonde people eat the larvae (Silow 1976). **Zimbabwe:** This species is consumed (Dube et al. 2013).

Bunaeopsis caffraria → *Bunaea alcinoe*

Bunaeopsis **sp.** A species belonging to this genus is considered as foodstuff in **DRC** and **Zambia** (Latham 2003, Hoare 2007).

Caio championi The larvae are eaten by the Nahuatl, Totonaco and Huasteco people in Veracruz State, **Mexico** (Ramos-Elorduy et al. 2011).

Caio richardsoni → *Arsenura richardsonii*

Callosamia promethea The larvae and pupae are eaten by the Tzeltzales Tzoltziles and Lacandones people in **Mexico** (Ramos-Elorduy and Pino 2002).

Cinabra hyperbius (banded emperor) **DRC:** The Kibemba name is *finkubala*. The larvae are consumed (Malaisse 1997). **Zambia:** People of the Cokwe, Lucazi, Luvale, and Nkangala tribes eat the larvae (Silow 1976). This species is also a food source in other countries of **South Central Africa** (Hoare 2007).

Cirina forda (pallid emperor) **DRC:** The Kibemba name is *mukosa*. The larvae are esteemed as good foodstuff (Malaisse 1997). **Nigeria:** The Nupe people desiccate the larvae after boiling for preservation. It is essential as an ingredient of soup (Fasoranti and Ajiboye 1993). **Zambia:** The Mbunda name is *kakomba*. People of the Ayisenga, Cokwe, Kangali, Lucazi, Lukolwe, Luvale, Mashasha, Mbunda, Ndembu, Nkangala, Nkoya, and Yauma tribes eat the larvae (Silow 1976). This species is also eaten in **Burkina Faso** (Latham 1999), **PRC** (DeFoliart 2002), **Namibia, RSA, Zimbabwe**, and other countries in **South Central Africa**.

Cirina forda butyrospermi In **Mali,** the Bambara and Koulikoro people boil the larvae, and then roast with butter of the butter-tree. This species is also consumed in other **western African countries** (Bergier 1941).

Cirina similis In **RSA,** the larvae are used to make soup (Junod 1927 and 1962).

Coloradia pandra (Pandra moths) The **USA:** Native Americans belonging to the Mono Lake Paiute (Aldrich 1921, Davis 1963), Owens Valley Paiute, Northern Paiute, Miwok tribes eat the larvae (Fowler 1986, Ikeda et al. 1993). People of the Klamath and Modoc tribes, however, ate the pupae but not the larvae (Patterson 1929).

Coscinocera anteus In West Papua State, **Indonesia,** this species is a wholesome food (Ramandey and van Mastrigt 2010).

Cyrtogonecana → Microgone cana

Dirphia **sp.** The larvae and pupae are considered good food by the Tukanoan people in **Colombia** (Dufour 1987).

Eacles **aff.** *ormondei yacatanensis* The larvae are food source by the Náhuatl, Yutoazteca and Otomi people in Veracruz State in **Mexico** (Ramos-Elorduy et al. 2011).

Eacles **sp.** The larvae, whose Mexican name is *gusanito,* are considered as food source by the Maya people in **Mexico** (Ramos-Elorduy et al. 2011).

Epiphora bauhiniae The larvae are consumed in **Africa** (Pagezy 1975).

Eriogyna pyretorum → Saturnia pyretorum

Euleucophaeus tolucensis The Mexican name is *zacamiche*. In **Mexico,** the larvae are eaten (Ramos-Elorduy et al. 1998).

Gonimbrasia alopia The larvae are consumed in central **Africa** (Latham 2002).

Gonimbrasia anthina The larvae are eaten in central **Africa** and southern central **Africa** (Latham 2002).

Gonimbrasia aurantiaca → Bunaeopsis aurantiaca

Gonimbrasia belina (mopani worm moths) **(PL VII-6) Malawi:** People eat the larvae as an important protein source (Munthali and Mughogho 1992). **Zambia:** The Mbunda name is *muyaya*. The larvae are relished (Mbata and Chidumayo 2003). People of the Cokwe, Lamba, Lucazi, Luvale, Mambwe, Namwang, Nkangala, and Yauma tribes eat the larvae (Silow 1976). This species is very popular and is widely used as food in various countries such as **Botswana** (DeFoliart 2002), **Namibia, RSA** (Bodenheimer 1951, Quin 1959), **Zimbabwe** (Chavanduka 1975), and central, southern areas of **Africa** (Oberprieler 1995).

Gonimbrasia cytherea The larvae are consumed in southern **Africa** (Malaisse 2005).

Gonimbrasia dione → Imbrasia dione

Gonimbrasia ertli → Imbrasia ertli

Gonimbrasia epimethea → Imbrasia epimethea

Gonimbrasia hecate **DRC:** The larvae are called *finakibobo* by the people of the Kibemba, and are eaten (Malaisse 1997). This species is also consumed in southern central areas of **Africa**.

Gonimbrasia macrothyris The larvae are considered as foodstuff in central and southern areas of **Africa** (Malaisse and Parent 1980).

Gonimbrasia melanops → Imbrasia melanops

Gonimbrasia rectilineata The larvae are used as food item in southern central areas of **Africa** (Malaisse and Parent 1980).

Gonimbrasia rhodina The larvae are considered as food in **Africa** (Latham 1999).

Goninbrasia richelmanni In **DRC**, this species is called *kisansapelebele* by the people of Kibemba, and the larvae are eaten (Malaisse 1997).

Gonimbrasia truncata → Imbrasia truncata

Gonimbrasia tyrrhea → Imbrasia tyrrhea

Gonimbrasia zambesina **DRC:** This species is called *finamiembe* by the people of Kibemba, and are usd as foodstuff (Malaisse 1997). **RSA:** The Thonga people eat the larvae (Junod 1927). **Zambia:** The Bemba name is *mumpa*. This insect is the most important species for the Bemba people as a foodstuff and as material for cashing (Junod 1962, Sugiyama 1997, DeFoliart 1999). **Zimbabwe:** This species is eaten (Dube et al. 2013).

Goodia kunzei **DRC:** The Kibemba name is *mitasondwa*. The larvae are eaten (Malaisse 1997). This species is also eaten in **Zimbabwe** and other countries of southern **Africa** (DeFoliart 2002).

Gynanisa ata → *Gynanisa maja*

Gynanisa maia → *Gynanisa maja*

Gynanisa maja **DRC:** The Kibemba name is *kawanatengo.* The larvae are esteemed as food item (Malaisse 1997). **Malawi:** People eat the larvae as an important protein source (Munthali and Mughogho 1992). **RSA:** People relish the larvae, and consequently the population density of this species has decreased markedly. Therefore the development of artificial rearing methods of this species is under way (Styles 1996, Ditlhogo 1996). **Zambia:** The Mbunda name is *likese.* The Bemba name is *Chipumi.* People of the Ayisenga, Cokwe, Kangali, Lucazi, Lukolwe, Lunda, Mashasha, Mbunda, Ndembu, Nkangala, Nkoya, and Yauma tribes eat the larvae (Silow 1976). The Bemba people are very fond of this species (Sugiyama 1997). **Zimbabwe:** People eat this species, and the dried larvae are sold at markets (Mbata et al. 2002). This species is also eaten in **Namibia** (Silow 1976), and in other countries of southern **Africa.**

Gynanisa maja ata → *Gynanisa maja*

Hemileuca tolucensis → *Euleucophaeus tolucensis*

Heniocha apollonia The larvae are consumed as food stuff in **Namibia,** and in southern **Africa** (Marais 1996).

Heniocha dyops The larvae are considered as a food source in **Namibia** and some other countries of southern **Africa** (Marais 1996). This species is also consumed as food in **Zimbabwe** (Dube et al. 2013).

Heniocha marnois The larvae are recognized as food in **Namibia** and some other countries located in southern **Africa** (Marais 1996, Malaisse 2005).

Holocerina agomensis This species is called *lindengola* by the Nkangala people. The larvae are regarded as a food item by the Nkangala people in **Zambia** (Silow 1976).

Hyalophora euryalus (ceanothus silk moth) In the **USA,** Californian Indians used the cocoon of this species as rattles in spiritualistic ceremonies and as musical instruments. Before making the rattles, people removed the pupae and ate them (Essig 1965). Peigler (1994), however, doubts that the pupae of this species were consumed, at least routinely "because of the power most groups associated with the rattle made from these cocoons".

Hyalophora **sp.** The larvae and the pupae are considered edible by the Tzeltzales Tzoltziles people in **Mexico** (Ramos-Elorduy and Pino 2002).

Hylesia coinopus The Mexican name is *mariposa de hilo grande.* The larvae are appreciated as food by people of many tribes in **Mexico** (Ramos-Elorduy et al. 2011).

Hylesia frigida The Mexican name is *mariposa del madroño*. The larvae are consumed in Chiapas and Oaxaca States, **Mexico** (Ramos-Elorduy de Conconi et al. 1984).

Hylesia sp. The larvae of a species belonging to this genus are considered edible by people of many tribes in Oaxaca State, **Mexico** (Ramos-Elorduy et al. 2011).

Imbrasia alopia The Kikongo name is *minsongo*. The larvae are regarded as foodstuff in **DRC** (Latham 2003).

Imbrasia anthina The Kicongo name is *Minsuka*. The larvae are food item in **DRC** (Latham 2003).

Imbrasia aurantiaca → *Bunaeopsis aurantiaca*

Imbrasia belina → *Gonimbrasia belina*

Imbrasia cytherea (pine tree emperor moth) The larvae are regarded as good for eating in **Zambia** (Silow 1976).

Imbrasia dione The Ngandu name is *lilangachike*. The Kicongo name is *finasepe*. **DRC:** The larvae are treated as good food by the Ngandu people (Takeda 1990).**Zambia** and some other countries of central and southern **Africa** (Malaisse and Parent 1980, Takeda 1990, Latham 2003).

Imbrasia eblis In **DRC,** the Kicongo name is *kwesu*. This species is eaten (Malaisse 1997).

Imbrasia epimethea In **DRC,** the Kicongo name is *mvinsu*. The larvae are fit for eating (Latham 2003). **Zambia:** The Mbunda name is *cuva*. People of the Ayisenga, Bisa, Cokwe, Lucazi, Lunda, Luvale, Mbunda, Nkangala, Nkoya, and Yauma tribes eat the larvae (Silow 1976, Mbata and Chidumayo 2003). This species is also consumed in **CAR** (Hoare 2009), **PRC** (Bani 1995), **Zimbabwe** (Gelfand 1971) and some other countries of central and southern **Africa**.

Imbrasia ertli The larvae are appreciated as good food item in **Angola** (Oliveira et al. 1976), **PRC** (Latham 1999), **Zimbabwe** (DeFoliart 2002) and some other countries of central and southern **Africa**.

Imbrasia forda → *Cirina forda*

Imbrasia hecate → *Goninbrasia hecate*

Imbrasia macrothyris The Kibemba name is *kawanatengo*. The larvae are esteemed as good food in **DRC** (Malaisse 1997).

Imbrasia melanops The Kicongo name is *minsendi*. **DRC:** The larvae are regarded as edible (Latham 2003). This species is also eaten in some other countries in central and northern **Africa**.

Imbrasia nictitans The larvae are relished in **DRC** (Malaisse 2005).

Imbrasia obscura The Kicongo name is *minsendi*. The larvae are considered as good food in **DRC** (Latham 2003). This species is also eaten in **CAR** (Latham 2003), **PRC** (Bani 1995) and some other countries of central and northern **Africa**.

Imbrasia oyemensis The Kicongo name is *makangu*. This species is consumed in **DRC** (Mapunzu 2002), and in **PRC** (Bani 1995).

Imbrasia petiveri The Kicongo name is *bisu*. The larvae are regarded as foodstuff in **DRC** (Latham 2003).

Imbrasia rectilineata The larvae are appreciated as a delicacy in **DRC** (Malaisse 1997, Malaisse and Parent 1980).

Imbrasia rhodina The larvae are considered as foodstuff in **DRC** (Mapunzu 2002, Hoare 2007).

Imbrasia richelmanni → *Gonimbrasia richelmanni*

Imbrasia rubra **DRC:** The Kibemba name is *kisukubia*. The larvae are used as food item (Mapunzu 2002). **Zambia:** The Bisa people relish the larvae (Mbata and Chidumayo 2003).

Imbrasia truncata The Lingala name is *likoto*. **DRC:** The larvae are used as foodstuff (Mapunzu 2002). This species is also eaten in following countries: **CAR** (Hoare 2007). **PRC** (Bani 1995) and some other countries of central and northern **Africa**.

Imbrasia tyrrhea (willow tree emperor) The larvae are considered edible in **Namibia** (Oberprieler 1995), **DRC** (DeFoliart 2002) and some other countries of southern and southern **Africa**.

Imbrasia whalbergii The Kicongo name is *minsendi noir*. The larvae are used as foodstuff in **DRC** (Mapunzu 2002).

Imbrasia zambesina → *Gonimbrasia zambesina*

Latebraria amphipyroides In **Mexico**, native people of the Chontale, Huasteco, Huave, Lacandone, Maya, Mestizo, Mixteco, Náhua, Tlapaneca, Totonaca, and Zapoteco tribes eat this species (Ramos-Elorduy and Pino 1989).

Lobobunaea christyi The larvae are consumed in **Zambia** (DeFoliart 2002).

Lobobunaea goodie The Ngandu name is *lingonju*. The larvae are considered edible by the Ngandu people in **DRC** (Takeda 1990).

Lobobunaea phaedusa **DRC:** The Kicongo name is *kaba*. The larvae are regarded as foodstuff (Latham 2003). In **PRC**, this species is also eaten (Hoare 2007).

Lobobunaea saturnus **DRC:** The Kibemba name is *finamuniangu*. The larvae are appreciated as food (Mapunzu 2002). In **Zambia** and some countries in southern central **Africa**, people also eat this species (DeFoliart 2002, Hoare 2007). This species is also recognized as food in **Zimbabwe** (Dube et al. 2013).

Melanocera menippe The larvae are considered good for eating in **Gabon** (Bergier 1941), and **South Africa** (DeFoliart 2002).

Melanocera nereis The larvae are eaten in central **Africa** (Mapunzu 2002).

Melanocera parva **DRC:** The larvae are consumed in southern central **Africa** (Mapunzu 2002). **Zambia:** The Mbunda name is *liung mundenba*. People of the Cokwe, Lucazi, Luvale, and Nkangala tribes eat the larvae (Silow 1976).

Micragone ansorgei **DRC:** The larvae are eaten (Malaisse 1997). **Zambia:** The Mbunda name is *cingoyi*. People of the Cokwe, Lucazi, Luvale, Mbunda, and Nkangala tribes eat the larvae (Silow 1976). In southern central Africa, this species is also consumed.

Microgone cana The larvae are used as a food item in **DRC** (Malaisse and Parent 1980), **RSA,** southern Africa (DeFoliart 2002), and southern central **Africa** (Mapunzu 2002).

Microgone herilla The Mu-Lang people in Ituri forests of **DRC** eat the larvae after removing the thick spines, which cover the larval cuticle (Bequaert 1921). This species is also considered as foodstuff in **Cameroon** (Bodenheimer 1951).

Microgone **sp.** A species belonging to this genus is used as food item in **Equatorial Africa** (Bergier 1941).

Nudaurelia authina **DRC:** This species is consumed (Toirambe Bamoninga 2007).

Nudaurelia dione → *Imbrasia dione*

Nudaurelia macrothyris → *Imbrasia macrothyris*

Nudaurelia oyemensis → *Imbrasia oyemensis*

Nudaurelia richelmanni In **tropical Africa,** this species is consumed as food (Malaisse and Lognay 2003).

Nyodes prasinodes **DRC:** This species is eaten (Mapunzu 2002).

Paradirphia fumosa The larvae are eaten by the Náhuatl and Totonaco people in Puebla State, **Mexico** (Ramos-Elorduy et al. 2011).

Paradirphia hoegei The larvae are regarded as foodstuff by the Náhuatl and Totonaco people in Puebla State, **Mexico** (Ramos-Elorduy et al. 2011).

Philosamia cynthia → *Samia cynthia*

Philosamia ricini → *Samia ricini*

Platisamia euryalus → *Hyalophora euryalus*

Pseudantheraea arnobia The Lingala name is *mbona*. This species is recognized as food in **DRC** (Mapunzu 2002).

Pseudantheraea discrepans **DRC:** The Ngandu name is *boona*. The larvae and pupae are eaten by the Ngandu people (Takeda 1990). **Zambia:** People of the Cokwe and Pende tribes call the larvae *mahondela*. The people eat the larvae (Silow 1976). This species is also eaten in **CAR** (Hoare 2007), **PRC** (DeFoliart 2002), and central and northern central **Africa**.

Pseudobunaea irius The larvae are recognized as a food item in **Namibia** (Marais 1996) and **Zimbabwe** (Dube et al. 2013).

Pseudodirphia mexicana The adults are considered as food resource by the Náhuatl, Yutoazteca and Otomi people in Veracruz State, **Mexico** (Ramos-Elorduy et al. 2011).

Rhodinia fugox The Japanese name is *Usutabiga*. **Japan:** The cocoons were used for treating whooping cough (Umemura 1943).

Rohaniella pygmaea The larvae are one of component of diet in **Namibia** (Malaisse 2005) and **southern Africa.**

Samia cynthia (one of Tusar oakworm moth) This species is considered as foodstuff by the Ao-Naga people in Nagaland State, **India** (Meyer-Rochow 2005), and in **Sri Lanka** (Nandasena et al. 2010).

Samia euryalus → Hyalophora euryalus

Samia ricini (eri silkworm moth) **India:** In north eastern areas, many people rear this silkworm. They prefer to eat the prepupae and to sell the cocoon at markets (Chowdhury 1982). The larvae and pupae are eaten cooked, baked or as a chutney and curry by the Karbi and Rengma Naga people in Assam State (Ronghang and Ahmed 2010). **Sri Lanka:** This species is considered as a food item (Nandasena et al. 2010).

Saturnia marchii This species is regarded as food in **Gabon** (Bergier 1941).

Saturnia pyretorum In southern **China**, particularly Hainan Island, people extract silk glands from matured larvae, and make strong fishing line from them. The larvae are sometimes fried and eaten after the silk glands are extracted (Peigler 1993).

Saturnia pyri **Mexico:** This species is raised as food (Ramos-Elorduy 1997).

Saturnia sp. A species belonging to this genus and called *lihakala* by the Ngandu people is considered as food item in **DRC** (Takeda 1990).

Synanthedon cardinalis The Mexican name is *gusanos remosos*. The larvae are consumed by the Tarasco, Náhuatl and Otomi people in Michoacán State, **Mexico** (Ramos-Elorduy et al. 2011).

Syntherata apicalis **Indonesia** (West Papua State)**:** The larvae are regarded as a food item (Ramandey and van Mastrigt 2010).

Syntherata weyneri **Indonesia** (West Papua State): The larvae are consumed (Ramandey and van Mastrigt 2010).

Tagoropsis flavinata The Kibemba name is *kisansapelebele*. The larvae are considered as food item in **DRC** (Mapunzu 2002).

Tagoropsis natalensis The larvae are appreciated as food in **DRC** (Malaisse 2005), and southern central **Africa.**

Tagoropsis **sp.** A species belonging to this genus is considered as foodstuff in **Madagascar** (DeFoliart 2002).

Urota sinope **DRC:** The Kicongo name is *finakisungwa*. The larvae are said good for eating (Mapunzu 2002). This species is also eaten in **Gabon** (Bergier 1941), **RSA,** and the southern central areas of **Africa** (DeFoliart 2002).

Usta terpsichore **Angola:** The larvae are appreciated as a food item (Womeni et al. 2009). **DRC:** The Kicongo name is *finachimpampa*. This species is consumed (Mapunzu 2002).

Usta wallengrenii The larvae are considered good for eating in **Namibia** and southern **Africa** (Oberprieler 1995).

Satyridae (Satyr butterflies) **(2)**

Magisto metaleuca In Mexico, people in Chiapas State eat the larvae (Ramos-Elorduy and Pino 2002).

Mycalesis gotoma The pupae are eaten in **China** (Hu and Zha 2009).

Sesiidae (Clearwings) **(3)**

Paranthrene regale → *Paranthrene regalis*

Paranthrene regalis (grape clearwing moth) **China:** The larvae and pupae are recognized as food item (Hu and Zha 2009). **Japan:** The Japanese name is *budō-sukashiba*. In the Chiba Prefecture, the larvae are roasted and then dipped into solution of sugar in soy sauce. They are also skewer-roasted (Miyake 1919). The larvae were medicinally used for treating convulsive fits in children, stomach disorders, diseases of the uterus, etc. (Hirase and Shitomi 1799).

Parathene tabaniformis The larvae and pupae are eaten in **China** (Hu and Zha 2009).

Synanthedron cardinalis The Mexican Name is *gusano cremosos*. The larvae are said good for eating by people of Tarasco, Náhuatl and Otomi tribes in **Mexico** (Ramos-Elorduy et al. 2011).

Sphingidae (Hawk moths, Sphinx moths) **(33)**

Acherontia achesis In **Indonesia** (West Papua State), people living in the Bime Valley eat the larvae (Tommaseo-Ponzetta and Paoletti 2005).

Acherontia atropos (dead's head hawk moth) The larvae are consumed in **DRC** and the central areas of **Africa** (Latham 2003).

Acherontia lechesis The Japanese name is *kuromengata-suzume*. In **Japan,** the larvae were used for treating convulsive fits in children, digestive disorders, tuberculosis, etc. (Watanabe 1982).

Acherontia styx crathis (death's head) The Japanese name is *mengata-suzume*. In **Japan,** the adults are fried with oil in the Nagano Prefecture (Takagi 1929g). The larvae were used for treating convulsive fits in children, digestive disorders, tuberculosis, etc. (Watanabe 1982).

Agrius convolvuli → *Herse convolvuli*

Clanis bilineata The larvae and pupae are salt-soaked and then fried in **China** (DeFoliart 2002).

Clanis bilineata tsingtauica The larvae are fried or roasted in **China** (Chen and Feng 1999).

Clanis deucalion This species is a host of a fungus *Cordyceps* sp., and used as a medicine in **China** (Chen and Feng 1999).

Cocytius antaeus The Mexican name is *gusano cornudo*. The larvae are fit for eating by people of many tribes in **Mexico** (Ramos-Elorduy et al. 2011).

Coelonia fulvinotata The larvae are treated as a food item in **CAR** (Malaisse 2005).

Coenotes eremophilae This species is consumed in **Australia** (Reim 1962).

Daphnis hypothous In **Indonesia** (West Papua State), people eat this species (Ramandey and Mastrigt 2008).

Deilephila elpenor lewisii → *Pergesa elpenor lewisii*

Erinnyis ello **Brazil:** The larvae are roasted (Araujo and Beserra 2007). **Venezuela:** The Yanomamo name is *opomoschi*: The Yanomamo people eat the larvae (Paoletti and Dufour 2005).

Herse convolbuli (sweet potato horn worms) **China:** The larvae and pupae are accepted as food (Hu and Zha 2009). **Japan:** The Japanese name is *ebigara-suzume*. The adults are fried with oil in the Nagano Prefecture (Takagi 1929g). The larvae are used for treating convulsive fits in children, digestive disorders, tuberculosis, etc. (Watanabe 1982). **Zambia:** The Mbunda name is *liung kandolo*. People of the Ayisenga, Cokwe, Luval, Nkangala, Nkoya, and Yauma tribes eat the larvae raw or cooked (Silow 1976). This species is also eaten in **Botswana** (Nonaka 1996), **RSA, Zimbabwe** (DeFoliart 2002), the southern areas of **Africa** (Quin 1959) and the southern central areas of **Africa**.

Hippotion celerio In **Botswana,** people living around the Kalahari Desert eat the larvae (Grivetti 1979). **Malaysia:** This species is eaten in Sabah State (Chung et al. 2002).

Hippotion eson The larvae are regarded as foodstuff in **DRC** (Malaisse 2005), and the southern central areas of **Africa**.

Hyles lineata (white lined sphinx) **Mexico:** The Seri people in northwestern Merxico, eat the larvae by cooking with oil after removing the heads and viscera (Felger and Moser 1985). The **USA:** Native Americans belonging to the following tribes eat this species; Pyramid Lake Paiute (Fowler and Fowler 1981), Navajo (Sutton 1988), Tohono O'Odham (Tarre 2003), Washoe, Northern Paiute, Utah Southern Paiute, Western Shoshone, Western Ute (Fowler 1986).

Hyles lineata livornicoides In **Australia**, aborigines starve the larvae for a day or two before roasting them. The roasted larvae can be stored for a long time (Low 1989).

Langia zenzeroides nawai The Japanese name is *ōshimohuri-suzume*. The larvae are used for treating convulsive fits in children, digestive disorders, tuberculosis, etc. in **Japan** (Watanabe 1982).

Lophostethus demolini The larvae are considered good for eating in **CAR** (Malaisse 2005) and western areas of **Africa**.

Macroglossum stellatarum (hummingbird hawk-moth) The Japanese name is *hōjaku*. The adults are fried with oil in the Nagano Prefecture, **Japan** (Takagi 1929g). The larvae are used for treating convulsive fits in children, digestive disorders, tuberculosis, etc. (Watanabe 1982).

Macrosila carolina → *Manduca sexta*

Manduca sexta (tobacco hornworm) The **USA:** Native Americans of the Yakut and Pimo tribes ate the larvae (Powers 1877, Simmonds 1885, Woodward 1934). **Mexico:** The Mexican name is *gusano del cuerno*. The larvae and adults are relished by people of the Náhuatl, Yotoazteca, and Otomi tribes (Ramos-Elorduy et al. 2011). This species is raised as food (Ramos-Elorduy 1997).

Manduca **sp.** The larvae and adults of a species whose Mexican name is *gusano grande verde*, are appreciated as food in Veracruz, **Mexico** (Ramos-Elorduy et al. 2011).

Marumba gaschkewitschii echephron (peach horn worms) The Japanese name is *momo-suzume*. The adults are fried with oil in the Nagano Prefecture, **Japan** (Takagi 1929g). The larvae are used for treating convulsive fits in children, digestive disorders, tuberculosis, etc. (Watanabe 1982).

Nephele comma The Mbunda name is *cikilakila*. This species is a common hawk moth in **Zambia**. People of the Ayisenga, Cokwe, Kangali, Lucazi, Lukolwe, Lunda, Luvale, Mashasha, Mubunda, Ndembu, Nkangala, Nkoya, and Yauma tribes eat the larvae and pupae raw or cooked. The larvae appear in October and eat a plant, *Diplorhynchus condylocarpon* (Silow 1976).

Oxyamblyx dohertyi This species is used as foodstuff in West Papua State, **Indonesia** (Ramandey and van Mastrigt 2010).

Pachilia ficus The larvae are accepted as food item by people of the Maya, Tzotzil, Tzeltal, Chol, Lacandon, and Tojolabal tribes in Chiapas State, **Mexico** (Ramos-Elorduy et al. 2011).

Pergesa elpenor lewisi (elephant hawk-moth) The Japanese name is *beni-suzume*. The adults are fried with oil in the Nagano Prefecture, **Japan** (Takagi 1929g). The larvae were used for treating stomach disorders, convulsive fits in children, tuberculosis, etc. (Umemura 1943).

Platysphinx stigmatica The larvae are used as foodstuff in the tropical areas of **Africa** (Malaisse 2005).

Platysphinx **sp.** A species belonging to this genus, whose Kicongo name is *munsona,* is eaten in **DRC** (Latham 2003).

Platysphinx **spp.** The larvae of a species belonging to this genus are considered as food item in the central areas of **Africa** (Latham 1999).

Psilogramma increta The Japanese name is *shimofuri-suzume*. The adults are fried with oil in the Nagano Prefecture, **Japan** (Takagi 1929g). The larvae were used for treating convulsive fits in children, digestive disorders, tuberculosis, etc. (Watanabe 1982).

Smerinthus planus (cherry horn worms) **China:** The pupae infected by a fungus, *Cordyceps militaris* are used as a medicine (Chen and Feng 1999). **Japan:** The Japanese name is *uchi-suzume*. The adults are fried with oil in the Nagano Prefecture (Takagi 1929g). The larvae are used for treating convulsive fits in children, digestive disorders, tuberculosis, etc. (Watanabe 1982).

Sphinx ludoviciana In the **USA**, Californian Indians ate this species (Powers 1877).

Sphinx **spp.** In **China**, some species belonging to this genus are considered edible roasted after soaking in salt water and cooked with noodles in the Shangdong, Henan, Hebei, Anhwei and Jiangsun regions (Zhi-Yi 2005).

Theretra japonica The Japanese name is *ko-suzume*. In **Japan,** the adults are fried with oil in the Nagano Prefecture (Takagi 1929g). The larvae are used for treating convulsive fits in children, digestive disorders, tuberculosis, etc. (Watanabe 1982).

Theretra nessus **Japan:** The Japanese name is *kiiro-suzume*. The adults were fried with oil in the Nagano Prefecture (Takagi 1929g). The larvae are used for treating convulsive fits in children, digestive disorders, tuberculosis, etc. (Watanabe 1982). **Indonesia:** In West Papua State, people eat this species (Ramandey and van Mastrigt 2010).

Theretra oldenlandiae (hawk moths) **China:** In olden times this species was used to enhance potency (Li 1596). **Japan:** The Japanese name is *sesuji-suzume*. The adults are fried with oil in the Nagano Prefecture (Takagi 1929g). The larvae are used for treating convulsive fits in children, digestive disorders, tuberculosis, etc. (Watanabe 1982).

Tortricidae (Leafroller moths) **(1)**

Leguminivora glycinivorella This species is well known as an injurious insects of beans in China. People eat the larvae fried (Chen and Feng 1999).

Uraniidae (Swallowtail moths) **(1)**

Nyctalemon patroclus goldiei **Indonesia** (West Papua State): The larvae are eaten (Ramandey and van Mastrigt 2010).

Xyloryctidae (Xyloryctid moths) **(1)**

Linoclostis gonatias This species is a host of a fungus *Cordyceps* sp., and the infected insects are used as a medicine in **China** (Chen and Feng 1999).

Zygaenidae (Burnets and Foresters) **(2)**

Zygaena ephialtes **Italy:** Around Tramonti de Sotto in the Friurli district, people eat the abdomens of the adult moths raw in summer (Paoletti and Dreon 2005).

Zygaena transalpine **Italy:** Around Tramonti de Sotto in the Friurli district, people eat the abdomens of the adult moths raw in summer (Paoletti and Dreon 2005).

18. HYMENOPTERA [385]

Agaonidae (Fig wasps) **(1)**

Blastoophaga pumilae The eggs, larvae, pupae and the adults are considered good for eating in **China** (Hu and Zha 2009).

Anthophoridae (0)

Anthophora sp. **Botswana:** The honey of a species belonging to this genus is eaten raw (Nonaka 1996). The **USA:** The honey of a species called digger bee is known as a food of the Tarahumare children (Hrdlicka 1908).

Apidae (Bees) **(155)**

Apis adansoni The larvae and the honey are eaten raw in **Cameroon** (Tessmann 1913 and 1914, Mbata 1995), **DRC** (Takeda 1990), **Tanzania** (Harris 1940) and **Zambia** (Mbata 1995).

Apis andreniformis In **Thailand,** people eat this species (Chen and Wongsiri 1995).

Apis cerana (oriental honey bee) **China:** This species is raised, and used for curing rheumatism, menstruation disorders and hepatitis (Zimian et al. 1997). **Indonesia** (Java Island): The larvae are treated as foodstuff as *botok* cuisine, and the honey is eaten raw (Matsuura 1998a). **Malaysia:** In Sabah State in the Borneo Island, people call this species *pomosuon,* and eat the larvae raw or drink together with honey. They also stir-fry or boil the larvae with porridge (Chung 2010). **Mexico:** This species is raised as food (Ramos-Elorduy 1997). The **Philippines:** The extra larvae and pupae in the process of harvesting honey are relished. The cultured brood is not eaten although this species have been adapted for domestic cultivation (Adalla and Cervancia 2010). **Sri Lanka:** The Vedda are accustomed to eating the larvae of this species (Wijesekara 1964). **Thailand:** People eat this species (Chen et al. 1998).

Apis cerana indica **India:** The larvae and adults are regarded as food item in Arunachal Pradesh State. People roast the bees to eat or use as a paste. When the adults are eaten, their wings and antennae are removed. The honey is used for coughs, fevers, stomach disorders, etc. (Chakravorty et al. 2011). The Karbi and Rengma Naga people in Assam State eat the larvae and pupae fried, baked or as a chutney (Ronghang and Ahmed 2010). **Malaysia** (Sabah State): The larvae and pupae are used as a foodstuff (Chung et al. 2002). The **Philippines:** The larvae are regarded as a food item (Gibbs et al. 1912). **Sri Lanka:** This species is called *mee massa.* The honey of this species is the sweetest among honey produced by five species of wild bee found in Sri Lanka (Spittel 1924). **Thailand:** The Laotian people collect the nest of this bee, and eat the honey stored in the nest and larvae (Bristowe 1932).

Apis cerana japonica **Japan:** The Japanese name is *nihon-mitsubachi.* Earlier the larvae and the honey stored in the nest were widely consumed. Sometimes people crushed the bee nests containing larvae and honey, ground it by adding water, decocted it, and squeezed it for medicinal uses (Hirase and Shitomi 1799). **North Korea:** The honey is boiled, and the larvae are pulverized. They are used for stomach disorders, constipation, etc. in the North P'yongan and South P'yongan Provinces (Okamoto and Muramatsu 1922). **South Korea:** The honey is boiled, and the larvae are pulverized. They are used for stomach disorders, constipation, etc. in the North Joella, South Chungcheong and Gyeonggi Provinces (Okamoto and Muramatsu 1922).

Apis dorsata **China:** This species is eaten especially in the Yunnan Province (Chen et al. 2009). **India:** The larvae and pupae are fried, baked or as a chutney by the Karbi and Rengma Naga people (Ronghang and Ahmed 2010). The Galo people in Arunachal Pradesh State eat the larvae and the honey raw (Kato and Gopi 2009). The Onge aboriginal people in Andaman Islands, eat the larvae and the honey stored in the nest (Crane 1967). **Indonesia:** On Sumatra Island, this species is consumed (Matsuura 2002). **Malaysia:** In Sarawak State, the Dayak people eat the honey stored in the nest (Beccari 1904). In Sabah State, people eat the larvae and pupae raw, boiled in porridge or rice, stir-fried or take it with diluted honey as a drink. Sometimes the brood together with the hive is squeezed to extract some liquid, which is then boiled. It becomes hard like fried eggs (Chung et al. 2002). **Mexico:** This species is raised (Ramos-Elorduy 1997). **Nepal:** People of the Gurung tribe eat the honey stored by this species (Strickland 1932). The **Philippines:** The extra larvae and pupae in the process of harvesting honey are eaten. The cultured brood is not eaten although this species have been adapted for domestic cultivation (Adalla and Cervancia 2010, Gibbs et al. 1912). **Sri Lanka:** The Vedda are accustomed to eating the larvae of this species (Wijesekara 1964). **Thailand:** The larvae and pupae are roasted, and the honey is eaten raw (Bristowe 1932). **Timor:** People collect the honey and wax (Brygoo 1946, Wallace 1869).

Apis florea **China:** This species is eaten especially in the Yunnan Province (Chen et al. 2009). **India:** The larvae, pupae and the honey are fried, baked or as a chutney by the Karbi and Rengma Naga people (Ronghang and Ahmed 2010). **Sri Lanka:** The Vedda are accustomed to eating the larvae of this species (Wijesekara 1964). **Thailand:** In the north eastern area of the country, the larvae, pupae and the adults are roasted (Hanboonsong et al. 2000). **Laos, Myanmar** and **Vietnam:** This species is considered as foodstuff (Yhoung-Aree and Viwatpanich 2005).

Apis indica → *Apis cerana indica*

Apis indica var. *japonica* → *Apis cerana japonica*

Apis laboriosa **Mexico:** This species is raised (Ramos-Elorduy 1997). **Nepal:** The larvae, prepupae and the pupae are heated and then eaten (Burgett 1990). **Sri Lanka:** This species is used as a food item (Nandasena et al. 2010).

Apis mellifera (honey bees) **China:** This species is used for curing rheumatism, menstruation disorders, hepatitis, lung cancer, etc. (Zimian et al. 1997). **Ecuador:** The larvae, pupae and the honey stored in the nests are eaten raw by the Negr. Esmerald people (Posey 1983a and b, Onore 1997, Paoletti and Dufour 2005). **India:** The larvae, pupae, eggs and honey are consumed as a chutney or baked (Ronghang and Ahmed 2010). **Japan:** The Japanese name is *seiyō-mitsubachi*. The male larvae are used like *Vespula flavipes lewisii* (Matsuura 1999). **Kenya:** The honey is eaten raw (Huntingford 1955). **Mexico:** The Mexican name is *abeja de la miel*. The eggs, larvae, pupae and honey collected by worker bees are consumed everywhere (Ramos-Elorduy de Conconi et al. 1984). **Nepal:** The larvae, prepupae and the pupae are heated and then eaten (Sokolov 1991), **Nigeria:** The larvae and pupae are consumed (Womeni et al. 2009). **Paraguay:** The larvae, pupae and the honey stored in the nests are eaten raw by the Ache people (Posey 1983a and b, Onore 1997, Hill et al. 1984). **Senegal:** This species is considered as foodstuff (Gessain and Kinzler 1975). **Tanzania:** The honey is eaten raw (Huntingford 1955). The **USA:** The larvae are fried (Huyghe 1992). **Vietnam:** The larvae are regarded edible (Tiêu 1928).

Apis mellifera capensis The honey is used as sweet food material raw in **Zambia** (Mbata 1995).

Apis mellifera scutellata **Brazil:** The larvae, pupae, honey and the pollen stored in the nest are consumed by the Guarani M'byá people of San Paulo State, the Pankararé people of Bahia State and the Nhambiquara people of Mato Grosso state (Costa-Neto 2003a and b). **Mexico:** In Chiapas State, the eggs, larvae, pupae and the honey collected by worker bees are regarded as foodstuff (Ramos-Elorduy and Pino 2002). This species is raised (Ramos-Elorduy 1997).

Apis mellifica ligustica **Mexico:** This species is raised (Ramos-Elorduy 1997).

Apis mellifica var. *adansoni* **Cameroon:** The Pangwe people relish the honey stored by this species (Tessmann 1913 and 1914, Bequaert 1921). **DRC:** The Ngandu people eat the honey (Takeda 1990). **Mexico:** People of Mazatecos, Nahuas, Popolocas, Totonacos eat this bee (Ramos-Elorduy 2009). **Tanzania:** The larvae and the honey are eaten raw (Ichikawa 1982).

Apis nigrocincta In Sarawak State, **Malaysia**, this species is called *nuang*. The honey stored by this species is not good in quality, but is produced by the bees in large amounts. The honey is, nevertheless, highly relished by the Dayak people (Beccari 1904).

Apis unicolor adansoni The larvae and honey are eaten raw in southern **Sahara** (Bodenheimer 1951).

Apis unicolor fasciata The larvae and honey are used as food raw in **Egypt** and **Arabia** (Bodenheimer 1951).

Apis unicolor unicolor The larvae and honey are consumed raw in **Bourbon, Madagascar,** and **Mauritius** (Bodenheimer 1951).

Apis zonata In the **Philippines**, the larvae are considered as foodstuff (Gibbs et al. 1912).

Apis **spp. India:** The adults of some species belonging to this genus are fried or used to make paste by people of the Galo tribe in Arunachal Pradesh State (Chakravorty et al. 2011). **Laos:** The Laotian name is *hang pheung*, or *nam pheung*. Several species belonging to this genus are considered edible (Boulidam 2008).

Apotrigona ref. *ferruginea* The larvae and honey are eaten raw in DRC (Ichikawa 1982).

Apotrigona ref. *komiensis* The larvae and honey are used as food raw in DRC (Ichikawa 1982).

Axestotrigona erythra interposita → *Trigona erythra interposita*

Axestotrigona richardsi → *Trigona richardsi*

Axestotrigona simpsoni The larvae and honey are eaten raw in DRC (Ichikawa 1982).

Cephalotrigona capitata In **Brazil**, the Guarani M'byá people in San Paulo State eat the honey collect by the worker bees (Rodrigues 2005).

Cephalotrigona femorata In **Brazil,** the Nahmbiquara people in Mato Grosso State eat the larvae, pupae, and the honey and pollen stored in the nest (Setz 1991).

Cephalotrigona zexmemiae The eggs, larvae, pupae and the honey collected by the bees are consumed in Chiapas State of **Mexico** (Ramos-Elorduy and Pino 2002).

Dactylurina staudingeri The larvae and honey are consumed as foodstuff raw in DRC (Ichikawa 1982).

Dolichotrigona schultessi → *Trigonisca schultessi*

Duckeola ghilianii In **Brazil,** the Nahmbiquara people in Mato Grosso State eat the honey collected by the bees (Setz 1991).

Friesella schrottkyi In **Brazil,** the Guarani M'byá people in San Paulo State consume the larvae, pupae, honey, bee-wax and the pollen in the nests (Rodrigues 2005).

Frieseomellita silvestrii In **Brazil**, the Pankararé People in Bahia State eat the raw pollen stored in the nests. It is also used as a medince (Costa Neto and de Melo 1998).

Frieseomellita varia In **Brazil**, the Kayapó people in Pará State eat the larvae and pupae raw (Posey 1983a and b). **Venezuela:** This species is consume by the Yanomamo people (Paoletti and Dufour 2005).

Friesesmellita nigra The eggs, larvae, pupae and the honey collected by the bees are relished in Oaxaca State of **Mexico** (Ramos-Elorduy et al. 1997).

Frieseomellita silvestrii → *Trigona silvestrii*

Frieseomellita sp. → *Tetragonisca angustula angustula*

Hypotrigona araujor The larvae and honey are eaten raw in DRC (Ichikawa 1982).

Hypotrigona gribodoi → *Trigona gribodoi*

Hypotrigona occidentalis → *Trigona occidentalis*

Hypotrigona ruspolii → *Trigona ruspolii*

Hypotrigona sp. → *Trigona* spp.

Lestrimelita limao **Brazil:** The Nhambiquara people of Mato Grosso State (Setz 1991), and the people in Bahia State (Costa Neto 2003b) eat the larvae, pupae, and the honey and pollen stored in the nests. **Mexico:** The eggs, larvae, pupae and the honey collected by the bees are consumed in Campeche, Chiapas and Yucatan States (Ramos-Elorduy de Conconi et al. 1984). The Maya people domesticated this bee (Bodenheimer 1951).

Lestrimelita niitkib The eggs, larvae, pupae and the honey collected by the bees are used as human food in Chiapas State of **Mexico** (Ramos-Elorduy and Pino 2002).

Lestrimelita sp. The eggs, larvae, pupae and the honey collected by the bees belonging to this genus are considered edible in Oaxaca and Yucatan of **Mexico** (Ramos-Elorduy de Conconi et al. 1984, Ramos-Elorduy et al. 1997).

Megapis dorsata → *Apis dorsata*

Megapis zonata → *Apis zonata*

Melipona asilvai In **Brazil,** people in Bahia eat the honey stored in the nests (Dias 2003).

Melipona atratula The honey stored in the nests is used as sweet material in **Brazil** (Bodenheimer 1951).

Melipona beechei The Mexican name is *abeja que no pica*. In **Mexico,** native people of the Lacandone and Maya tribes eat this species (Ramos-Elorduy and Pino 1989). The Maya people domesticated this bee (Bodenheimer 1951). The eggs, larvae, pupae and the honey collected by the bees are consumed in Chiapas State (Ramos-Elorduy and Pino 2002). This species is used also in Campeche, Hidalgo, Oaxaca, Quintana Roo and Veracruz States (Ramos-Elorduy 2009).

Melipona beechei fulvipes **Cuba:** The honey stored in the nests is used as sweetening raw (Schwarz 1948).

Melipona bicolor In **Brazil,** the Guarani M'byá people in San Paulo State eat the honey stored in the nests (Rodrigues 2005).

Melipona bilineata The honey stored in the nests is used as sweetening in **Brazil** (Bodenheimer 1951).

Melipona bleckei The dried larvae and pupae are considered as food for preservation in Yasotato of **Mexico** (Ramos-Elorduy and Pino 1990).

Melipona compressipes The honey collected by bees is eaten in **Brazil** (Crane 1992).

Melipona compressipes fasciculata In **Brazil,** the Kayapó people in Pará State eat the larvae, pupae and the honey stored in the nests raw (Posey and Camargo 1985).

Melipona crinita In **Brazil,** the Ashaninka people in Acre State eat the honey stored in the nests (Brilhante and Mitoso 2005).

Melipona dorsalis The honey stored in the nests is used as sweetening in **Brazil** (Bodenheimer 1951).

Melipona ebumea fuscopilosa In **Brazil,** the Ashaninka people in Acre State eat the honey stored in the nests (Brilhante and Mitoso 2005).

Melipona fasciata The Mexican name is *abeja chica.* The eggs, larvae, pupae and the honey collected by the bees are consumed in Chiapas State of **Mexico** (Ramos-Elorduy and Pino 2002).

Melipona fasciata guerreroensis The eggs, larvae and the pupae are regarded as foodstuff in Chiapas and Guerrero States of **Mexico** (Ramos-Elorduy de Conconi et al. 1984). Native people of the Maya tribe domesticated this bee (Bodenheimer 1951).

Melipona fasciata scutellaris The honey stored in the nests is good for sweetening of food material in **Brazil.** This species has been domesticated in northern Brazil (Bodenheimer 1951).

Melipona favosa The larvae and pupae are considered edible in Bali Island, **Indonesia** (Césard 2006).

Melipona grandis **Brazil:** The larvae, pupae, and the honey stored in the nests are eaten raw by the Ashaninka people in Acre State (Brilhante and Mitoso 2005), and the Surui people (Coimbra 1984). **Mexico:** The eggs, larvae, pupae and the honey collected by the bees are treated as foodstuff in Oaxaca State (Ramos-Elorduy et al. 1997).

Melipona interrupta In **Mexico,** the eggs, larvae, pupae and the honey collected by the bees are consumed in Oaxaca State (Ramos-Elorduy et al. 1997).

Melipona interrupta grandis In **Brazil,** the Nhambiquara people in Mato Grosso State eat the larvae, pupae, honey and the pollen in the nests (Setz 1991).

Melipona mandacaia In **Brazil,** people in Bahia State eat the honey collected by the bees (Cost Neto 2003b).

Melipona marginata In **Brazil,** the Guarani M'byá people in San Paulo State eat the honey collected by the bees (Rodrigues 2005).

Melipona melanoventer The larvae, pupae and the honey stored in the nests are used as food item in **Brazil** (Posey and Camargo 1985).

Melipona minuta The larvae, and pupae are roasted, and the honey is eaten raw in **Indonesia** (van der Burg 1904).

Melipona nigra The honey collected by the bees are used as sweetening in **Brazil** (Crane 1992).

Melipona pseudocentris The honey collected by the bees are used as sweetening in **Brazil** (Crane 1992).

Melipona quadrifasciata In **Brazil,** the Guarani M'byá people in San Paulo State eat the honey stored in the nests (Rodrigues 2005).

Melipona rufiventris In **Brazil,** the Ashaninka people in Acré State eat the honey stored in the nests (Brilhante and Mitoso 2005). The Kayapó people in Pará State eat the larvae and pupae raw (Posey 1983a and b).

Melipona rufiventris flavolineata The larvae and pupae are eaten raw by the Kayapó people in Pará State, **Brazil** (Posey 1983a and b).

Melipona schencki picadensis The honey collected by the bees is used as sweetening in **Brazil** (Crane 1992).

Melipona schencki schencki **Brazil:** The honey collected by the bees is used to sweeten other food material in **Brazil** (Crane 1992).

Melipona schwarzi In **Brazil,** the Surui people in Pará State eat the larvae, pupae and the honey collected by the bees (Coimbra 1984).

Melipona scutellaris In Bahia State of **Brazil,** the honey stored in the nests is roasted (Costa-Noto 2003b). The honey of this species is also used as a medicine for epidemic mumps (Marques and Costa-Neto 1997).

Melipona seminigra merrillae The honey collected by the bees is used as sweetening in **Brazil** (Crane 1992).

Melipona seminigra **cf.** *pernigra* **Brazil:** The Kayapó people in Pará State eat the larvae and pupae raw (Posey 1983a and b). The honey is eaten raw to cure impotence (Cost-Neto 1996).

Melipona solani The eggs, larvae, pupae and the honey collected by the bees are consumed in Chiapas State of **Mexico** (Ramos-Elorduy and Pino 2002).

Melipona vidua The larvae, and pupae are roasted, and the honey is eaten raw in **Indonesia** (van der Burg 1904).

Melipona **spp. Africa:** The larvae and honey are used as foodstuff raw (Bodenheimer 1951). **Mexico:** In Chiapas, the eggs, larvae, pupae and the honey collected by the bees belonging to this genus are considered as food source (Ramos-Elorduy and Pino 2002). **Paraguay:** The honey stored in the nests are fit for eating raw (Bodenheimer 1951).

Meliponula bocandei This species is recognized as food in **Africa** (Ichikawa 1982).

Meliponula ferruginea → *Trigona ferruginea*

Meliponula lendliana → *Trigona lendliana*

Meliponula richardsi → *Trigona richardsi*

Micrapis florae → *Apis florae*

Nannotrigona bipunctata polystica **Brazil:** The Surui people in Pará State eat the larvae, pupae, and the honey stored in the nests raw (Coimbra 1984).

Nannotrigona perilampoides The eggs, larvae, pupae and the honey collected by the bees are consumed in Chiapas State of **Mexico** (Ramos-Elorduy and Pino 2002).

Nannotrigona testaceicornis The eggs, larvae, pupae and the honey collected by the bees are recognized as food item in Chiapas State of **Mexico** (Ramos-Elorduy and Pino 2002).

Nannotrigona xanthotricha **Brazil:** The Surui people in Pará State eat the larvae, pupae and the honey stored in the nests raw (Coimbra 1984).

Nannotrigona **spp. Brazil:** The Surui people in Pará State eat the larvae, pupae, and the honey stored in the nests raw (Coimbra 1984). **Mexico:** The eggs, larvae, pupae and the honey collected by the bees are fit for eating in Chiapas state of Mexico (Ramos-Elorduy and Pino 2002).

Oxytrigona obscura In **Brazil,** the Nhambiquara people in Mato Grosso State eat the larvae, pupae, honey and the pollen stored in the nests (Setz 1991).

Oxytrigona tataira In **Brazil,** the Kayapó people in Pará State and the Nhambiquara people in Mato Grosso State eat the larvae, pupae, honey and the pollen stored in the nests raw (Posey 1983a).

Oxytrigona **spp. Brazil:** The Kayapó people in Pará State and the Nhambiquara people in Mato Grosso state eat the larvae, pupae and the honey stored in the nests of several species belonging to this genus raw (Posey 1983a).

Paratrigona **sp.** In **Brazil,** the Guarani M'byá people in San Paulo State eat the larvae, pupae and the honey collected by the bees of a species belonging to this genus (Costa Neto and Ramos-Elorduy 2006).

Partamona bilineata The eggs, larvae, pupae and the honey collected by the bees are consumed in Chiapas State of **Mexico** (Ramos-Elorduy and Pino 2002).

Partamona **cf.** *cupira* In **Brazil,** the Nhambiquara people in Mato Grosso State (Setz 1991), the Pankararé people in Bahia State (Costa Neto 1994 and 1996) and people in Algoas State (Cost Neto 1994) eat the larvae, pupae and the honey and pollen stored in the nests raw.

Partamona orizabaensis The eggs, larvae, pupae and the honey collected by the bees are used as food item in Chiapas State of **Mexico** (Ramos-Elorduy and Pino 2002).

Partamona testacea This species is relished by the Yanomamo people in **Venezuela** (Paoletti and Dufour 2005).

Partamona **sp. Brazil:** The Guarani M'byá people in San Paulo State eat the larvae, pupae and the honey stored by the bees raw (Rodrigues 2005, Posey 1983a and b). **Mexico:** In Campeche and Yucatan States the eggs, larvae and the pupae are good for eating (Ramos-Elorduy de Conconi et al. 1984). Maya people domesticated this bee (Bodenheimer 1951).

Plebeia emerina In **Brazil,** the Pankararé people in Bahia State eat the larvae and pupae roasted (Crane 1992). They also use the raw honey as a remedy for a sore throat (Costa-Neto 1996).

Plebeia frontalis The eggs, larvae, pupae and the honey collected by the bees are regarded as foodstuff in Chiapas State of **Mexico** (Ramos-Elorduy and Pino 2002).

Plebeia mexica The honey collected by this bee is consumed in **Mexico** (Acuña et al. 2011).

Plebeia mosquito In **Brazil**, the people of Alagoas State eat the honey stored in the nests (Lima 2000).

Plebeia remota The honey stored by the bees is eaten in **Brazil** (Crane 1992).

Plebeia **spp. Brazil:** The Guarani M'byá people in San Paulo State (Rodrigues 2005), and the Surui people (Coimbra 1984) eat the larvae, pupae and the honey of some species belonging to this genus raw. **Mexico:** The eggs, larvae, pupae and the honey collected by several species of bees belonging to this genus are considered as foodstuff in Chiapas State (Ramos-Elorduy and Pino 2002).

Plebeiella lendliana → *Trigona lendliana*

Plebina denoita (mopani bee) The honey collected by this small bee is used as sweetening by herd boys in **Botswana** (Mojeremane and Lumbile 2005).

Ptilotrigona lurida The larvae, pupae and the honey collected by the bees are consumed by the Surui people in **Brazil** (Coimbra 1984).

Scaptotrigona mexicana The Mexican name is *mosca de la virgin*. In Chiapas and Oaxaca States, **Mexico,** the eggs, larvae, pupae and the honey collected by the bees are recognized as food source (Ramos-Elorduy et al. 1997). This species can be raised (Ramos-Elorduy 1997).

Scaptotrigona nigrohirta In **Brazil**, native people of Kayapó in Pará State (Posey and Camargo 1985) and people of the Nhambiquara tribe in Mato

Grosso State (Setz 1991) eat the larvae, pupae, honey and the pollen stored in the nests.

Scaptotrigona pectoralis The eggs, larvae, pupae and the honey collected by the bees are treated as foodstuff in Chiapas and Oaxaca States, **Mexico** (Ramos-Elorduy et al. 1997).

Scaptotrigona polysticta In **Brazil**, native people of Kayapó in Pará State (Posey and Camargo 1985) and people of the Nhambiquara tribe in Mato Grosso State (Setz 1991) eat the larvae, pupae, honey and the pollen stored in the nests.

Scaptotrigona postica The honey stored in the nests are used as sweetening in **Brazil** (Crane 1992).

Scaptotrigona tubiba The people in Alagoas State, **Brazil** eat the honey stored in the nests (Costa-Neto 1994).

Scaptotrigona xanthotrica **Brazil:** The Nhambiquara people in Mato Grosso State eat the larvae, pupae, honey and the pollen stored in the nests (Setz 1991). This species is also relished by the Yanomamo in **Venezuela** (Paoletti and Dufour 2005).

Scaptotrigona **spp.** The eggs, larvae, pupae and the honey collected by the bees belonging to this genus are considered as food item in Chiapas State of **Mexico** (Ramos-Elorduy and Pino 2002).

Scaura longula In **Brazil,** the Kayapó people in Pará State eat the larvae and pupae raw (Posey 1983b).

Tetragona benneri **Brazil:** The Kayapó people in Pará State eat the larvae and pupae raw (Posey and Camargo 1985).

Tetragona fulviventris In **Brazil,** the Kayapó people in Pará State eat the larvae and pupae raw (Posey 1983a and b).

Tetragona goettei In **Brazil,** the Kayapó people in Pará State eat the larvae and pupae raw (Posey 1983a and b).

Tetragonanigra → *Friesesmelittanigra*

Tetragona quadrangular In **Brazil,** the Kayapó people in Pará State eat the larvae and pupae raw (Posey 1983a and b).

Tetragona spinnipes In **Brazil,** the Kayapó people in Pará State consumed the larvae and pupae raw (Posey 1983a and b).

Tetragonisca angustula angustula **Brazil:** The Kayapó people in Pará State eat the larvae and pupae raw (Posey 1983a and b). People of the Pankarare tribe in Bahia State and those in Alsgoas State eat the pupae and honey stored in the nests raw (Costa-Neto 1994 and 1996). The Suri people also eat this species raw (Coimbra 1985). **Ecuador:** The larvae mixed with honey and pollen is relished by the Canari people in the Valladolid area (Onore 2005, Paoletti and Dufour 2005).

Tetragonisca branneri In **Brazil,** the larvae, pupae and the honey stored in the nests are eaten raw by the Suri people (Posey 1983a and b).

Tetragonisca dorsalis In **Brazil,** the larvae, pupae and the honey stored in the nests are consumed raw by the Suri people (Coimbra 1984).

Trigona amalthea **Brazil:** The Kayapó people in Pará State eat the larvae and pupae raw (Coimbra 1984). **Venezuela:** The larvae, pupae and the honey are eaten by people of the Yukpa tribe (Ruddle 1973).

Trigona angustula In **Brazil,** the Kayapó people in Pará State (Posey 1987), the Guarani M'byá people in San Paulo State (Rodrigues 2005) and the people in Bahia State (Costa Neto 1996) eat the larvae, pupae, honey and the pollen stored in the nests.

Trigona apicalis The Malay name is *clulut*. In Sarawak State, **Malaysia,** the Dayak people eat the stored honey, which is slightly sour (Beccari 1904).

Trigona biroi The extra larvae and pupae in the process of harvesting honey are consumed in the **Philippines**. The cultured brood is not eaten although this species are adapted for domestic cultivation (Adalla and Cervancia 2010).

Trigona branneri In **Brazil,** the Nhambiquara people in Mato Grosso State (Setz 1991), and the Surui people in Pará State (Coimbra 1984) eat the larvae, pupae and the honey stored in the nests.

Trigona (=Hypotrigona) braunsi The larvae and honey are eaten raw in **DRC** (Parent et al. 1978, Ichikawa 1987)

Trigona carbonaria In **Australia,** the larvae, pupae, adults, honey, and the nests are used as food item (Hockings 1884).

Trigona cassiae In Queensland State of **Australia**, the larvae, pupae, adults, honey, and the nests are appreciated as food (Hockings 1884).

Trigona chanchamayoensis In **Brazil,** the Kayapó people in Pará State eat the larvae, pupae, and the honey stored in the nests raw (Posey 1987).

Trigona cilipes pellusida In **Brazil,** the Nhambiquara people in Mato Grosso State eat the larvae, pupae, and the honey and pollen stored in the nests raw (Setz 1991).

Trigona clavipes In **Brazil,** the honey stored in the nests are used as sweetening (Crane 1992).

Trigona (Tetragona) clavipes **Colombia:** The Yukupa people roast the larvae, and eat the honey stored in the nests raw (Posey 1983a and b). **Venezuela:** The larvae and pupae are roasted, and the honey stored in the nests is consumed raw by Yukpa people (Posey 1983a and b).

Trigona clypeata The larvae and honey are eaten raw in **Africa** (Bodenheimer 1951).

Trigona (*Trigona*) *corvina* The eggs, larvae, pupae and the honey collected by bees are esteemed as delicacy in Chiapas State, **Mexico** (Ramos-Elorduy and Pino 2002).

Trigona dallatorreana In **Brazil,** the Kayapó people in Pará State (Posey and Camargo 1985), and the Nhambiquara people in Mato Grosso State (Setz 1991) eat the larvae, pupae, and the honey and pollen stored in the nests raw.

Trigona dorsalis In **Brazil,** the Surui people in Pará State eat the larvae, pupae and the honey stored in the nests raw (Coimbra 1984).

Trigona (=*Axestotrigona*) *erythra interposita* The larvae and honey are consumed raw in the Shaba area of **DRC** (Parent et al. 1978).

Trigona erythra togoensis The larvae and honey are eaten raw in **Tanzania** (Bodenheimer 1951).

Trigona ferruginea gambiensis The larvae and honey are used as food item raw in **Senegal** (Gessain and Kinzler 1975).

Trigona flaveola The honey collected by the bees is used as sweetening in **Brazil** (Bodenheimer 1951).

Trigona fulviventris The eggs, larvae, pupae and the honey collected by the bees are eaten in Chiapas State in **Mexico** (Ramos-Elorduy and Pino 2002). The Maya people have domesticated this bee (Bodenheimer 1951).

Trigona (*Trigona*) *fuscipennis* **Brazil:** The Kayapó people in Pará State eat the larvae and pupae raw (Coimbra 1984). **Mexico:** The eggs, larvae, pupae and the honey collected by the bees are regarded as food item in Chiapas State (Ramos-Elorduy and Pino 2002).

Trigona geniculata The honey stored in the nests are eaten in **Brazil** (Bodenheimer 1951).

Trigona ghilianii In **Brazil,** the Nhambiquara people in Mato Grosso State eat the larvae, pupae, honey and the pollen stored in the nests (Setz 1991).

Trigona gribodoi The Ngandu name is *mbolo*. The honey is used as sweetening in **DRC** (Takeda 1990).

Trigona hypogea In **Brazil,** the Nhambiquara people in Mato Grosso State eat the larvae, pupae, honey and the pollen stored in the nests (Setz 1991).

Trigona jati **Bolivia-Brazil:** The larvae, pupae, and the honey stored in the nests are eaten raw (Posey 1983a and b). Native people of the Paressi tribe raise the bees in gourds to get honey (Gilmore 1963). The Maya people domesticated these bees (Bodenheimer 1951). **Mexico:** The pupae are considered good for eating in Oaxaca and Tabasco States (Ramos-Elorduy de Conconi et al. 1984).

Trigona leucogaster In **Brazil,** the Nhambiquara people in Mato Grosso State eat the larvae, pupae, honey and the pollen stored in the nests (Setz 1991).

Trigona lendliana The larvae and honey are eaten raw in **DRC** (Parent et al. 1978, Ichikawa 1982).

Trigona lurida In **Brazil,** the Surui people in Pará State eat the larvae, pupae and the honey stored in the nests (Coimbra 1984).

Trigona madecassa The larvae and honey are eaten raw in **Madagascar** (Simmonds 1885).

Trigona mexicana This species regarded as food item by the people of Chol, Maya, Tzetzal tribes in Yucatán State, **Mexico** (Ramos-Elorduy 2009).

Trigona mombuca The honey stored in the nests is used as sweetening in **Brazil** (Crane 1992).

Trigona muscaria The honey stored in the nests is used to sweeten foodstuff in **Brazil** (Bodenheimer 1951).

Trigona nigra nigra In **Mexico,** the eggs, larvae, pupae and the honey collected by the bees are consumed in Campeche, Chiapas, Oaxaca, Tabasco and Yucatan States (Ramos-Elorduy de Conconi et al. 1984). The Maya people domesticated these bees (Bodenheimer 1951).

Trigona occidentalis **DRC:** The larvae and honey are eaten raw in the Shaba area (Parent et al. 1978). **Senegal:** The larvae and honey are eaten raw (Gessain and Kinzler 1975).

Trigona pallens In **Brazil,** the Kayapó people in Pará State eat the larvae and pupae raw (Gilmore 1963).

Trigona pectoralis **Mexico:** Native people of the Maya tribe domesticated these bees (Bodenheimer 1951).

Trigona postica **Mexico:** This species is raised (Ramos-Elorduy 1997).

Trigona recursa The honey stored in the nests is used for sweetening of dish components in **Brazil** (Ott 1998).

Trigona richardsi The larvae and honey are consumed in the Shaba area of **DRC** (Parent et al. 1978).

Trigona ruspolii The larvae and honey are relished as raw food in **Senegal** (Gessain and Kinzler 1975).

Trigona schmidti The larvae and honey are consumed raw in **Africa** (Bodenheimer 1951).

Trigona senegalensis The larvae and honey are eaten raw in **Senegal** (Gessain and Kinzler 1975).

Trigona silvestrii The Nhambiquara people in Mato Grosso State of **Brazil** eat the larvae, pupae, honey and the pollen stored in the nests (Setz 1991).

Trigona spinipes In **Brazil**, the Guarani M'byá people in San Paulo State eat the larvae, pupae, honey and the pollen stored in the nests raw (Rodrigues 2005). The Pankararé people living in northeast Brazil make a syrup for treating coughs from scutella and sugar (Costa-Neto 1999).

Trigona togoensis **var.** *junodi* The larvae and honey are used as raw food in **RSA** (Junod 1927 and 1962).

Trigona trinidadensis **Colombia:** The Yukupa people roast the larvae, and eat the honey in the nests raw (Posey 1983a and b). **Venezuela:** This species is used as foodstuff in this country (Ruddle 1973).

Trigona vidua → Melipona vidua

Trigona **(***Hypotrigona***) spp. Africa:** The larvae and honey are eaten raw (Bodenheimer 1951). **Australia:** In the central areas of the country, the Walbiri and Pintupi tribes of aborigines eat several species belonging to this genus as highly prized food. About six *Trigona* species are known here (Bodenheimer 1951, Meyer-Rochow and Changkija 1997). **Mexico:** In Chiapas and Oaxaca States, the eggs, larvae, pupae and the honey collected by the bees of several species belonging to this genus are considered as food item (Ramos-Elorduy et al. 1997, Ramos-Elorduy and Pino 2002). **DRC:** The larvae and honey are eaten raw (Ichikawa 1982).

Trigonisca **(=***Dolichotrigona***)** *schultessi* The eggs, larvae, pupae and the honey collected by the bees are regarded as foodstuff in Chiapas State of **Mexico** (Ramos-Elorduy and Pino 2002).

Trigonisca jaty In Campeche, Oaxaca, Tabasco and Yucatan States of **Mexico**, the eggs, larvae, pupae and the honey collected by the bees are relished (Ramos-Elorduy et al. 1997).

Trigonisca pipioli **Mexico:** In Chiapas State, the eggs, larvae, pupae and the honey collected by the bees are treated as foodstuff (Ramos-Elorduy and Pino 2002).

Trigonisca **(***Drichotrigona***)** *schultessi* The eggs, larvae, pupae and the honey collected by the bees are used as food item by the Mames people in Hidalgo State, **Mexico** (Ramos-Elorduy and Pino 2002).

Argidae (Argid sawflies) (1)

Dielocerus formosus The pupae are used as food item by the Nhambiquara people in Rondonia State, **Brazil** (Carrer 1992).

Bombidae (16)

Bombinae **spp.** The honey collected by several species belonging to *Bombinae* is eaten in **Zambia** (Mbata 1995).

Bombus appositus (white-shouldered bumble bee) The larvae and pupae are recognized as food item by Californian Indians in the **USA** (Sutton 1988).

Bombus atratus The larvae are considered good for eating by the Canari and Otavalo people in **Ecuador** (Onore 1997, Paoletti and Dufour 2005).

Bombus diligens The eggs, larvae, pupae, adults and the honey collected by the bees are used as foodstuff in Chiapas, Puebla, Tabasco States and DF of **Mexico** (DeFoliart 2002, Ramos-Elorduy 2009, Ramos-Elorduy and Pino 2002).

Bombus diversus tersatus The Japanese name is *ezotora-maruhanabachi*. In the Nagano, Aichi, and Yamanashi Prefectures, **Japan**, the larvae or pupae are roasted with a bit of salt, cooked with sugar and soy sauce, or prepared for bee-rice (Takagi 1929e and h). In the Nagano Prefecture, the honey in the nests are consumed. The honey stored by this species is sweeter than that of *Apis mellifera*, however it is often fermented, and intoxicates the drinker (Mukaiyama 1987). The roasted larvae were used for treating convulsive fits in the Niigata Prefecture (Miyake 1919).

Bombus ecuadorius The larvae are used as foodstuff in **Ecuador** (Onore 1997).

Bombus ephippiatus → *Bombus formosus*

Bombus formosus The Mexican name is *jicote gordo*. The larvae, pupae, adults and the honey are regarded as good foodstuff in Jalisco and Quintana Roo States, **Mexico** (Ramos-Elorduy 2009, Ramos-Elorduy et al. 1998, DeFoliart 2002).

Bombus funebris The larvae are relished by the Canari and Otavalo people in **Ecuador** (Onore 1997, Paoletti and Dufour 2005).

Bombus ignitus The Japanese name is *kuro-maruhabachi*. In the Nagano, Aichi, and Yamanashi Prefectures, **Japan,** the larvae or pupae are roasted with a bit of salt, cooked with sugar and soy sauce, or prepared for bee-rice (Takagi 1929e and h). In the Nagano Prefecture, the honey in the nests is consumed. The honey stored by this species is sweeter than that of *Apis mellifera*, however it is often fermented, and intoxicates the drinker (Mukaiyama 1987). The roasted larvae were used for treating convulsive fits in the Niigata Prefecture (Miyake 1919).

Bombus kainowskyi The Japanese name is *ki-maruhanabachi*. In the Nagano, Aichi, and Yamanashi Prefectures, **Japan**, the larvae or pupae are roasted with a bit of salt, cooked with sugar and soy sauce, or prepared for bee-rice (Takagi 1929e and h). In the Nagano Prefecture, the honey in the nests is consumed. The honey stored by this species is sweeter than that of *Apis mellifera*, however it is often fermented, and intoxicates the drinker (Mukaiyam 1987). The roasted larvae were used for treating convulsive fits in the Niigata Prefecture (Miyake 1919).

Bombus medius The eggs, larvae, pupae, adults and the honey collected by the bees are considered good for eating in Chiapas, Puebla, Tabasco States and DF of **Mexico** (DeFoliart 2002, Ramos-Elorduy 2009, Ramos-Elorduy and Pino 2002).

Bombus nevadensis (Nevada bumble bee) The larvae and pupae are treated as food item by Californian Indians in the **USA** (Sutton 1988).

Bombus robustus The larvae are eaten in **Ecuador** (Onore 1997).

Bombus rufocinctus The eggs, larvae, pupae and the honey collected by the bees are considered as foodstuff in Chiapas State of **Mexico** (Ramos-Elorduy and Pino 2002).

Bombus speciosus **China:** The larvae and pupae are relished (Hu and Zha 2009). **Japan:** The Japanese name is *maruhanabachi*. The larvae were used for treating convulsive fit (Shiraki 1958).

Bombus terricola occidentalis (yellowbanded bumble bee) In the **USA,** the larvae and pupae are appreciated as food by Californian Indians (Sutton 1988).

Bombus vosnesenskii (Vosnesensky bumble bee) The larvae and pupae are considered good for eating by Californian Indians in the **USA** (Sutton 1988).

Bombus **spp. China:** Some species belonging to this genus were stir fried in olden times (Li 1596). **Mexico:** A species whose Mexican name is *jicote* is eaten. In Chiapas State, the eggs, larvae, pupae and the honey collected by some species belonging to this genus are used as food item (Ramos-Elorduy and Pino 2002). Native people of the Lacandone, Otomi, Tepehuane Zoque, and Tzottzile tribes eat some species belonging to this genus (Ramos-Elorduy and Pino 1989). This species can be raised (Ramos-Elorduy 1997).

Braconidae (Braconid wasps) **(1)**

Euurobracon penetrator **(=*E. yokohamae*)** The Japanese name is *umanoobachi*. The roasted larvae were used for treating convulsive fit in the Nara Prefecture in **Japan** (Miyake 1919).

Crabronidae (0)

Philanthus **sp.** In **Mexico,** native people of the Hñähñu tribe living in Vallée du Mezquital, Hidalgo State eat a species belonging to this genus (Aldasoro Maya 2003).

Trypoxylon **sp.** In **Brazil,** a species belonging to this genus is eaten in Bahia State (Costa-Neto and Pacheco 2005).

Cynipidae (Gall wasps) **(4)**

Aulacides levantina **Greece:** People in Crete ate the galls made by this species on *Salvia* spp. (Bodenheimer 1951). **Turkey:** The galls of sage made by this species were esteemed in the **Levant** for their aromatic and acid flavor, especially when prepared with honey and sugar, and formed a considerable article of commerce from Scio to Istanbul, where they were regularly sold in the markets (Olivier 1813).

Aulax glechomae The galls made on ground-ivy, *Glechoma hederacea*, by this wasp are consumed with the larvae inside in **France** (Lorraine district) and **Sweden** (Fagan 1918).

Aulax latreille → *Aulax glechomae*

Aulax **sp.** The galls made by this wasp are considered a delicacy with the larvae inside in **Greece** (Olivier 1801–1807).

Cynips rosae **Levant:** The galls formed on rose trees are used as a remedy for treating diarrhea, dysentery, etc. (Ealand 1915).

Hedickiana levantina → *Aulacides glechomae*

Liposthenus glechomae → *Aulacides glechomae*

Trichagalma serratae (quercus gall wasp) The Japanese name is *kunugiedaiga-tamabachi*. In **Japan,** the gall made by this wasp contains a high concentration of tannin, and is used as a remedy for diarrhea (Hayashi 1903).

Diprionidae (Pine sawflies) **(2)**

Neodiprion guilletei The prepupae are considered good for eating by people living around Sierra Tarasca of **Mexico** (Ramos-Elorduy and Pino 1990).

Zadiprion vallicale The larvae are eaten by people living around Sierra Tarasca State of **Mexico** (Ramos-Elorduy and Pino 1990).

Eumenidae (Potter wasps) **(8)**

Eumenes canariculata **Venezuela:** This species is used as food by the Yukpa people (Ruddle 1973).

Eumenes petiolata In **Thailand,** this species is called *mang taan*. People in Hua Hin district, eat the larvae fried (Bristowe 1932).

Eumenes **sp. India:** The larvae of a species belonging to this genus are eaten raw in Arunachal Pradesh State, whereas the pupae are boiled or used to make a paste (Chakravorty et al. 2011). **Mexico:** The larvae and pupae of a species belonging to this genus are considered edible in Chiapas State (Ramos-Elorduy and Pino 2002).

Euodynerus **sp.** The larvae and pupae of a species belonging to this genus are consumed in Chiapas State of **Mexico** (Ramos-Elorduy and Pino 2002).

Montezumia dimidiata **Ecuador:** The adults are relished in this country (Onore 1997).

Nectarinia lecheguana → *Brachygastra lecheguana*

Synagris cornuta flavofasciata **Africa:** In certain countries or among some tribes, the mud, from which the nest is formed is relished by pregnant women (Hunter 1984).

Synagris **sp.** In **DRC,** this species is eaten raw (Adriaens 1951).

Formicidae (Ants) **(78)**

Acromyrmex octospinosus The reproductive adults are said good for eating in Chiapas State of **Mexico** (Ramos-Elorduy and Pino 2002).

Acromyrmex rugosus The reproductive adults are considered good as foodstuff in Chiapas State of **Mexico** (Ramos-Elorduy and Pino 2002).

Anomma nigricans This species is consumed in **Cameroon** (van Huis 2005).

Atta bisphaerica The winged adults are relished in **Brazil** (Laredo 2004).

Atta capiguara The winged adults are appreciated as delicacy in **Brazil** (Laredo 2004).

Atta cephalotes **Amazonia:** The Indios living in this area eat the abdomens of adult ants raw or roasted with salt or smoked with salt (Wallace 1852/53). **Mexico:** The Mexican name is *hormiga arriera.* The reproductive adults are consumed in Chiapas, Guerrero, and Oaxaca States (Ramos-Elorduy de Conconi et al. 1984). Native people of the Chontale, Hñähñu, Huasteco, Lacandone, Mestizo, Náhua, Totonaca, Tzetzale, Tzottzile, and Zapoteco tribes also eat this species (Ramos-Elorduy and Pino 1989, Aldasoro Maya 2003). This species is raised by different tribes in many states (Ramos-Elorduy 2009). **Brazil:** The winged adults are regarded as a luxury food by the Kayapó people of Pará State (Posey 1987), and the Enawenê-Nawê people of Mato Grosso State (Dufour 1987). **Colombia:** The adults are eaten raw, roasted, or smoked by the Tukanoan people (Wallace 1852/1953, Dufour 1987). **Ecuador:** The adults are eaten raw, roasted or smoked by the Ashuara, Awa, Cofane, Huaorani, Negr. Esmeraldo, Quichua, Secoya, Shuara, Siona and Siona Secoya people (Wallace 1852/1853, Dufour 1987, Onore 1997, Paoletti and Dufour 2005). **Guyana:** The roasted females of winged ants were considered a favorite delicacy by the native people (Schomburgk 1847/1848). **Honduras and Nicaragua:** Native people of the Miskito and Sumu tribes eat only the abdomens of the alate females roasted after removing their wings, legs and heads (Conzemius 1932, Clausen 1954). **Venezuela:** Some people eat only the heads and thorax of the ants, whereas others eat whole bodies of the ants (Spruce 1908). The Ye'kuana people living around Alto Orinoco eat the adults raw (Araujo and Beserra 2007, Paoletti and Dufour 2005). **Paraguay:** The aboriginal people ate the adults by putting butter on the abdomens of ants and then roasted or fried them, and ate them after adding syrup (Rengger 1835).

Atta laevigata **Brazil:** The winged adults are eaten by the Alto and Xingu people living near Xingu River (Carvalho 1951), and by the Nhambiquara people of Mato Grosso State (Setz 1991). **Colombia:** The Colombian name is *hormigas culonas*. The Tukanoan people eat the soldiers and alates (Dufour 1987). **Venezuela:** The soldiers and the queens are appreciated as a delicacy (Paoletti et al. 2000).

Atta mexicana The Mexican name is *chicatanas*. In **Mexico**, the reproductive adults are esteemed as a luxury in Chiapas, Guanajuato, Puebla, Oaxaca, Veracruz States (Ramos-Elorduy de Conconi et al. 1984). Native people of the Mazahua, Mestizo, Mixteco, Náhua, Tarasco and Zapoteco tribes also eat this species (Ramos-Elorduy and Pino 1989). This species is raised in Oaxaca and Hidalgo States (Ramos-Elorduy 2009).

Atta opacipes The adults are considered as food item in **Brazil** (Laredo 2004).

Atta sexdens **Brazil:** Native people of the Mave, Nhambiquara, Sateré-Maué, Tukana, Tupinambá, and Kayapó tribes eat the winged adults raw or roasted (Posey 1978 and 1987, Setz 1991). **Colombia:** The Tukanoan people eat the soldiers and alates raw or roasted (Schomburgk 1847–48, Posey 1978, Dufour 1987). **Colombia-Brazil-Venezuela:** Native people of the Carib tribe ate the female adults by roasting and making a paste with casaba powder (Posey 1978). **Ecuador:** The adults are regarded as good food by the Ashuara, Awa, Cofane, Huaorani, Negr. Esmerald, Quichuas, Secoya, Shuara and Siona people (Onore 1997). **Guyana:** This species is a speciality of the aboriginal people, and they eat the abdomens of the adult ants raw or roasted (Schomburgk 1847/48). **Venezuela:** The adults are consumed raw, or roasted (Posey 1978, Dufour 1987, Paoletti et al. 2000).

Atta **spp. Bolivia:** The Lecos people eat a species belonging to this genus (Stefano Vanzin 2003). **Brazil:** The adults of several species belonging to this genus are eaten raw, smothered or skewer-roasted by the Kayapo people (Posey 1987), Sateré-Maué people (Lenko and Papavero 1979), and Mundurucu people (Santos 1957) in Pará State, the Desâna people (Ribeiro and Kenhiri 1987, Balée 2000), Tariano people (Noice 1939), Wanana people (Lenko and Papavero 1979), Tuyuka people (Lenko and Papavero 1979) and the Tucano people (Lenko and Papavero 1979) in Amazonas State, the Surui people (Coimbra 1984) in Rondonia State, and the Ye'Kuana people (Paoletti and Dreon 2005) along Orinoco River. **Colombia:** The Yukpa people eat the adults of several species belonging to this genus raw, smothered or skewer-roasted (Costa-Neto and Ramos-Elorduy 2006, Ruddle 1973). **Mexico:** This species is raised as food (Ramos-Elorduy 1997). **Peru:** Native people of the Campa tribe eat alates of at least five or six species of this genus. The adults are eaten raw, smothered or spit-roasted (Denevan 1971).

Venezuela: The Miskito people and Sumu people roast the adult ants after removing their wings, legs and heads, and eat only their abdomens (Conzemius 1932, Clausen 1954). The Piaroa and Guajibo people eat the alates or soldiers of some species belonging to this genus (Paoletti and Dufour 2005). The Yanomamo, Ye'kuana and Yukpa people eat soldiers or alates of some species belonging to this genus (Ruddle 1973, Paoletti and Dufour 2005). **Honduras:** The Miskito people and Sumu people roast the adult ants after removing their wings, legs and heads, and eat only their abdomens (Conzemius 1932, Clausen 1954).

Bothroponera rufipes The whole body of the adults are used as a remedy for high blood pressure, malaria, etc. in **India** (Chakravorty et al. 2011).

Camponotus consobrinus In Victoria State of **Australia,** the pupae are used to make a paste (Smyth 1878). In Tasmania Island, the pupae are eaten (Noetling 1910).

Camponotus (Terraernimex) dumetorum The adults are relished in Chiapas State of **Mexico** (Ramos-Elorduy and Pino 2002).

Camponotus fulvopilosus In **Africa**, the adults are used to get the sour flavor from formic acid of this species (Green 1998).

Camponotus gigas In Saba State of **Malaysia**, this species is a food item (Chung 2008).

Camponotus inflatus (one of honey-pot ant species) In the central area of **Australia**, the repletes of this species are collected by the Walbiri people and Pintupi people to get honey (Conway 1991, Meyer-Rochow 2005).

Camponotus japonicus The eggs, larvae, pupae and the adults of this species are used as food and a medicine in **China** (Chen and Feng 1999).

Camponotus maculates The **USA:** The Shoshone people in California State cook the larvae, pupae and the adults (Sutton 1988).

Camponotus pennsylvanicus (black carpenter ant) **Canada:** Woodmen living in the South eastern areas eat the adult ants (Fladung 1924).

Camponotus **spp. Botswana** and **Cameroon:** People used the adults of a species belonging to this genus to get a sour flavor from the formic acid of the ants (De Colombel 2003). **Mexico:** A species belonging to this genus is considered as foodstuff in Hidalgo State (Ramos-Elorduy 2006). The **Philippines:** The eggs of a species, whose vernacular name is *karakara*, are a delicacy in the northern Philippines (Ilocos region). The eggs are cooked as spicy adobo or sautéed with garlic and onions and a small amount of pepper (Adalla and Cervancia 2010). The **USA:** In California State, the Digger Indians are fond of both the larvae and adults of a species (ca. 3/4 inch long, jet-black, woodboring ant). They bite off and reject the head and eat the flavor acid body with relish (Muir 1911).

Carebara castanea: The larvae are appreciated as food item in North eastern areas of **Thailand** (Hanboonsong et al. 2000).

Carebara junodi The alate females are consumed in **Africa** (Silow 1983).

Carebara lignata **China:** The eggs, lavae, pupae and the adults of this species are used as food or as a medicine. The eggs are highly esteemed (Chen and Feng 1999). **Southern area of Africa:** This species is considered good for eating (van Huis 2003). **Thailand:** People of Hua Hin district, which is located south west of Bangkok, ate the larvae after adding them to a curry (Bristowe 1932).

Carebara vidua **DRC:** The large winged queens, which emerge from termite mounds during some seasons are highly prized as delicacies. People eat only the abdomen raw, fried, or roasted (Bequaert 1922). **Malawi:** People eat the adults fried with a little salt, but no fat. The dishes are served hot or cold as a relish (Shaxson et al. 1985). **RSA:** The Pedi people, one of the Sotho tribes, eat the adults raw or roasted, and consider them as a very appetizing dish served with grain-meal porridge. They prefer the females having large egg sacks with a much higher fat content than the males (Quin 1959). **Zambia:** People collect the winged adults, and remove their wings. The ants are then fried in their own fat. They are eaten as snacks or used as a relish. The ants may also be boiled or eaten raw (Mbata 1995). **Zimbabwe:** The winged adults emerging from anthills, are collected and used as a relish. Sometime, people may eat the ants raw (Gelfand 1971, Chavanduka 1975).

Colobopsis grasseri In Tasmania Island of **Australia**, the pupae are esteemed as delicacy (Noetling 1910).

Crematogaster dohni (red ant) This species is regarded as good food by the Ao-Naga people in Nagaland State, **India** (Meyer-Rochow 2005).

Crematogaster **sp. Indonesia:** In Kalimantan State, the larvae of a species belonging to this genus are occasionally eaten (Chung 2010). **Thailand:** The larvae of a species belonging to this genus are used for a curry in the Hua Hin district (DeFoliart 2002).

Ecton burchelli The adults are eaten raw by the Yanomamo and Yekuana people in Alto Orinoco area, **Venezuela** (Araujo and Beserra 2007).

Eciton **sp.** The adults of a species belonging to this genus are considered edible in Chiapas, Hidalgo, Puebla States of **Mexico** (Ramos-Elorduy and Pino 2002).

Formica aquilonia This species is used as food or as a medicine in **China** (Chen and Feng 1999).

Formica beijingensis The larvae and adults are used as food item in **China** (Chen and Feng 1999).

Formica consobrina → *Camponotus consobrinus*

Formica edulis **Guyana:** The adults are regarded as foodstuff (Barrère 1741).

Formica exectoides **USA:** This species is recommended by the US army to use as a dressing for wild grass salade, when soldiers encounter an emergency (Akre 1992).

Formica fusca The larvae and adults are consumed in **China** (Hu and Zha 2009).

Formica japonica The larvae and adults are recognized as food item in **China** (Hu and Zha 2009).

Formica major **Guyana:** The adults are fit for eating (Barrère 1741). **Europe:** The alcoholic infusion of this ant was used for treating apoplexy, giddiness, etc. (Yasumatsu 1948).

Formica melliger → *Myrmecosistus melliger*

Formica minor **Europa:** This species was used for treating gout, paralysis, etc. (Ealand 1915).

Formica obscuripes **USA:** This species is recommended by US army to use as a dressing for wild grass salade, when soldiers encounter an emergency (Akre 1992).

Formica pensylvanica **Canada:** Children and woodmen eat the adults collected from old stumps. The ants are a favorite food for them (Provancher 1882).

Formica rufa (wood ant) **Austria:** In the Alpine areas, people crushed the ants on bread, and ate the bread absorbed with ant body fluid (Mayr 1855). **China:** The larvae and adults are consumed (Hu and Zha 2009). **North Korea:** The adults are pulverized, and the powder is drunk with hot water or as a liqueur in the North P'yongan, North Hamgyong and South Hamgyong Provinces (Okamoto and Muramatsu 1922). The **USA:** In the state of California American Indians eat the larvae, pupae, and the adults (Sutton 1988).

Formica sanguinea (slave maker ant) **China:** The larvae and adults are eaten (Hu and Zha 2009). **Japan:** The Japanese name is *aka-yamaari*. The adults were collected in the Nagano Prefecture, and were processed to chocolate covered ants. The products were exported to the USA (Yasumatsu 1965).

Formica uralensis The larvae and adults are considered as food item in **China** (Hu and Zha 2009).

Formica volans The adults are regarded as food item in Guyana (Barrère 1741).

Formica yessensis **China:** The larvae and adults are consumed as food (Hu and Zha 2009). **Japan:** The Japanese name is *yezoaka-yamaari*. The adults were used for as a remedy for lumbago (Shiraki 1958).

Lasius fliginosus (field ant) The Japanese name is *kurokusa-ari*. The adults were used for treating pleuritis, peritonitis, gonorrhea, etc. in **Japan** (Koizumi 1935).

Lasius flavus The larvae and adults are used as foodstuff in **China** (Hu and Zha 2009).

Lasius niger In the state of California the **USA**, the larvae, pupae and the adults are consumed as food (Sutton 1988).

Leptomyrmex varians (one of honey-pot ant species) In Queensland State of **Australia**, the repletes of this species are collected to get honey (Conway 1991).

Liometopum apiculatum The Mexican name is *escamol de obreras*. In **Mexico,** the eggs, larvae and the pupae are eaten in Chiapas, Hidalgo, Oaxaca, Puebla, Tlaxcala, Zacatecas States and Mexico DF. In Hidalgo State, native people of the Hñähñu tribe eat this species (Ramos-Elorduy de Conconi et al. 1984, Esparza-Farausto et al. 2008, Aldasoro Maya 2003). Native people of the Lacandone, Mestizo, Náhua, and, Otomi tribes eat this species (Ramos-Elorduy and Pino 1989). This species is raised by different tribes in many states (Ramos-Elorduy 1997 and 2009).

Liometopum occidentale **var.** *luctuosum* The Mexican name is *escamol de reproductores*. The eggs, larvae and the pupae are eaten in Michoacan, Puebla and Zacatecas States of **Mexico** (Ramos-Elorduy de Conconi et al. 1984). Native people of the Mazahua, Mestizo, and Tarasco tribes eat this species (Ramos-Elorduy and Pino 1989). This species is raised by different tribes in many states (Ramos-Elorduy 1997 and 2009).

Melophorus bagoti (one of honey-pot ant species) In **Australia,** the repletes of this species are collected to get honey (Conway 1991).

Melophorus cowleyi (one of honey-pot ant species) In **Australia,** the repletes of this species are collected to get honey (Conway 1991).

Melophorus inflatus → *Camponotus inflatus*

Melophorus midas (one of honey-pot ant species) In the central area of **Australia**, the repletes of this species are collected to get honey (Conway 1991).

Messor aegyptiacus This species itself is not eaten. Instead, the seeds of plants, which are collected by the ants are eaten in **Africa** (Gast 2000).

Monomorium nipponensis The Japanese name is *aka-ari*. The larvae were said to be effective for treating gonorrhea (Kuwan 1930).

Monomorium pharaonis (Pharaoh ant) The **USA:** This species was accidentally discovered and was said to have a good taste (anonymous 1893).

Myrmecia pyriformis (black bulldog ants) In Victoria State of **Australia,** the pupae are used to make a paste (Smyth 1878).

Myrmecia sanguinea (red bulldog ants) In Victoria State of **Australia,** the pupae are used to make a paste (Smyth 1878).

Myrmecocystus depilis (one of honey-pot ant species) In California, Arizona, and Colorado States of the **USA,** people eat the honey stored in the repletes (Conway 1986).

Myrmecosystus melliger (one of honey-pot ant species) **Mexico:** The Mexican name is *hormiga mielera.* In Tamaulipas, Hidalgo, Mexico D.F., Puebla, Oaxaca States, the honey stored in the repletes is eaten (Ramos-Elorduy de Conconi et al. 1984). **USA:** In California, Arizona, and Colorado States, people eat the honey stored in the repletes (Conway 1986).

Myrmecosistus melligera → *Myrmecosistus melliger*

Myrmecosistus mexicanus (one of honey-pot ant species) The Mexican name is *hormiga mielera.* In **Mexico,** native people of the Hñähñu tribe living in Hidalgo State eat this species (Aldasoro Maya 2003). The honey stored in the repletes is relished as delicacy in Chiapas, Hidalgo, Mexico D.F., Puebla, Tamauripas States (Ramos-Elorduy 2006).

Myrmecocystus mexicanus hortideorum (one of honey-pot ant species) In the **USA,** people in California, Arizona, and Colorado States, eat the honey stored in the repletes (Conway 1986).

Myrmecocystus mimicus (one of honey-pot ant species) The **USA:** In California, Arizona, and Colorado States of the **USA,** people eat the honey stored in the repletes (Conway 1986).

Myrmecocystus testaceus (one of honey-pot ant species) In California, Arizona, and Colorado States of the **USA,** people eat the honey stored in the repletes (Conway 1986).

Myrmecocystus **spp.** In **Mexico,** native people of the Chinanteco, Mazateco, Mestizo, Náhua, and Otomi tribes eat several species belonging to this genus (Ramos-Elorduy and Pino 1989).

Myrmica rubra In **India,** the larvae and pupae are cooked or baked by the Karbi and Rengma Naga people in Assam State (Ronghang and Ahmed 2010).

Oecodoma cephalotes → *Atta cephalotes*

Oecodoma laevigata → *Atta laevigata*

Oecodoma sexdens → *Atta sexdens*

Oecophylla longinoda → *Oecophylla smargdina longinoda*

Oecophylla smaragdina (red ant, red tree ant, green tree ant, weaver ant) **Africa:** The larvae and adults are consumed in large areas of Africa (Chinn 1945). **Australia:** Aborigines living northern Queensland State collect the larvae and pupae, and make a type of cake by crushing and squeezing them. They eat the cake without further processing, or dip the cake in water, which is drunk as beverage later (Campbell 1926). **China:** Natives of the Thai tribe make a type of paste from the eggs (Shu 1989). Some other minorities make vinegar from the boiled ants (Chen and Feng 1999). **India:** The Murrie people living around Bastar district, central India, eat these ants routinely (Long 1901). The adults and larvae are consumed raw by people of the Nishi and Galo tribes in Arunachal Pradesh State. The entire body of the adults is used as a remedy for stomach disorders and dysentry (Chakravorty et al. 2011). The Karbi and Rengma Naga people in Assam State eat the adults and eggs baked, as a chutney or curry (Ronghang and Ahmed 2010). **Indonesia:** In Kalimantan State, people eat the adults by mixing with rice (Beccari 1904). In Bali Island this species is also eaten (Césard 2006). **Laos:** The Laotian name is *khai mot dieng*. The larvae, prepupae, and the pupae are fried (Nonaka 1999a). **Malaysia:** The local name of this species is *sumut kassu*. In Sarawak State, the adults are used to add taste to rice (Beccari 1904). In Sabah State, people often mix the adults with chili and salt, and serve as condiments, while the larvae are eaten raw or cooked with porridge or rice (Chung 2010). The Dayaku people used to make a kind of condiment from this species (Curran 1939). **Myanmar:** The nests of this species are collected and the adults and larvae are killed by smoke. They are then made into a paste which turns sour and is called *khagyin*. It is used as a medicine. The larvae are consumed raw, roasted or cooked (Ghosh 1924). **PNG:** The Kiriwina people (Trobriand Islanders) and Onabasulu people (Southern Highlanders) eat the adults raw (Meyer-Rochow 1973 and 2005). The **Philippines:** In Luzon Island, the larvae are mostly stir-fried (Starr 1991). In Palawan Island, the larvae are eaten raw, roasted or cooked. Also, it is used to make a paste or sate (Revel 2003). **Sri Lanka:** This species is treated as food item (Nandasena et al. 2010). **Thailand:** The Thai name is *mod duan*. People living in the North eastern area, use this species to give some foods a sour taste. The eggs and larvae are often eaten raw. Sometimes they are cooked and put in a Thai salad (*yam*), fried with eggs, put in bamboo shoot stew (*gaeng noremai*), curried fish wrapped in banana leaves (*hore mohk*) or pickled with some herbs (Jonjuapsong 1996, Hanboonsong et al. 2000).

Oecophylla smaragdina virescens: In Queensland State, **Australia**, the pupae are eaten raw, or are used to make a type of soft drink (Campbell 1926).

Oecophylla smargdina longinoda **DRC:** The larvae and adults are crushed into a paste, and then steamed (Chinn 1945). **Tanzania:** The nests of this species is used as a remedy for coughs (van Huis 2003).

Oecophylla **spp. Cameroon:** A species belonging to this genus is considered edible (Bani 1995). **PRC:** The adults of a species belonging to this genus are used as foodstuff raw (Bani 1995).

Oecophylla virescens In **PNG**, people living along Fly River eat this species and another species belonging to the same genus (Ohtsuka et al. 1984).

Pachycondyla tarsata **CAR:** The Gbaya children eat the ants raw. People also collect the cocoons, and eat them boiled. Generally people eat the larvae, adults, and the nests raw (Roulon-Doko 2003).

Pachycondyla **sp.** People eat plant seeds stored in the nests of the ants belonging to this genus in **Africa** (Seignobos et al. 1996).

Pheiodole **sp. Mozambique**, **Zambia** and **Zimbabwe:** The adults often invade various food stuffs such as sugar, beer, etc. People, however, eat the ants believing that this species is good for gastrointestinal diseases (van Huis 1996).

Plagiolepis **sp.** (one of honey-pot ant species) **Australia:** The repletes of this species are collected to get honey (Conway 1991).

Pogonomyrmex barbatus (red harvester ant) The Mexican name is *hormiga colorada.* The larvae, pupae, the worker-ants are appreciated as food in Hidalgo State, **Mexico** (Ramos-Elorduy et al. 1998).

Pogonomyrmex californicus (California harvester ant) In the **USA,** the larvae, pupae, and the adults are treated as foodstuff by Californian Indians (Sutton 1988).

Pogonomyrmex desertorum In the **USA,** the larvae, pupae, and the adults are relished by Californian Indians (Sutton 1988).

Pogonomyrmex occidentalis (western harvester ant) The larvae, pupae, and the adults are used to make soup by Californian Indians in **USA** (Sutton 1988).

Pogonomyrmex owyheei In the **USA,** the larvae, pupae, and the adults are used to make soup by Californian Indians such as the Snake tribe (Sutton 1988).

Pogonomyrmex **sp. Mexico:** The larvae and pupae of a species called *hormiga colorada* are eaten (Ramos-Elorduy 2006). The **USA:** In olden times, American Indians seemed to eat a species belonging to this genus, as ants of this genus were detected in the storage of foods found in the ruins of Eden-Farson in Wyoming State (Frison 1971).

Polyrhachis dives In some restaurants in Guangzhou, **China**, a dish using this species is served. The adults are sprinkled on noodles (Umeya 1994).

Polyrhachis illaudata The larvae and adults are eaten in **China** (Hu and Zha 2009).

Polyrhachis lamellidens The eggs, larvae, pupae and the adults are used as food item in **China** (Chen and Feng 1999).

Polyrhachis mayri This species is eaten in **China** (Chen and Feng 1999).

Polyrhachis vicina This species is raised for medicinal uses in **China** (Zimian et al. 1997). It has medicinal effects on various diseases (Hayashi 2005).

Sternotornis **sp.** In **DRC**, a species belonging to this genus is considered as food item (Adriaens 1951).

Tetramorium caespitum The eggs, larvae, pupae and the adults are consumed in **China** (Chen and Feng 1999).

Halictidae (Sweat bees) **(0)**

Nomia **spp.** In **Thailand**, the larvae, honey and the wax are consumed (Bristowe 1932).

Trichona **spp.** In **Thailand**, the honey is used as sweetener (Bristowe 1932).

Megachilidae (Leafcutting bees) **(0)**

Anthophora **sp.** The **USA:** The honey stored by a species belonging to this genus is eaten raw by Native Americans of the Pima tribe in California (Hrdlicka 1908, Essig 1934 and 1965).

Chalicodema **sp. Kenya and Tanzania:** The Dorobo people eat the honey stored by the bees raw (Huntingford 1955).

Halictus **sp. Kenya and Tanzania:** The Dorobo people used the raw honey stored by bees as a sweetening (Huntingford 1955).

Megachile **spp. China:** The honey stored in bamboo stems was used as a remedy for canker (Li 1596). **Botswana:** The honey collected by a species belonging to this genus is eaten raw (Nonaka 1996). **Ecuador:** The larvae of a species belonging to this genus are regarded as foodstuff by the Otavalo, Pilahuine, Quichuas, Salazakas and Saraguro people (Onore 1997).

Melissodes **sp.** The **USA:** The honey stored by a species belonging to this genus is used as a sweetner by native Americans of the Pima tribe in California (Hrdlicka 1908, Essig 1934 and 1965).

Osmia **sp. Kenya and Tanzania:** The Dorobo people eat the honey stored by the bees raw (Huntingford 1955).

Mutillidae (Velvet ants) **(1)**

Dasymutilla occidentalis **Mexico:** The Hñähñu people in Hidalgo State use this species as a remedy for measles (Aldasoro Maya 2003).

Perilampidae (Perilampids) **(0)**

Trachilogastir **sp.** In **Australia,** the larvae in the galls are considered edible (Cleland 1966).

Polybiidae (34)

Agelaia aereata The eggs, larvae, pupae and the honey collected by the wasps are regarded as foodstuff in Chiapas State of **Mexico** (Ramos-Elorduy and Pino 2002).

Agelaia angulata The larvae and pupae are consumed raw or roasted by the Tukanoan people in **Colombia** (Dufour 1987).

Agelaia baezae The larvae are treated as a food item by the Canari people in **Ecuador** (Onore 1997).

Agelaia corneliana The larvae are fit for eating in **Ecuador** (Onore 2005).

Agelaia lobipleura The larvae are accepted as food in **Ecuador** (Onore 2005).

Agelaia ornata The larvae are recognized as a foodstuff in **Ecuador** (Onore 2005).

Agelaia pallidiventris In **Venezuela,** the larvae and pupae are eaten raw, or roasted by the Yanomamo people (Paoletti and Dufour 2005).

Agelaia panamensis The eggs, larvae, pupae and the honey collected by the wasps are esteemed as food in Chiapas State of **Mexico** (Ramos-Elorduy and Pino 2002).

Agelaia **sp.** The larvae of a species belonging to this genus are consumed raw by the Yanomamo people in Alto Orinoco area, **Venezuela** (Araujo and Beserra 2007).

Angiopolybia paraensis The larvae and pupae are treated as food by the Ashuara and Shuara people in **Ecuador** (Onore 1997).

Angiopolybia **sp.** The larvae of a species belonging to this genus are eaten raw by the Yanomamo people in Alto Orinoco area, **Venezuela** (Araujo and Beserra 2007).

Myrametra occidentalis nigratella → *Polybia occidentalis nigratella*

Polybia aequatorialis The larvae and pupae are fit for eating by the Shuara people in **Ecuador** (Onore 1997).

Polybia diguetana The eggs, larvae and the pupae are appreciated as food in **Mexico** (Ramos-Elorduy and Pino 1989).

Polybia dimidiata **Brazil:** The larvae and pupae are regarded as delicacy by the Nhambiquara people of Mato Grosso State (Setz 1991). **Ecuador:** The larvae and pupae are used as foodstuff by the Shuara people (Onore 1997).

Polybia emaciata The larvae and pupae are recognized as foodstuff by the Shuara people in **Ecuador** (Onore 1997).

Polybia flavifrons The larvae and pupae are preferred edible insects by the Shuara people in **Ecuador** (Onore 1997).

Polybia ignobilis The larvae are roasted by the Yukpa people in **Colombia-Venezuela** (Ruddle 1973).

Polybia (Trichothorax) micans The roasted larvae are favorite food the Yanomamo people in Alto Orinoco area, **Venezuela** (Araujo and Beserra 2007).

Polybia nigratella **Mexico:** This species is used for treating diseases of the urinary organs (Ramos-Elorduy de Conconi and Pino 1988).

Polybia occidentalis **Brazil:** The larvae, pupae and the honey collected by the wasps are consumed raw by the Nhambiquara people in Mato Grosso State (Bodenheimer 1951, Setz 1991). **Mexico:** This species is used for treating diseases of the urinary organs (Ramos-Elorduy de Conconi and Pino 1988). **Venezuela:** The larvae are roasted by the Yanomamo people in Alto Orinoco area (Araujo and Beserra 2007).

Polybia occidentalis bohemani The Mexican name is *avispa raya amarilla*. Native people of the Chinanteco, Mazahua, Mixe, Mixteco, Náhua, Popolaca, and Tarasco tribes in **Mexico** eat the eggs, larvae and the pupae (Ramos-Elorduy de Conconi et al. 1984, Ramos-Elorduy and Pino 1989). This species is raised by different tribes in many states (Ramos-Elorduy 1997 and 2009).

Polybia (Myrametra) occidentalis nigratella The Mexican name is *avispa negra*. Native people of the Chontale, Lacandone, Maya, Mestizo, Náhua, Otomi, and Zapoteco tribes eat this species in **Mexico** (Ramos-Elorduy and Pino 1989). The eggs, larvae, pupae and the honey collected by the wasps are eaten in Chiapas, Hidalgo, Michoacan, Mexico D.F., Oaxaca, Puebla and Veracruz States (Ramos-Elorduy de Conconi et al. 1984). This species is raised by many tribes in various states (Ramos-Elorduy 1997 and 2009).

Polybia (Myrametra) parvulina Native people of the Chontale, Huasteco, Lacandone, Maya, Mestizo, Náhua, Otomi, Tzetzale, Tzottzile, and Zapoteco tribes in **Mexico** eat this species (Ramos-Elorduy and Pino 1989). The eggs, larvae, pupae and the honey collected by the wasps are used as foodstuff in Chiapas and Oaxaca States (Ramos-Elorduy et al. 1997). This species is raised by many tribes in various states (Ramos-Elorduy 1997 and 2009).

Polybia pygmaea The eggs, larvae, pupae and the honey collected by the wasps are regarded as foodstuff in Chiapas State of **Mexico** (Ramos-Elorduy and Pino 2002).

Polybia rejecta The Tatujo people eat the larvae in **Colombia** (Dufour 1987).

Polybia scrobalis The larvae and pupae are appreciated as food item in Puebla State of **Mexico** (Ramos-Elorduy de Conconi 1991).

Polybia sericea The larvae and pupae are used as food item by the Pankararé people of Bahia State in **Brazil** (Costa-Neto 2003b).

Polybia striata The eggs, larvae, pupae and the honey collected by the wasps are consumed as food in Chiapas State of **Mexico** (Ramos-Elorduy and Pino 2002).

Polybia testaceicolor The larvae and pupae are recognized as food item by the Shuara people in **Ecuador** (Onore 1997).

Polybia **spp.** In **Amazonia**, the larvae, pupae and the honey collected by the wasps of several species belonging to this genus are eaten raw by the Pankararé people of Bahia State, **Brazil** (Costa Neto 1996). **Colombia:** The Yukpa people eat some species belonging to this genus (Ruddle 1973). **Ecuador:** The adults of some species belonging to this genus are regarded as foodstuff (Onore 1997 and 2005). **Mexico:** The eggs, larvae, pupae and honey collected by the wasps, whose Mexican name is *avispa colita amarilla*, are consumed in Chiapas, Michoacan, Oaxaca, Puebla States (Ramos-Elorduy et al. 1997). **Paraguay:** The Chaco people like to eat the larvae of some wasps belonging to this genus (Clastres 1972). **Venezuela:** People eat the larvae of some species belonging to this genus (Ruddle 1973).

Protopolybia exigua exigua The nest of the wasps are used as inhalation for medical treatment by the Pankararé people of Bahia State in **Brazil** (Costa-Neto 1998).

Pseudopolybia vespiceps The larvae and pupae are consumed by the Nhambiquara people of Mato Grosso State in **Brazil** (Setz 1991).

Stelopolybia aereata → *Agelaia aereata*

Stelopolybia angulata → *Agelaia angulata*

Stelopolybia panamensis → *Agelaia panamensis*

Stelopolibia **sp.** → *Agelaia* **sp.**

Stenopolybia baezae → *Agelaia baezae*

Stenopolybia corneliana → *Agelaia corneliana*

Stenopolybia lobipleura → *Agelaia lobipleura*

Stenopolybia ornata → *Agelaia ornata*

Synoeca corneliana The larvae are used as food source in **Ecuador** (Onore 1997).

Synoeca lobipleura The larvae are considered edible in **Ecuador** (Onore 1997).

Synoeca ornata The larvae are fit for eating in **Ecuador** (Onore 1997).

Synoeca surinama **Mexico:** The larvae and pupae are treated as food item in Chiapas State (Ramos-Elorduy and Pino 2002). **Brazil:** The nests of this species are used as a medicine for treating asthma, tonsillitis, giddiness, etc. (Marques and Costa-Neto 1997).

Synoeca virginia The adults are eaten in **Ecuador** (Onore 1997).

Polistidae (42)

Apoica pallens **Brazil:** The larvae and pupae are appreciated as food by the Nhambiquara people of Mato Grosso State (Setz 1991), Tapirapé people of Mato Grosso State (Lenko and Papavero 1996), Desâna people of Amazonas State (Ribeiro and Kenhiri 1987), and the Yanomamo people of Amazonas State (Lizot 1977). The smoke from burning nests are said to be good for heart diseases (Costa-Neto and de Melo 1998). **Ecuador:** The larvae and pupae are eaten by the Ashuara and Shuara people (Onore 1997).

Apoica pallida The larvae and pupae are regarded as food by the Ashuara and Shuara people in **Ecuador** (Onore 1997).

Apoica strigata The larvae and pupae are treated as food item by the Ashuara and Shuara people in **Ecuador** (Onore 1997).

Apoica thoracica **Colombia:** The Tatujo people eat the larvae and pupae (Dufour 1987, Onore 1997). **Ecuador:** The larvae and pupae are eaten by the Ashuara and Shuara people (Onore 1997). **Venezuela:** The larvae are roasted (Araujo and Beserra 2007).

Apoica sp. The eggs, larvae and the pupae are considered as foodstuff in Chiapas State of **Mexico** (Ramos-Elorduy and Pino 2002).

Epipona quadrituberculata In **Brazil**, the larvae and pupae are consumed by people of the Nhambiquara and Tapirapé tribes in Mato Grosso State, and people of the Desâna and Yanomamo tribes in Amazonas State (Setz 1991).

Epipona sp. A species belonging to this genus is regarded as food by people of the Chol, Lacandon, Mame, Tojolabal, Tzetzal, Tzotzil and Zoque tribes in Chiapas State, **Mexico** (Ramos-Elorduy 2009).

Euchromia lethe The eggs are one of food item in **Cameroon** (De Colombel 2003).

Mischocyttarus **(Kappa)** *cubensis* The eggs, larvae and the pupae are accepted as a food item in Chiapas State of **Mexico** (Ramos-Elorduy and Pino 2002).

Mischocyttarus (Mischocytarus) basimaculata The eggs, larvae, pupae and the honey collected by the wasps are considered as foodstuff in Chiapas, Oaxaca and Puebla States of **Mexico** (Ramos-Elorduy et al. 1997). This species is raised by many tribes in various states (Ramos-Elorduy 2009).

Mischocyttarus (Mischocyttarus) pallidipectus The eggs, larvae and the pupae are considered as foodstuff in Chiapas State of **Mexico** (Ramos-Elorduy and Pino 2002).

Mischocyttarus rotundicollis The adults are used as a food item in **Ecuador** (Onore 2005).

Mischocyttarus tomentosus The adults are consumed as a food in **Ecuador** (Onore 2005).

Mischocyttarus **spp. Mexico:** The eggs, larvae, pupae and the honey collected by the wasps belonging to this genus are considered as foodstuff in Chiapas and Oaxaca States (Ramos-Elorduy et al. 1997). **Colombia-Venezuela:** The Yukupa people roast the larvae (Ruddle 1973).

Montezmia dimidiata The larvae and pupae are used as food by the Ashuara and Shuara people in **Ecuador** (Onore 1997).

Polisotiusmajor → *Polistes major*

Polistes apicalis The eggs, larvae, pupae and the adults are accepted as foodstuff in Chiapas State of **Mexico** (Ramos-Elorduy and Pino 2002).

Polistes bicolor The larvae and pupae are considered delicacy by the Ashuara and Shuara people in **Ecuador** (Onore 1997).

Polistes canadensis The Mexican name is *huaricho chico.* The eggs, larvae, pupae and the adults are used as foodstuff in Chiapas and Oaxaca States, **Mexico** (Ramos-Elorduy et al. 1997). This species is raised by many tribes in various states (Ramos-Elorduy 1997 and 2009).

Polistes canadensis erythrocephalus **Colombia:** The Yukupa people roast the larvae (Ruddle 1973). **Venezuela:** The larvae are roasted (Ruddle 1973).

Polistes carnifex The eggs, larvae, pupae and the adults are eaten in Chiapas State of **Mexico** (Ramos-Elorduy and Pino 2002).

Polistes chinensis (long-legged wasp) **China:** This species is said to be effective against lung cancer (Zimian et al. 1997).

Polistes chinensis antennalis (Japanese paper wasp) The Japanese name is *futamon-ashinagabachi.* The larvae and the pupae are cook-roasted, or cooked with sugar and soy sauce in the Nagano Prefecture, **Japan** (Katagiri and

Awatsuhara 1996). The larvae were used for treating convulsive fits or as a tonic (Umemura 1943).

Polistes deceptor The larvae and pupae are eaten by the Asuara and Shuara people in **Ecuador** (Onore 1997).

Polistes dorsalis The eggs, larvae, pupae and the adults are regarded as food item in Chiapas State of **Mexico** (Ramos-Elorduy and Pino 2002).

Polistes erythrocephalus → Polistes canadensis erythrocephalus

Polistes fadwigae The Japanese name is *seguro-ashinagabachi*. In **Japan,** the larvae were used for treating convulsive fits or as a tonic (Umemura 1943).

Polistes gigas The larvae and pupae are accepted as food item in **China** (Hu and Zha 2009).

Polistes hebraeus **China:** The larvae and pupae are used as foodstuff (Hu and Zha 2009). **Japan:** The Japanese name is *ashinagabachi*. In many prefectures (Tochigi, Saitama, Shizuoka, Ishikawa, Nara, Okayama, Tottori, Shimane, Ehime, Kōchi, Nagasaki, Miyazaki), the larvae and pupae are consumed. They are fried, roasted adding a touch of soy sauce, dipped in soy sauce or miso paste, or cooked with soy sauce and sugar. People also make wasp rice from boiled rice and cooked wasps (Miyake 1919, Mizuno and Ibaraki 1983, Tsurufuji 1985). The roasted larvae and pupae, or the infusion of the larvae and pupae were used for treating convulsive fits (Miyake 1919). **Madagascar:** The larvae are relished (Womeni et al. 2009).

Polistes ignobilis **Colombia:** The Yukpa people, who live in the northern part of Colombia, use the larvae of this species to make special food for new born babies. The food is called *kosés*, and made of corn flour with the larvae in it (Ruddle 1973).

Polistes instabilis The Mexican name is *avispa guitarrilla*. The eggs, larvae, pupae and the adults are recognized as food in Hidalgo, San Luis Potosi, Chiapas, Oaxaca, States and **Mexico** DF (Ramos-Elorduy et al. 1997). The roasted or boiled adults are eaten as a remedy for nervous prostration (Ramos-Elorduy de Conconi and Pino 1988). This species has be raised by many tribes in various states (Ramos-Elorduy 1997 and 2009).

Polistes jadwigae jadwigae (Japanese paper wasp) **Japan:** The Japanese name is *seguro-ashinagabachi*. The larvae and pupae are eaten raw or roasted adding a bit of soy sauce on them in the Nagano Prefecture (Matsuura 2002).

Polistes **(Aphanilopterus)** *kaibabensis* The eggs, larvae, pupae and the adults are consumed as foodstuff in Chiapas State of **Mexico** (Ramos-Elorduy and Pino 2002).

Polistes major The Mexican name is *huaricho*. The eggs, larvae, pupae and the adults are appreciated as food source by the people of Lacandones and Tajolabals in Chiapas State, **Mexico** (Ramos-Elorduy de Conconi et al. 1984,

Ramos-Elorduy and Pino 2002). The Hñähñu people in Hidalgo State use this species as a remedy for rheumatism (Aldasoro Maya 2003). This species can be raised (Ramos-Elorduy 1997).

Polistes mandarinus **China:** The larvae and pupae are eaten (Hu and Zha 2009). **Japan:** The Japanese name is *kiboshi-ashinagabachi*. The larvae were used for treating convulsive fits or as a tonic (Umemura 1943).

Polistes mexicanus **Mexico:** The Hñähñu people in Hidalgo State use this species as a remedy for rheumatism (Aldasoro Maya 2003).

Polistes modestus → *Polistes pacificus modestus*

Polistes occipitalis The larvae and pupae are considered as food item by the Asuara and Shuara people in **Ecuador** (Onore 1997).

Polistes pacificus → *Polistes pacificus modestus*

Polistes pacificus modestus **Colombia:** The larvae are recognized as food item (Ruddle 1973). **Venezuela:** The Yukpa people eat the larvae roasted (Ruddle 1973).

Polistes parvulina In **Mexico,** this species is consumed as food in Hidalgo, San Luis Potosi, States and **Mexico** D.F. (Ramos-Elorduy 2009).

Polystes rothney The Japanese name is *ki-ashinagabachi*. In **Japan,** the larvae were used for treating diarrhea, convulsive fits, etc. (Umemura 1943, Shiraki 1958).

Polistes sagittarius The larvae and pupae are fried in **China**, especially by the minority people in the Yunnan Province (Chen and Feng 1999).

Polistes salcatus The larvae and pupae are treated as foodstuff in **China** (Hu and Zha 2009).

Polistes snelleni The Japanese name is *ko-ashinagabachi*. In **Japan,** the larvae and pupae are relished as foodstuff in many prefectures by frying or roasting with a bit of soy sauce on them (Takagi 1929h). The larvae were medicinally used for treating convulsive fits or as a tonic (Umemura 1943).

Polistes stigmata **India:** The larvae, pupae and the eggs are cooked, baked or as a chutney or curry by the Karbi and Rengma Naga People in Assam State (Ronghang and Ahmed 2010). This species is also eaten in **Laos, Myanmar, Thailand** and **Vietnam** (Yhoung-Aree and Viwatpanich 2005).

Polistes sulcatus The larvae and pupae are fried in **China**, especially by the minority people in the Yunnan Province (Chen and Feng 1999).

Polistes testaceicolor The larvae and pupae are appreciated as food source by the Asuara people in **Ecuador** (Onore 1997).

Polistes versicolor **Colombia:** The larvae are good for eating (Ruddle 1973). **Venezuela:** The Yukpa people roast the larvae (Ruddle 1973).

Polistes yokohamae → *Polistes snelleni*

Polistes **spp. Guatemala:** Native people of the Chuh tribe eat the larvae and pupae of several species belonging to this genus (Lenko and Papavero 1979). **India:** The adults of some species belonging to this genus are fried in Arunachal Pradesh State. The muscle is chewed, after the wings are removed. The larvae and pupae are also consumed (Chakravorty et al. 2011). **Mexico:** The eggs, larvae, pupae, adults and the honey collected by wasps belonging to this genus are consumed in Chiapas, Michoacan and Puebla States (Ramos-Elorduy de Conconi et al. 1984).

Sphecidae (Digger wasps) **(1)**

Ammophila infesta **China:** This species was used to induce perspiration (Li 1596). **Japan:** The Japanese name is *jigabachi*. The larvae and adults were used for treating acute peritonitis (Shiraki 1958). **North Korea:** In North Hwanghae Province, people ate the larvae by roasting with salt (Okamura and Muramatsu 1922).

Ammophila **sp.** The eggs, larvae and the pupae are accepted as article of food in Oaxaca State, **Mexico** (Ramos-Elorduy de Conconi et al. 1984).

Pelopoeus **sp.** → *Sceliphron* **sp.**

Sceliphron **sp.** A species belonging to this genus is considered fit for eating in **DRC** (Adriaens 1951).

Vespidae (Honets) **(32)**

Brachygastra azteca The Mexican name is *avispa cola amarilla*. The eggs, larvae, pupae and the honey collected by the wasps are regarded as food item in Chiapas and Oaxaca states, **Mexico** (Ramos-Elorduy et al. 1997).

Brachygastra lecheguana **Brazil:** The larvae, pupae and the honey collected by the wasps are recognized as food by the Guarani M'byá people in San Paulo state (Costa Neto and Ramos-Elorduy 2006). In San Paulo of Brazil, the larvae, pupae and the honey collected by wasps are roasted by the Guarani M'byá people (Rodrigues 2005). **Ecuador:** The larvae and the honey are accepted as food (Onore 2005). **Mexico:** The eggs, larvae, pupae and the honey collected by the wasps are consumed in Chiapas and Michoacan states (Ramos-Elorduy de Conconi et al. 1984).

Brachygastra mellifica **Brazil:** The Native people of the Pankararé tribe eat the honey stored in the nests, which is often fermented and has high alcoholic content (Gunther 1931). **Mexico:** The Mexican name is *avispa, panalde castilla*. Native people of the Huasteco, Mestizo, Mixe, Mixteco, Náhua, Popolaca, Tarasco, Tlapaneca, and Zapoteco tribes eat the eggs, larvae and the pupae (Ramos-Elorduy et al. 1997, Ramos-Elorduy and Pino 1989).

Brachygastra sp. In **Mexico**, the eggs, larvae, pupae and the honey collected by a species of wasps belonging to this genus are consumed in Chiapas state (Ramos-Elorduy and Pino 2002).

Brachymenes wagnerianus The larvae and pupae are considered as foodstuff by the Ashuara and Shuara people in **Ecuador** (Onore 1997).

Dolichovespula flora In **China**, the people in the Yunnan Province, eat the larvae, pupae and the adults by immersing them in an alcoholic beverage (Matsuura 2002).

Dolichovespula **spp.** The larvae and pupae of several species belonging to this genus are considered as food item by Californian Indians in the **USA** (Sutton 1988).

Icaria artifex The larvae are accepted as foodstuff by the Ao-Naga people in Nagaland State, **India** (Meyer-Rochow 2005).

Parachartegus apicalis The Mexican name is *avispa ala blanca*. The eggs, larvae and the pupae are used as food item in Chiapas and Oaxaca States of **Mexico** (Ramos-Elorduy et al. 1997).

Provespa anomala In Kalimantan State of **Indonesia**, the larvae are relished raw or boiled with rice (Chung 2010).

Provespa barthelemyi The larvae and pupae are highly esteemed by the minority people in the Yunnan Province, **China**. They can be purchased in markets (Chen and Feng 1999).

Provespa **sp. Indonesia:** In the northern areas of Sumatra Island, the Batak people eat the pupae and the adults of a species belonging to this genus by frizzling them with palm oil, shallots, salt, pepper and other spices (Matsuura 1998b).

Ropalidia **spp. Indonesia:** In Kalimantan State, the larvae are consumed raw or boiled with rice (Chung 2010). **Malaysia:** In Sabah State a species called *tampiperes* in the local language is eaten. People eat the larvae raw, stir-fried or boiled with porridge (Chung 2010).

Synagris **sp.** A species belonging to this genus is considered as foodstuff in **DRC** (DeFoliart 2002).

Vespa affinis (lesser hornet) **Taiwan:** This species is eaten (Chen et al. 1998). **Japan:** The Japanese name is *tsumaguro-suzumebachi*. The larvae and pupae are cooked with sugar and soy sauce in the Nagano and Gifu Prefectures (Matsuura 2002). **Laos:** The larvae and pupae are fried (Matsuura 2002). **Malaysia:** The larvae and pupae are regarded as food source in Sabah State (Chung et al. 2002).

Vespa affinis indosinensis This species is appreciated as food item by the people living in the North eastern areas of **Thailand** (Hanboonsong et al. 2000).

Vespa analis (golden wasp) **China:** The larvae and pupae are highly esteemed by the minority people in the Yunnan Province (Chen and Feng 1999). In **Taiwan,** the larvae and pupae are fried (Matsuura 2002). **Japan:** The Japanese name is *kogata-suzumebachi*. The larvae and pupae are cooked with sugar and soy sauce in the Nagano and Gifu Prefectures (Matsuura 2002). In Java Island of **Indonesia,** people eat the larvae by botok cuisine (Matsuura 1998a).

Vespa auraria **Myanmar:** The larvae and pupae are accepted as food (Ghosh 1924). **Thailand:** This species is an article of food (Leksawasdi 2008).

Vespa basalis **China:** The larvae and pupae are highly esteemed by the minority people in the Yunnan Province (Chen and Feng 1999). **Taiwan:** The larvae and pupae are fried (Matsuura 2002).

Vespa bicolor bicolor **China:** The larvae and pupae are highly esteemed by the minority people in the Yunnan Province. They can be purchased in markets (Chen and Feng 1999). **India:** The larvae are eaten roasted by the Galo people in Arunachal Pradesh State (Kato and Gopi 2009).

Vespa cincta The larvae and adults are eaten fried in **Thailand** (Bristowe 1932).

Vespa crabro (hornet) **China:** The larvae and pupae are eaten (Hu and Zha 2009). **Japan:** The Japanese name is *mon-suzumebachi*. The larvae and pupae are cooked with sugar and soy sauce in the Nagano and Gifu Prefectures (Matsuura 2002).

Vespa ducalis (small yellow jacket) **China:** The larvae and pupae are eaten (Hu and Zha 2009). This species is also used for curing rheumatism, menstruation disorders and hepatitis (Zimian et al. 1997). **Indonesia:** In the northern areas of Sumatra Island, the Batak people eat the pupae and adults by frizzling them with palm oil, shallots, salt, pepper and other spices (Matsuura 1998b).

Vespa magnifica This species is eaten in the northern areas of **Thailand** (Leksawasdi 2008) and in **China** (Chen et al. 2009).

Vespa manchurica In **South Korea,** the larvae are eaten, and also are used as medicine (Pemberton 1994).

Vespa mandarinia (yellow jacket) **(PL VIII-1) China:** The larvae and pupae are highly esteemed by the minority people in the Yunnan Province (Chen and Feng 1999). **Taiwan:** The larvae and pupae are fried (Matsuura 2002). **India:** The larvae are eaten raw by the Galo people in Arunachal Pradesh State (Kato and Gopi 2009). **Japan:** The Japanese name is *ō-suzumebachi*. The larvae and pupae are consumed in the Tochigi, Saitama, Nagano, Oita, Miyazaki Prefectures. People eat them raw, or by cooking, roasting and dipping into soy sauce or miso paste, or cooking with sugar and soy sauce. In August, around Takachiho city in Miyazaki Prefecture, the combs containing the larvae and pupae are sold in fruit and vegetable markets or supermarkets (Miyake 1919, Katagiri and Awatsuhara 1996). In the Miyazaki Prefecture, some people make soup stock from the salt water used to boil the larvae (Matsuura 2002). The infusion of the nest was used for treating various diseases such as heart diseases, fever, convulsive fits, etc. (Miyake 1919).

Vespa orientalis (giant hornet) The larvae are highly esteemed in Arunachal Pradesh State, **India**. The larvae are smoked along with the nests. When eating the adults, the wings are removed before preparation (Chakravorty et al. 2011).

Vespa simillima The Japanese name is *kiiro-suzumebachi*. In **Japan**, the larvae and pupae are cooked-and-roasted, or cooked with sugar and soy sauce in the Nagano Prefecture (Katagiri and Awatsuhara 1996). In the Ishikawa Prefecture, the larvae and pupae are eaten by boiling with salt water, or frizzling (Mizuno and Ibaraki 1983).

Vespa soror **China:** The fried larvae of this species are a popular food item in the Yunnan Provice (Matsuura 2002). **Laos:** The larvae, pupae and the newly emerged adults are fried (Matsuura 2002) **(PL VIII-2)**. **Thailand:** This species is eaten also (Matsuura 2002).

Vespa tropica (banded hornet) **China:** This species is eaten (Chen and Feng 1999). **India:** The larvae are eaten roasted by the Galo people in Arunachal Pradesh State (Kato and Gopi 2009). **Indonesia:** In Java Island, people eat the larvae by botok cuisine (Matsuura 1998a). **Laos:** The larvae and pupae are fried (Matsuura 2002). **Malaysia:** The larvae and pupae are eaten in Sabah State (Chung et al. 2002). **Thailand:** This species is eaten (Chen et al. 1998).

Vespa variabilis The larvae and pupae are highly esteemed by the minority people in the Yunnan Province, **China** (Chen and Feng 1999).

Vespa velutina **China:** The larvae and pupae are highly esteemed by the minority people in the Yunnan Province (Chen and Feng 1999). **Taiwan:** The larvae and pupae are fried (Matsuura 2002). **Indonesia:** In Java Island, people eat the larvae by botok cuisine (Matsuura 1998a). **Thailand:** People eat this species (Chen et al. 1998).

Vespa velutina auraria **China:** The larvae and pupae are highly esteemed by the minority people in the Yunnan Province (Chen and Feng 1999). **Myanmar:** The larvae, pupae and the adults nesting underground are eaten by the Shan people (Ghosh 1924).

Vespa xanthoptera → *Vespa simillima*

Vespa **spp. China:** The larvae and pupae of some species belonging to this genus are eaten (Hu and Zha 2009). **India:** The adults of a species belonging to this genus are eaten fried. The fresh insects are chewed after

Plate VIII. 1. *Vespa mandarinia*. A worker wasp collecting tree sap in Tokyo, Japan. Body length: 30 mm (cf. p.215); 2. Fried larvae, pupae and newly emerged adults of *Vespa soror* served at a restaurant in Yunnan, China (cf. p.216); 3–6. *Vespula flavipes lewisii*. Body length: queen: 16 mm, worker: 12 mm. 3. Fried larvae, pupae and newly emerged adults of wasps (cf. p.218). 4. Susi with cooked wasp larvae, pupae and adults (cf. p.218). 5. Release of a worker wasp after marking (arrow). Insert: let a wasp hold small meat ball with a trigonal pyramid made with polystyrene foam in Gifu, Japan (cf. p.218). 6. A miniature house for a nest of *V. flavipes lowisii*. The house is about 1 meter tall (cf. p.218).

removing the wings by the Galo people in Arunachal Pradesh State. The larvae and pupae are also consumed by the Ao-Naga people in Nagaland State (Meyer-Rochow 2005, Chakravorty et al. 2011). **Indonesia:** The larvae of wasps belonging to this genus are often eaten raw or boiled with rice (Chung 2010). **Laos:** Several species belonging to this genus called *tor* in Rao are eaten (Boulidam 2008). **Malaysia:** The Sabah name is *surun*. In Sabah State in Borneo, people eat the larvae raw, stir-fried or boiled with porridge (Chung 2010). **Thailand:** A species belonging to this genus is eaten (Jamjanya et al. 2001).

Vespula arenaria → *Vespula diabolica*

Vespula diabolica In the **USA,** the larvae and pupae are eaten by Californian Indians (Sutton 1988), such as Washoe, Northern Paiute, Western Shoshone, Northern Shoshone, Southern Ute (Fowler 1986).

Vespula (Vespula) pennsylvanica (western yellow jacket) In the **USA,** the larvae and pupae are eaten by Californian Indians (Sutton 1988), such as Washoe, Northern Paiute, Western Shoshone, Northern Shoshone, Southern Ute (Fowler 1986).

Vespula flavipes lewisii The Japanese name is *kuro-suzumebach*. In **Japan,** this species is one of the most popular insect foods. The larvae and pupae are consumed in many prefectures (Iwate, Tochigi, Gunma, Shizuoka, Yamanashi, Nagano, Gifu, Aichi, Nara, Shimane, Okayama, Miyazaki). People eat the wasps raw, or by roasting, cooking, frying, etc. **(PL VIII-3).** Among them, *hachinoko meshi* (wasp rice) and *hachi susi* (wasp *susi*) are famous (Miyake 1919) **(PL VIII-4).** This wasp makes a nest underground. People can find the nests by marking the wasps with a small white substance such as a piece of cotton, and pursuing the marked wasps when they fly back to their nests **(PL VIII-5).** Recently, many professional wasp gatherers caught large amounts of wasps, and consequently the population of the wasp has decreased greatly. People in some areas are trying to protect them **(PL VIII-6).** The demand for the wasp surpasses the supply, and wasps are imported from Korea and New Zealand, although the species are *Vespula koreensis* and *V. germanica* respectively, instead of *V. flavipes lewisii* (Mitsuhashi 1997, 2003, 2005 and 2012). The infusion of the nests was used as a diuretic, or antifebrile (Sekiya 1972).

Vespula germanica (German wasp, German yellow jacket) In **New Zealand,** the larvae and pupae are collected, and exported to Japan as food material (Matsuura 1999).

Vespula koreensis **South Korea:** The larvae and pupae are boiled with soy sauce. A part of the product is exported to Japan (Matsuura 1999).

Vespula rufa (Red wasp) **Japan:** In the Nagano Prefecture, eating of this species was recorded, it is however, not popular (Aruga 1983).

Vespula shidai **Japan:** The larvae and pupae are cooked with sugar and soy sauce in the Nagano and Gifu Prefecturs (Matsuura 2002).

Vespula squamosa (southern yellowjacket) The Mexican name is *avispa panal de tierra*. In **Mexico**, the eggs, larvae, pupae and the honey collected by the wasps are eaten in Chiapas, Hidalgo and Oaxaca States (Ramos-Elorduy de Conconi et al. 1984). Native people of the Mestizo, Mixe, Mixteco, Náhua, Popolaca, Tarasco, Tlapaneca, and Zapoteco tribes eat this species (Ramos-Elorduy and Pino 1989).

Vespula vulgaris The larvae are eaten roasted by the Nishi people in Arunachal Pradesh State, **India** (Singh et al. 2007).

Vespula **spp.** In **South Korea**, the people in the provinces of South Chungcheong, South Gyeongsang, and North Gyeongsang, eat a species belonging to this genus after roasting (Nonaka 1991). **Mexico:** The larvae, pupae and the worker wasps of a species of this genus, which is called *panal de tierra* in Mexican, are eaten (Ramos-Elorduy et al. 1998). The **USA:** The larvae and pupae of several species belonging to this genus are eaten (DeFoliart 2002).

Xylocopidae (Carpenter bees) **(8)**

Xylocopa aestuans → *Xylocopa confusa*

Xylocopa appendiculata circumvolans The Japanese name is *kumabachi* or *kimune-kumabachi*. In **Japan,** the larvae are grilled or cooked with sugar and soy sauce in the Nara and Fukuoka Prefectures (Miyake 1919). The larvae were used for treating convulsive fits, tuberculosis, rheumatism, etc. (Koizumi 1935, Umemura 1943, Shiraki 1958).

Xylocopa californica The larvae and pupae are eaten by Californian Indians in the **USA** (Sutton 1988).

Xylocopa confusa (carpenter bee) The adults are eaten raw around Bangkok, **Thailand** (Bristowe 1932).

Xylocopa dissimilis **Japan:** The adults imported from Hong Kong or Taiwan were used as a remedy for various diseases (Nanba 1980).

Xylocopa latipes **Indonesia:** The Javanese name is *tawon endas*. The larvae and pupae are fried with butter or with onions and salt (Lukiwati 2010). **Malaysia:** The larvae and pupae are eaten in Sabah State (Chung et al. 2002). **Thailand:** The adults are eaten raw around Bangkok (Bristowe 1932).

Xylocopa orpifex The larvae and pupae are eaten by Californian Indians in the **USA** (Sutton 1988).

Xylocopa pictifrons The Japanese name is *Taiwan-takekumabachi*. In **Japan,** the adults imported from Hong Kong or Taiwan are used as a remedy for various diseases (Nanba 1980).

Xylocopa violacea In **China,** this species is sold for medicinal purposes by Chinese drug stores in Shanghai City (Nanba 1980). **Japan:** The imported adults are used as a remedy for various diseases (Nanba 1980).

Xylocopa sp. **India:** The larvae and adults of a species belonging to this genus are eaten. People remove the wings at first, then boil the wasps in Arunachal Pradesh State (Chakravorty et al. 2011). **Malaysia:** A species belonging to this genus is eaten in Sabah State (Chung et al. 2002). **Mexico:** The Seri people greatly relish the sweet beebread (pollen plus nectar) made by the adults of a species belonging to this genus (Felger and Moser 1985). **Sri Lanka:** A species belonging to this genus is eaten (Knox 1817).

19. Numbers of Identified Species in Each Order and Family [2127]

Numbers in [ ] show total numbers of orders or suborders.

THYSANURA	[2]	Tettigoniidae	38	Dactylopiidae	4
Lepismatidae	2	Tetrigidae	0	Eriosomatidae	4
				Flatidae	3
EPHEMEROPTERA	[7]	PHASMIDA	[7]	Fulgoridae	2
Baetidae	2	Phasmatidae	6	Lacciferidae	1
Caenidae	2	Necrosciidae	1	Membracidae	5
Ephemerellidae	1			Pseudococcidae	6
Ephemeridae	2	ISOPTERA	[67]	Psyllidae	3
		Kalotermitidae	1	Heteroptera	[141]
ODONATA	[40]	Rhinotermitidae	4	Alydidae	2
Aeschnidae	8	Termitidae	59	Aphelochiridae	1
Agrionidae	1	Hodotermitidae	3	Belostomatidae	13
Calopterygidae	0			Cimicidae	1
Coenagrionidae	0	BLATTARIA	[12]	Coreidae	14
Cordulegasteridae	1	Blaberidae	0	Corixidae	13
Corduliidae	2	Cryptocercidae	0	Gerridae	2
Gomphidae	3	Panesthiidae	1	Naucoridae	4
Lestidae	1	Blattellidae	3	Nepidae	6
Libellulidae	24	Blattidae	5	Notonectidae	8
Megapodagrionidae	0	Corydiidae	2	Pentatomidae	75
		Phyllodromiidae	1	Pyrrhocoridae	1
PLECOPTERA	[8]			Rhopalidae	1
Nemouridae	0	MANTODEA	[20]	Scutelleridae	0
Perlidae	4	Hymenopodidae	1		
Perlodidae	0	Mantidae	19	COLEOPTERA	[578]
Pteronarcidae	4			Anobiidae	2
		ANOPLURA	[4]	Bostrichidae	1
ORTHOPTERA	[305]	Haematopinidae	1	Bruchidae	5
Acrididae	210	Pediculidae	2	Buprestidae	21
Catantopidae	10	Haematopinidae	1	Carabidae	1
Gryllacrididae	2			Cerambycidae	112
Gryllidae	32	HEMIPTERA	[239]	Chrysomelidae	7
Gryllotalpidae	7	Homoptera	[98]	*Coccinellidae*	1
Hemiacrididae	3	Aphididae	2	Cicindelidae	4
Raphidophoridae	0	Asterolecaniidae	2	Curculionidae	48
Schizodactylidae	2	Cicadellidae	2	Dermestidae	1
Stenopelmatidae	1	Cicadidae	62		
		Coccidae	2		

Dynastidae	32	Hyppoboscidae	1	Lymantriidae	1
Dytiscidae	43	Muscidae	3	Megalopygidae	1
Elateridae	6	Oestridae	2	Noctuidae	27
Gyrinidae	6	Piophilidae	1	Notodontidae	19
Helmidae	2	Rhagionidae	0	Nymphalidae	13
Haliplidae	3	Simuliidae	2	Papilionidae	7
Histeridae	0	Stratiomyidae	3	Pieridae	16
Hydrophilidae	16	Syrphidae	3	Psychidae	4
Ipidae	3	Tabanidae	12	Pyralidae	20
Lampyridae	2	Tachinidae	1	Saturniidae	104
Lariidae	3	Tephritidae	2	Satyridae	2
Lucanidae	20	Tipulidae	6	Sesiidae	3
Lyctidae	1			Sphingidae	33
Meloidae	12	**TRICHOPTERA**	[10]	Tortricidae	1
Mordellidae	1	Calamoceratidae	0	Uraniidae	1
Nitidulidae	0	Hydropsychidae	4	Xyloryctidae	1
Noteridae	0	Leptoceridae	2	Zygaenidae	2
Passalidae	9	Limnephilidae	0		
Psephenidae	0	Odontoceridae	0	**HYMENOPTERA**	[385]
Ptinidae	0	Phryganeidae	3	Agaonidae	1
Scarabaeidae	202	Rhyacophilidae	1	Anthophoridae	0
Scolitidae	3			Apidae	155
Silvaniidae	0	**LEPIDOPTERA**	[386]	Argidae	1
Tenebrionidae	11	Arctiidae	6	Bombidae	16
Trictenotomidae	0	Bombycidae	4	Braconidae	1
		Brahmaeidae	2	Crabronidae	0
NEUROPTERA	[7]	Carposinidae	1	Cynipidae	4
Corydalidae	6	Castniidae	6	Diprionidae	2
Myrmeleontidae	1	Cossidae	20	Eumenidae	8
		Ctenuchidae	1	Evaniidae	1
SIPHONAPTERA	[1]	Danaidae	3	Formicidae	78
Pulicidae	1	Eupterotidae	2	Halictidae	0
		Gelechiidae	1	Megachilidae	0
DIPTERA	[52]	Geometridae	7	Mutillidae	1
Calliphoridae	2	Gracillaridae	1	Perilampidae	0
Cecidomyiidae	1	Hepialidae	47	Polybiidae	34
Chaoburidae	5	Hesperidae	7	Polistidae	42
Drosophilidae	1	Hyblaeidae	1	Sphecidae	1
Dryomyzidae	1	Lasiocampidae	17	Vespidae	32
Ephydridae	4	Limacodidae	4	Xylocopidae	8
Hypodermatidae	2	Lycaenidae	1		

PART II

LIST OF EDIBLE INSECTS BY COUNTRIES OR DISTRICTS

Abbreviation

CAR: Central Africa Republic
DRC: Democratic Republic of Congo
PNG: Papua New Guinea
PRC: People's Republic of the Congo
RSA: Republic of South Africa
USA: United States of America

A. Countries

1. Europe

Austria
Hymenoptera Formicidae: *Formica rufa.*

Czech Republic
Coleoptera Scarabaeidae: *Melolontha melolontha.*

France
Hemiptera-Homoptera Cicadidae: *Cicada plebeja.*
Coleoptera Cerambycidae: *Cerambyx heros.*
Diptera Piophilidae: *Piophila (=Tyrophagus) casei.*
Lepidoptera Noctuidae: *Plusia gamma.*
Hymenoptera Cynipidae: *Aulax glechomae.*

Greece

Hemiptera-Homoptera Cicadidae: *Tettigonia antiquorum.*
Coleoptera Anobiidae: *Stegobium paniceum.* **Bostrichidae:** *Rhizopertha dominica.* **Chrysomelidae:** *Bruchus rufipes.* **Curculionidae:** *Sitophilus granaries.* **Silvaniidae:** *Oryzaephilus* sp.
Hymenoptera Cynipidae: *Aulacides levantina, Aulax* sp.

Italy

Hemitera-Homoptera Cicadidae: *Cicada orni.* **Dactylopiidae:** *Dactylopius coccus.*
Coleoptera Lucanidae: *Lucanus cervus.* **Scarabaeidae:** *Melolontha aprilina, Rhizotrogus assimilis.*
Diptera Piophilidae: *Piophila (=Tyrophagus) casei.*
Lepidoptera Ctenuchidae: *Syntomis phegea.* **Zygaenidae:** *Zygaena ephialtes, Z.* transalpine.

Romania

Coleoptera Cerambycidae: *Prionus coriarius,* **Scarabaeidae:** *Rhizotrogus pini.*

Rome Antiqua

Hemiptera-Heteroptera Cimicidae: *Cimex lectularis.*
Coleoptera Cerambycidae: *Prionus coriarius,* **Meloidae:** *Lytta vesicatoria.*
Lepidoptera Cossidae: *Cossus ligniperda.*

Russia

Blattaria Blattidae: *Blatta orientalis.*

Sweden

Hymenoptera Cynipidae: *Aulax glechomae.*

Ukraine

Orthoptera Acrididae: *Locusta migratoria.*

United Kingdom

Diptera Piophilidae: *Piophila (=Tyrophagus) casei.*

2. North America

Canada

Orthoptera Gryllidae: *Acheta domesticus.*
Anoplura Haematopinidae: *Haematopinus trichechi.*

Siphonaptera Pulicidae: *Pulex irritans.*

Diptera Hypodermatidae: *Hypoderma bovis, Oedemagena tarandi.*
Oestridae: *Cephenemyia phobifer, C. trompe.*

Hymenoptera Formicidae: *Camponotus pennsylvanicus, Formica pensylvanica.*

United States of America (USA)

Odonata Aeschnidae: *Aeschna multicolor.*

Plecoptera Perlodidae: *Isoperla* sp. **Pteronarcidae:** *Pteronarcys californica, P. dorsata, P. princes.*

Orthoptera Acrididae: *Acridium perigrinum, Arphia pseudonietana, Calopterus italicus, Camnula pellucida, Melanoplus atlanis, M. bivittatus, M. devastator, M. differentialis, M. femurrubrum, M. sanguinipes, Oedaleonotus enigma, Oedipoda migratoria, Schistocerca venusta.* **Gryllacrididae:** *Stenopelmatus fuscus.* **Gryllidae:** *Acheta domesticus, Gryllus assimilis.* **Stenopelmatidae:** *Stenopelmatus fuscus.* **Tettigoniidae:** *Anabrus simplex.*

Isoptera Rhinotermitidae: *Reticulitermes tibialis.* **Hodotermitidae:** *Zootermopsis angusticollis.*

Anoplura Pediculidae: *Pediculus humanus, P. humanus corporis.*

Hemiptera-Homoptera Aphididae: *Hyalopterus pruni.* **Asterolecaniidae:** *Cerococcus quercus.* **Cicadidae:** *Diceroprocla apache, D.* spp., *Magicicada cassini, M. septendecim, M. septendecula, M. tredecim, Okanagana bella, O. cruentifera, O.* spp., *Platypedia areolata, P. lutea.*

Hemiptera-Heteroptera Belostomatidae: *Lethocerus americanus.*

Coleoptera Cerambycidae: *Ergates spiculatus, Monochamus maculosus, M. scutellatus, Neoclytus conjunctus, Prionus californicus, Rhagium lineatum, Xylotrechus nauticus.* **Curculionidae:** *Rhynchophorus cruentatus.* **Dermestidae:** *Anthrenus* sp. **Ditiscide:** *Cybister ellipticus, C. explanatus.* **Lariidae:** *Algarobius* spp., *Neltumis* spp. **Ptinidae:** *Ptinus* sp. **Scarabaeidae:** *Cyclocephala borealis, C. dimidiate, Lachnosterna* sp., *Phyllophaga crinite, P. fusca, Polyphylla crinite.* **Tenebrionidae:** *Tenebrio molitor.*

Diptera Ephydridae: *Ephydra cinerea, E. macellaria, Hydropyrus hians.* **Oestridae:** *Cephenemyia trompe.* (**Rhagionidae:** *Atherix* sp.; cf. Plecoptera-Pteronarcidae-*Pteronarcys californica*). **Tiulidae:** *Holorusia rubiginosa, Tipula derbyi, T. quaylii, T. simplex.*

Lepidoptera Arctiidae: *Arctia caja americana.* **Danaidae:** *Danaus plexippus.* **Hesperiidae:** *Megathymus yuccae.* **Lasiocampidae:** *Malacosoma* spp. **Noctuidae:** *Helicoverpa zea, Homoncocnemis fortis, Spodoptera frugiperda.* **Notodontidae:** *Homonococnemis fortis.* **Pyralidae:** *Galleria mellonella.* **Saturniidae:** *Coloradia pandra, Hyalophora euryalus.* **Sphingidae:** *Hyles lineata, Manduca sexta, Sphinx ludoviciana.*

Hymenoptera Anthophoridae: *Anthophora* sp. **Apidae:** *Apis mellifera.* **Bombidae:** *Bombus appositus, B. nevadensis, B. terricola occidentalis, Bombus vosnesenskii.* **Formicidae:** *Camponotus maculates, Camponotus* spp., *Formica exectoides, F. obscuripes, F. rufa, Lasius niger, Monomorium pharaonis, Myrmecia sanguinea, Myrmecocystus depilis, M. melliger, M. mexicanus hortideorum, M. mimicus, M. testaceus, Pogonomyrmex barbatus, P. desertorum, P. occidentalis, P. owyheei, P.* sp. **Megachilidae:** *Anthophora* sp. **Megachilidae:** *Melissodes* sp. **Vespidae:** *Dolichovespula* spp., *Vespula diabolica, V. (Vespula) pennsylvanica, V.* spp. **Xylocopidae:** *Xylocopa californica, X. orpifex.*

3. Central area of America

Guatemala

Hymenoptera Polistidae: *Polistes* spp.

Honduras

Hymenoptera Formicidae: *Atta cephalotes, A.* spp. **Poybiidae:** *Polybia* spp.

Mexico

Ephemeroptera Baetidae: *Baetis* sp. **Ephemeridae:** *Ephemera* sp.

Odonata Aeschnidae: *Aeschna multicolor, A.* spp., *Anax* sp.

Orthoptera Acrididae: *Arphia fallax, Boopedon flaviventris, B.* sp. affin. *flaviventris, Encoptolophus herbaceous, Homocoryphus prasimus, Locusta migratoria, Melanoplus femurrubrum, M. mexicanus, M. sumichastri, M.* spp., *Ochrotettix cer salinus, Opeia* sp., *Orphula azteca, Orthochtha venosa, Osmilia fravolineata, Rhammatocerus maturius, R. viatorius, Schistocerca gregaria, S. paranensis, S. vaga vaga,* **S. spp.**, *Spharagemon aequale, Sphenarium histrio (=S. bolivari), S. magnum, S. mexicanum, S. purpurascens, S.* spp., *Taeniopoda auricornis, T.* sp., *Trimerotropis pallidipennis, T.* sp., *Tropinotus mexicanus, Xanthipus corallipes zapotecus.* **Gryllacrididae:** *Stenopelmatus* sp. **Gryllidae:** *Acheta domesticus, Gryllus assimilis.* **Gryllotalpidae:** *Gryllotalpa* spp. **Stenopelmatidae:** *Stenopelmatus fuscus.* **Tettigoniidae:** *Conocephalus triops, Microcentrum* sp., *Petaloptera zandala, Pyrgocorypha* sp., *Romalea colorata, R.* sp., *Stilpnochlora azteca, Stilpnochlora toracica.*

Phasmida Phasmtidae: *Extatosoma* spp.

Blattaria Blattellidae: *Blatella germanica, Pseudomops* sp. **Blattidae:** *Periplaneta americana, P. australasiae, Periplaneta* sp.

Anoplura Pediculidae: *Pediculus humanus corpor.*

Hemiptera-Homoptera Cicadidae: *Cicada montezuma, C. nigriventris, C. pruinosa, C.* spp., *Proarna* sp., *Quesada gigas.* **Coccidae:** *Coccus (Llavea) axin.* **Dactylopiidae:** *Dactylopius coccus, D. confusus, D. indicus, D. tomentosus.*

Membracidae: *Anthiante expansa, Hoplophorion monograma, Umbonia reclinata. U.* sp.

Hemiptera-Heteroptera Belostomatidae: *Abedus dilatatus, A. ovatus, A.* sp., *Lethocerus* spp. **Coreidae:** *Acantocephala declivis, A. luctuosa, A.* sp., *Mamurius mopsus, Pachilis gigas, Sephina vinula.* **Corixidae:** *Corisella edulis, C. mercenaria, C. texcocana, C.* spp., *Corixa femorale, Graptocorixa abdominalis, G. bimaculata, G.* sp., *Hesperocorixa laevignata, Krizousacorixa azteca, K. femorata, K.* sp. **Notonectidae:** *Notonecta fasciata, N. unifasciata, N.* sp. **Pentatomidae:** *Atizies sufultus, Banasa subrufescens, B.* sp., *Brochymena tenebrosa, B.* sp., *Chlorocoris distinctus, C. irroratus, C.* sp., *Edessa championi, E. conspersa, E. cordifera, E. discors, E. fuscidorsata, E. helix, E. indigena, E. lepida, E. mexicana, E. montezumae, E. petersii, E. reticula, E. rufomarginatus, E.* sp., *Euschistus bifibulus, E. biformis, E. comptus, E. crenator orbiculator, E. egglestoni, E. integer, E. lineatus, E. rugifer, E. schaffneri, E. spurculus, E. stali, E. strenuous, E. sufultus, E. sulcacitus, E. taxcoensis, E.* sp., *Mormidea (Mormidea) notulata, M.* sp., *Moromorpha tetra, Nezara (=Acrostemum) majuscule (=Chinavia montivaga), Nezara viridula, Oebalus mexicana, O. pugnax, Padaeus trivittatus, P. viduus, Pellaea stictica, Pharypia fasciata, Proxys puntalatus, Proxys* sp.

Coleoptera Buprestidae: *Chalcophora* sp., *Chrysobothris basalis, Euchroma gigantean, Thrincopyge alacris.* **Cerambycidae:** *Acrocinus longimanus, Aplagiognathus spinosus, A.* sp., *Arhopalus* sp., *Arophalus* afin *rusticus, A. rusticus montanus, Callipogon barbatus, Cerambyx* sp., *Derobrachus procerus, D.* sp., *Eburia stigmatica, Lagocheirus rogersii, Polyrhaphis* sp., *Prosopocera (Prosopocera)* sp., *Stenodontes* cer. *maxillosus, S.* cer. *molaria, Trichoderes pini.* **Chrysomelidae:** *Leptinotarsa decemlineata.* **Cicindellidae:** *Cicindela curvata, Cicindela roseiventris.* **Curculionidae:** *Metamasius spinolae, Rhynchophorus palmarum, R.* spp., *Scyphophorus acupunctatus, S.* sp. **Dermestidae:** *Thylodrias contractus.* **Dynastidae:** *Dynastes Hyllus, D.* sp., *Golofa (Golofa) imperialis, G. pusilla, G. (Golofa) tersander, Strategus aloeus aloeus, S. julianus, Xylotryctes* spp. **Ditiscidae:** *Cybister explanatus, C. flavocinctus, C. frimbiolatus, C. occidentalis, Dytiscus marginicollis, D.* sp., *Hololepta (Hololepta) guidnis, H.* sp., *Laccophilus apicalis, L.* sp., *Megadytes gigantean, M. giganteus, M.* sp., *Rhantus atricolor, R. consimilis, R.* sp., *Thermonectes bsilaria, T. marmoratus, T.* sp. **Elateridae:** *Chalcolepidius laforgei, Chalcolepidius rugatus, Pyrophorus mexicanus, P. pellucens.* **Gyrinidae:** *Gyrinus parcus, G. (Oreogyrinus) plicatus.* **Haliplidae:** *Haliplus punctatus, H.* sp., *Peltodytes mexicanus, P. ovalis.* **Helmidae:** *Austrelmis condimentarius.* **Histeridae:** *Homolepta* sp. **Hydrophilidae:** *Berosus* sp., *Dilobodeus* sp., *Dibolocelus* sp., *Tropisternus mexicanus, Tropisternus sublaevis, T. tinctus, T.* sp. **Lariidae:** *Algarobius* spp. **Lucanidae:** *Lucanus* sp. **Meloidae:** *Meloe dugesi, M. laevis, M. nebulosus.* **Noteridae:** *Suphisellus* sp. **Passalidae:** *Oileus reinator, O.* sp., *Passalus* sp., *Passalus (Passalus) interstitialis, P. (Pa.) punctiger, P. (Pa.)*

sp., *Passalus (Pertinax) punctarostriatus, Paxillus leachi.* **Scarabaeidae:** *Chanton humectus hidalguensis, Cotinis mutabilis var. oblicua, Cyclocephala fasciolata, Macrodactylus lineaticollis, Melolontha* sp., *Phyllophaga mexicana, P. rubella, P. rugipennis, P.* spp., *Strategus aloeus, S.* sp., *Xyloryctes teuthras, X. thestalus, X. ensifer, X.* spp. **Tenebrionidae:** *Asida rughosissima, Eleodes* sp., *Stenomorpha* sp., *Tenebrio molitor, Tribolium castaneum, Tribolium confusum, Zopherus jourdani, Zophobas morio, Z.* spp.

Neuroptera Corydalidae: *Corydalus cornutus.*

Diptera Drosophilidae: *Drosophila melanogaster.* **Ephydridae:** *Gymnopa tibialis, Hydropyrus hians.* **Muscidae:** *Musca domestica.* **Piophilidae:** *Piophila (=Tyrophagus) casei.* **Stratiomyidae:** *Helmetia illucens, Campylostoma* sp., *Copestylum anna, Copestylum haggi.* **Syrphidae:** *Copestylum haaggii, Ornidia obesa (=obescens ?).* **Tephritidae:** *Anastrepha ludens.*

Trichoptera Hydrosychidae: *Leptonema* spp., *Oecetis disjunta.* **Leptoceridae:** *Stenopsyche sauteri.*

Lepidoptera Arctiidae: *Amastus ochraceator, Elysius superba, Estigmene acrea, Pelochyta cervina.* **Bombycidae:** *Bombyx mori.* **Castniidae:** *Castnia chelone, C. synparamides chelone.* **Cossidae:** *Comadia redtenbacheri.* **Danaidae:** *Danaus gilippus thersippus, D. plexippus, Hamadryas* sp. **Geometridae:** *Achlyodes pallida, Acronyctodes mexicanaria, Panthera pardalaria, Pantherodes pardalaria, Synopsia mexicanaria.* **Hepialidae:** *Phassus trajesa, P. triangularis, P.* spp. **Hesperiidae:** *Achlyodes pallida, Aegiale hesperiaris.* **Lasiocampidae:** *Dendrolimus superans.* **Noctuidae:** *Ascalapha odorata, Gerra sevorsa, Helicoverpa zea, Latebraria amphipyrioides, Spodoptera exigua, S. frugiperda, S.* sp., *Thysania agrippina.* **Nymphalidae:** *Anartia fatima, Caligo memtion, Charaxes jasius, Chlosyne lacinia lacinia, Junonia lavinia, Morpho* sp., *Nymphalis antiopa, Pareuptychia metaleuca, Vanessa annabella, V. virginiensis.* **Papilionidae:** *Papilio polyxenes, Protographium philolaus philolaus, Pterourus multicaudata multicaudata.* **Pieridae:** *Catasticta flisa flisa, Catasticta nimbice nimbice, Catasticta teutila teutila, Eucheira socialis socialis, E. socialis westwoodi, Eurema lisa, E. salome jamapa, Leptophobia aripa elodia, Phoebis agarithe agarithe, P. philea philea, P. sennae macellina, Pieris brassicae, P.* sp., *Pontia protodice, Synchloe callidice.* **Pyralidae:** *Laniifera cyclades.* **Saturniidae:** *Actias luna, A. truncatipennis, Antheraea pernyi, A. Polyphemus, A. polyphemus mexicana, A.* sp., *Arsenura armida, A. polyodonta, Arsenura richardsonii, Caio championi, Callosamia promethea, Eacles* aff. *ormondei yacatanensis, E.* sp., *Euleucophaeus (=Hemileuca) tolucensis, Hemileuca* sp., *Hyalophora* sp., *Hylesia coinopus, H. frigida, H.* sp., *Latebraria amphipyroides, Paradirphia fumosa, P. hoegei, Pseudodirphia mexicana, Saturnia pyri, Synanthedon cardinalis.* **Satyridae:** *Magisto metaleuca.* **Sesiidae:** *Synanthedron cardinalis.* **Sphingidae:** *Cocytius antaeus, Hyles lineata, Manduca sexta, M.* sp., *Pachilia ficus.*

Hymenoptera Apidae: *Apis cerana, A. dorsata, A. laboriosa, A. mellifera, A. mellifera scutellata, A. mellifica ligustica, A. mellifica* var. *adansoni, Cephalotrigona zexmemiae, Friesesmellita nigra, Lestrimelita limao, L. niitkib, L.* sp.*, Melipona bleckei, M. fasciata, M. fasciata guerreroensis, M. grandis, M. interrupta, M. solani, M.* spp.*, Nannotrigona perilampoides, N. testaceicornis, N.* spp.*, Partamona bilineata, P. orizabaensis, Partamona* sp.*, Plebeia frontalis, P. mexica, P.* spp.*, Scaptotrigona mexicana, S. pectoralis, S.* spp.*, Trigona (Trigona) corvina, T. fulviventris, T. (Trigona) fuscipennis, T. jati, T. mexicana, T. nigra nigra, T. pectoralis, T. postica, T. (Hypotrigona)* spp.*, Trigonisca (Dolichotrigona) schultessi, T. jaty, T. pipioli, T. (Dolichotrigona) schultessi.* **Bombidae:** *Bombus diligens, B. formosus, B. medius, B. rufocinctus. B.* spp. **Crabronidae:** *Philanthus* sp. **Doprionidae:** *Neodiprion guilletei, Zadiprion vallicale.* **Eumenidae:** *B. lecheguana, Eumenes* sp.*, Euodynerus* sp. **Evaniidae:** *Brachgaster lecheguana.* **Formicidae:** *Acromyrmex octospinosus, A. rugosus, Atta cephalotes, A. mexicana, A.* spp.*, Camponotus (Terraernimex) dumetorum, Camponotus* spp.*, Eciton* sp.*, Liometopum apiculatum, L. occidentale* var. *luctuosum, Myrmecosystus melliger, M. mexicanus, M.* spp.*, Pogonomyrmex barbatus, Pogonomyrmex owyheei, Pogonomyrmex* sp. **Mutillidae:** *Dasymutilla occidentalis.* **Polybiide:** *Agelaia aereata, A. panamensis, Polybia diguetana, Polybia nigratella, P. occidentalis, P. occidentalis bohemani, P. (Myrametra) occidentalis nigratella, P. (Myrametra) parvulina, P. pygmaea, P. scrobalis, P. striata, Polybia* spp.*, Synoeca virginia, Synoeca surinama.* **Polistidae:** *Apoica* sp.*, Epipona* sp.*, Mischocyttarus (Kappa) cubensis, M. (Mischocytarus) basimaculata, M. (Mischocyttarus) pallidipectus, M.* spp.*, Polistes apicalis, P. canadensis, P. carnifex, P. dorsalis, P. instabilis, P.(Aphanilopterus) kaibabensis, P. major, P. mexicanus, P. parvulina, P.* spp. **Sphecidae:** *Ammophila* sp. **Vespidae:** *Brachygaster azuteca, B. lecheguana, B. mellifica, B.* sp.*, Parachartegus apicalis, Vespula squamosa, V.* spp. **Xylocopidae:** *Xylocopa* sp.

Nicaragua

Hymenoptera Formicidae: *Atta cephalotes.* **Polybiidae:** *Polybia* spp.

Panama

Coleoptera Hydrophilidae: *Tropisternus mexicanus.*

4. West Indies

Cuba

Hymenoptera Apidae: *Melipona beechei fulvipes.*

Dominica and Haiti (Hispaniola Island)

Coleoptera Elateridae: *Pyrophorus* sp.

Jamaica

Orthoptera Gryllidae: *Gryllus campestris.*

Martinique
Coleoptera Curculionoidae: *Rhynchophorus palmarum*.

Trinidad and Tobago
Coleoptera Curculionidae: *Rhynchophorus palmarum*.

5. South America

Argentina
Coleoptera Curculionidae: *Rhynchophorus palmarum*.

Barbados
Coleoptera Curculionidae: *Rhynchophorus palmarum*.

Bolivia
Coleoptera Curculionidae: *Rhynchophorus palmarum*. Isoptera Termicidae: *Nasutitermes corniger*.
Hymenoptera Formicidae: *Atta* spp.

Brazil
Orthoptera Acrididae: *Lophacris* sp., *Rhammatocerus schistocercoides*, *R.* sp., *Schistocerca cancellata cancellata, S. cancellata paranensis, S.* sp. , *Titanacris albipes,* **Tropidacris collaris. Gryllidae:** *Gryllus assimilis.* **Tettigoniidae:** *Lophacris* sp.

Isoptera Kalotermitidae: *Kalotermes flavicollis.* **Termitidae:** *Cornitermes* sp., *Labiotermes labralis, Nasutitermes* sp., *Syntermes aculeosus, S.* pr. *Spinosus, S.* spp.

Blattaria Blattallidae: *Eurycotis manni.* **Blattidae:** *Periplaneta Americana.*

Anoplura Pediculidae: *Pediculus humanus, Pediculus* sp.

Hemiptera-Homoptera Membracidae: *Umbonia spinosa, U.* sp.

Coleoptera Bruchidae: *Caryobruchus* sp., *Pachymerus cardo, P. nucleorum, P.* sp. **Buprestidae:** *Euchroma gigantean.* **Cerambycidae:** *Macrodontia cervicornis, Stenodontes damicornis.* **Curculionidae:** *Rhinostomus barbirostris, Rhynchophorus palmarum.* **Dynastidae:** *Dynastes hercules, Megaceras* sp., *Megasoma actaeon, M. anubis, Strategus* sp. **Lariidae:** *Caryobruchus* spp., *Pachymerus cardo, P. nucleorum, P.* sp. **Scarabaeidae:** *Geniatosoma nignum.*

Lepidoptera Megalopygidae: *Trosia* sp. **Nymphalidae:** *Brassolis sophorae.* **Pieridae:** *Oiketicus* spp. **Pyralidae:** *Myelobia (Morpheis) smenintha.* **Sphingidae:** *Erinnyis ello.*

Hymenoptera Apidae: *A. mellifera scutellata, Cephalotrigona capitata, C. femorata, Duckeola ghilianii, Friesella schrottkyi, Frieseomellita silvestrii, F. varia, Lestrimelita limao, Melipona asilvai, M. atratula, M. beecheii, M. bicolor, M. bilineata, M. compressipes, M. compressipes fasciculate, M. crinite, M. dorsalis, M. ebumea fuscopilosa, M. fasciata scutellaris, M. grandis, M. interrupta grandis, M. mandacaia, M. marginata, M. melanoventer, M. nigra, M. pseudocentris, M. quadrifasciata, M. rufiventris, M. rufiventris flavolineata, M. schencki picadensis, M. schencki schenski, M. schwarzi, M. scutellaris, M. seminigra merrillae, M. seminigra* cf. *pernigra. Nannotrigona bipunctata polystica, N. xanthotricha, N.* spp., *Oxytrigona obscura, O. tataira, O.* spp., *Paratrigona* sp., *Partamona* cf. *cupira, P.* sp., *Plebeia emerina, P. mosquito, P. remota, P.* spp., *Ptilotrigona lurida, Scaptotrigona mexicana, S. nigrohirta, S. polysticta, S. postica, S. tubiba, S. xanthotrica, Scaura longula, Tetragona benneri, T. fulviventris, T. goettei, T. quadrangular, T. spinnipes, Tetragonisca angustula angustula, T. branneri, T. dorsalis, Trigona amalthea, T. angustula, T. branneri, T. chanchamayoensis, T. cilipes pellusida, T. clavipes, T. dallatorreana, T. dorsalis, T. flaveola, T. (Trigona) fuscipennis, T. geniculata, T. ghilianii, T. hypogea, T. leucogaster, T. lurida, T. mombuca, T. muscaria, T. pallens, T. recursa, T. silvestrii, T. spinipes, Trigonisca pipioli.* **Argidae:** *Dielocerus formosus.* **Crabronidae:** *Trypoxylon* sp. **Eumenidae:** *Brachygastra lecheguana.* **Evaniidae:** *Brachygaster mellifica.* **Formicidae:** *Atta bisphaerica, A. capiguara, A. cephalotes, A. laevigata, A. opacipes, A. sexdens, A.* spp. **Polybiidae:** *Polybia dimidiata, P. occidentalis, P. sericea, P.* spp., *Protopolybia exigua exigua, Pseudopolybia vespiceps, Synoeca surinama.* **Polistidae:** *Apoica pallens, Epipona quadrituberculata.*

Chile

Coleoptera Helmidae: *Austrelmis chilensis, A. condimentarius.*

Colombia

Orthoptera Arididae: *Aidemona azteca, Orphulella* spp.,*Orthochtha venosa, Osmilia flavolineata, O.* spp., *Schistocerca cancellata cancellata, S. cancellata paranensis, S. paranensis, S.* spp., *Tropidacris* c. *cristata.* **Tettigoniidae:** *Conocephalus angustifrons, Neoconocephalus* sp.

Isoptera Termitidae: *Labiotermes labralis, Macrotermes* spp., *Syntermes aculeosus, S. parallelus, S. spinosus, S. synderi, S. tanygnathus.*

Anoplura Pediculidae: *Pediculus* sp.

Hemiptera-Homoptera Membracidae: *Umbonia spinosa.*

Coleoptera Bruchidae: *Caryobruchus* sp. **Buprestidae:** *Euchroma gigantean.* **Cermbycidae:** *Acrocinus longimanus, Macrodontia cervicornis.* **Curculionidae:** *Rhynchophorus palmarum.* **Dynastidae:** *Ancognatha* sp., *Dynastes hercules, Megaceras crassum.* **Helmidae:** *Austrelmis condimentarius.* **Passalidae:** *Passalus interruptus, Veturius sinuosus.* **Scarabaeidae:** *Ancognatha* sp.

Diptera Stratiomyidae: *Chryschlorina* spp.

Lepidoptera Noctuidae: *Mocis repanda, Spodoptera frugiperda.* **Saturniidae:** *Automeris* sp., *Dirphia* sp. **Sphingidae:** *Erinnyis ello.*

Hymenoptera Apidae: *Trigona (Tetragona) clavipes, T. trinidadensis.* **Formicidae:** *Atta cephalotes, A. laevigata, A. sexdens, A.* spp. **Polybiidae:** *Agelaia angulata, Polybia rejecta, P.* spp. **Polistidae:** *Apoica thoracica, Polistes canadensis erythrocephalus, P. ignobilis, P. pacificus modestus, P. versicolor.* **Vespidae:** *Brachygaster lecheguana, B. mellifica.*

Ecuador

Odonata Aeschnidae: *Aeschna brevifrons, A. marchali, A. peralta, Coryphaeschna adnexa.*

Orthoptera Acrididae: *Schistocerca* spp.

Isoptera Termitidae: *Termes destructor.*

Anoplura Pediculidae: *Pediculus humanus.*

Hemiptera–Hooptera Cicadidae: *Carineta fimbriata.* **Membracidae:** *Umbonia spinosa.*

Coleoptera Cerambycidae: *Macrodontia cervicornis, Oncideres* sp., *Pexteuso atys, Psalidognrathus atys, P. cacicus, P. erithrocerus, P. modestus.* **Curculionidae:** *Cosmopolites sordida, Dynamis borassi, D. nitidula, Metamasius cinnamominus, M. dimidiatipennis, M. hemipterus, M. sericeus, Psalidognathus atys, P. cacicus, P. erithrocerus, P. modestus, Rhinostomus barbirostris, Rhynchophorus palmarum.* **Dynastidae:** *Ancognatha jamesoni, A. vulgaris, Dynastes hercules, Golopha aeacus, G. aegeon, Praogolofa unicolor.* **Lucanidae:** *Sphaenognathus feisthamelii, S. lindenii, S. metallifer.* **Scarabaeidae:** *Ancognatha castanea, A. jamesoni, A. vulgaris, Clavipalpus antisanae, Coelosis biloba, Democrates burmeisteri, Golopha aegeon, Heterogomphus bourcieri, Leucopelaea albescens, Pelidnota nigricauda, Platycoelia forcipalis, P. lutescens, P. parva, P. rufosignata, Proagolofa unicolor.*

Diptera Tephritidae: *Anastrepha* spp., *Ceratitis capitata.*

Lepidoptera Castniidae: *Castnia daedalus, C. licoides, C. licus, Eupalamides cyparissias.* **Hepialidae:** *Hepialus* sp., **Nymphalidae:** *Brassolis astyra, B. sophorae, Caligo* spp., *Panacea prola.*

Hymenoptera Apidae: *Apis mellifera, Tetragonisca angustula angustula.* **Bombidae:** *Bombus atratus, B. ecuadorius, B. funebris, B. robustus.* **Eumenidae:** *Brachymenes lecheguana, Montezumia dimidiata.* **Formicidae:** *Atta cephalotes, A. sexdens.* **Megachilidae:** *Megachile* spp. **Polybiidae:** *Agelaia baezae, A. corneliana, A. lobipleura, A. ornata, Angiopolybia paraensis, Polybia aequatorialis, P. dimidiate, P. flavifrons, P. testaceicolor, P.* spp., *Synoeca corneliana, S. lobipleura, S. ornata, S. surinama.* **Polistidae:** *Apoica pallens, A. pallida, A. strigata, A. thoracica, Mischocyttarus rotundicollis, M. tomentosus, Montezmia* *dimidiate, Polistes bicolor, Polistes deceptor, P. occipitalis, P. testaceicolor, Brachygaster lecheguana, B. wagnerianus.*

Guyana

Orthoptera Acrididae: *Schistocerca cancellata cancellata, S. paranensis.*

Isoptera Termitidae: *Termes destructor.*

Coleoptera Cerambycidae: *Macrodontia cervicornis, Stenodontes damicornis.*

Hymenoptera Formicidae: *Atta cephalotes, A. sexdens, Formica edulis, F. major, F. Volans.*

Paraguay

Coleoptera Cerambycidae: *Macrodontia cervicornis.* **Crculionidae:** *Rhynchophorus palmarum.* **Passalidae:** *Passalus interruptus.*

Hymenoptera Apidae: *A. mellifera, Melipona* spp. **Formicidae:** *Atta cephalotes.* **Polybiidae:** *Polybia* spp.

Peru

Anoplura Pediculidae: *Pediculus humanus.*

Coleoptera Cerambycidae: *Stenodontes damicornis.* **Crculionidae:** *Rhynchophorus palmarum.* **Helmidae:** *Austrelmis chilensis, A. condimentarius.*

Neuroptera Corydalidae: *Corydalus armatus, C. peruvianus.*

Hymenoptera Formicidae: *Atta* spp.

Suriname

Orthoptera Acrididae: *Schistocerca cancellata cancellata, S. paranensis.*

Coleoptera Cerambycidae: *Stenodontes damicornis.* **Crculionidae:** *Rhynchophorus palmarum.* **Passalidae:** *Passalus interruptus.*

Venezuela

Odonata Calopterygidae: *Hetaerina* sp., *Mnesarete* sp.; **Coenagrionidae:** *Argia* sp. **Corduliidae:** *Lauromacromia dubitalis.* **Gomphidae:** *Agriogomphus* sp., *Progomphus* sp., *Zonophora* sp.; **Libellulidae:** *Brechmorhoga* sp., *Dasythemis* spp.; **Megapodagrionidae:** *Oxystigma* sp.

Orthoptera Acrididae: *Aidemona azteca, Orphulella* spp., *Orthochtha venosa, Osmilia flavolineata, O.* spp., *Rhammatocerus* sp., *Schistocerca* spp., *Tropidacris c. cristata.* **Tettigoniidae:** *Conocephalus angustifrons.*

Isoptera Termitidae: *Labiotermes labralis, Nasutitermes corniger, N. ephratae, N. ephrateae, N. macrocephalus, N. surinamensis, N.* sp., *Syntermes aculeosus, S. spinosus, S. tanygnathus, S. territus, S.* spp.

Anoplura Pediculidae: *Pediculus humanus, P.* sp.

Hemiptera-Heteroptera Belostomatidae: *Lepthocerus micantulum,*
L. spp. **Naucoridae:** *Ambrysus stali, A.* sp.*, Limnocoris* cf. *minutes.*
Coleoptera Bruchidae: *Caryobruchus* sp. **Curculionidae:** *Metamasius*
cinnamominus, M. sp.*, Rhinostomus barbirostris, Rhynchophorus palmarum.*
Helmidae: *Austrelmis condimentarius.* **Larridae:** *Caryobruchus* spp.
Scarabaeidae: *Pelidnota nigricauda, P. (Chalcoplethis)* sp. **Tenebrionidae:**
Zophobas spp.

Diptera Stratiomyidae: *Chryschlorina* spp.

Trichoptera Calamoceratidae: *Phylloicus* sp. **Leptoceridae:** *Triplectides* sp.
Odontoceridae: *Marilia* sp.

Lepidoptera Noctuidae: *Mocis repanda, Spodoptera frugiperda.* **Sphingidae:**
Erinnyis ello.

Hymenoptera Apidae: *Frieseomylitta varia, Partamona testacea, Trigona*
amaltea, T. (Tetragona) clavipes, T. trinidadensis. **Eumenidae:** *Eumenes*
canariculata. **Formicidae:** *Atta cephalotes, A. laevigata, A. sexdens, A.* spp.*,*
Ecton burchelli. **Polybiide:** *Agelaia pallidiventris, A.* sp.*, Angiopolybia* sp.*,*
Polybia (Trichothorax) micans, P. occidentalis, P. spp. **Polistidae:** *Apoica thoracica,*
Polistes canadensis erythrocephalus, P. pacificus modestus, P. versicolor.

6. Africa

Angola
Orthoptera Acrididae: *Schistocerca peregrinatoria.* **Gryllidae:** *Brachytrupes*
membranaceus.

Isoptera Termitidae: *Macrotermes natalensis, M. subhyalinus,*
M. vitrialatus, Pseudacanthotermes militalis, P. spiniger.

Coleoptera Buprestidae: *Chrysobothris fatalis, Psiloptera wellmani, Sterapsis*
amplipennis, Sternocera feldspathica. **Carabidae:** *Scarites* sp. **Curculionidae:**
Rhynchophorus phoenicis. **Dynastidae:** *Camenta* sp. **Scarabaeidae:** *Tricholespis*
sp.

Lepidoptera Sturniidae: *Bunaea alcinoe, Imbrasia ertli, Usta Terpsichore.*

Benin
Orthoptera Gryllidae: *Brachytrupes membranaceus.*

Isoptera Termitidae: *Macrotermes falciger.*

Coleoptera Curculionidae: *Rhynchophorus phoenicis.* **Dynastidae:** *Oryctes*
spp.

Botswana
Orthoptera Acrididae: *Cyrtacanthacris tatarica, Lamarckiana cucullata.*
Gryllidae: *Acanthoplus* sp.

Isoptera Hodotermitidae: *Hodotermes mossambicus.*

Hemiptera-Homoptera Cicadidae: *Monomatapa insignis, Orapa* sp. **Psyllidae:** *Arytaira mopane.*

Coleoptera Buprestidqe: *Sternocera orissa.* **Cerambycidae:** *Macrotoma natala.*

Lepidoptera Saturniidae: *Gonimbrasia belina.* **Sphingidae:** *Herse convolbuli, Hippotion celerio.*

Hymenoptera Anthophoridae: *Anthophora* sp. **Apidae:** *Plebina denoita.* **Formicidae:** *Camponotus* spp. **Megachilidae:** *Megachile* spp.

Bourbon

Hymenoptera Apidae: *Apis unicolor unicolor.*

Burkina Faso

Lepidoptera Saturniidae: *Cirina forda.*

Cameroon

Hemiptera-Heteroptera Belostomatidae: *Lethocerus cordofanus.*

Orthoptera Acrididae: *Acorypha glaucopsis, A. picta, Acrida bicolor, A. cinerea, A. turrita, Anacridium hurri, A. melanorhodon, Bruchycrotaphus lryxalicerus, Cataloipus cymbiferus, Chrotogonus senegalensis, Diabolocatantops axillaris, Exopropacris modica, Gastrimargus africanus, G. proceus, Harpezocatantops stylifer, Homoxyrrhepes punctipennis, Krausella amabile, Kraussaria angulifera, Locusta migratoria migratorioides, Oedaleus nigeriensis, Orthochtha venosu, Oxycantatops spissus, Paracinema tricolor, Pyrgomorpha cognata, Schistocerca gregaria, Truxalis johnstoni, Tylotropidius gracilipes, Zonocerus variegates.* **Gryllidae:** *Brachytrupes membranaceus.* **Tettigoniidae:** *Ruspolia* sp.

Isoptera Termitidae: *Macrotermes natalensis, M. subhyalinus, Odontotermes magdalense.*

Blattaria Blaberidae: *Gyna* sp.

Mantodea Mantidae: *Hoplocorypha garuana, Tarachodes saussurei.*

Hemiptera-Heteroptera Coreidae: *Anoplocnemis curvipes.* **Pentatomidae:** *Acrosternum millieri, Anoplocnemis curvipes, Aspongopus viduatus, Basicrryptus* sp., *Carbula pedalis, Diploxys cordofana, D.* sp., *Nezara viridula.*

Coleoptera Buprestidae: *Steraspis speciosa, S.* sp., *Sternocera castairea* subsp. *Irregularis, Sternocerca interrupta* subsp. *immaculata.* **Curculionidae:** *Rhynchophorus phoenicis.* **Dynastidae:** *Augosoma centaurus.* **Ditiscidae:** *Cybister occidentalis, Dytiscus* sp. **Scarabaeidae:** *Brachylepis bennigseri, Diplognatha gagates, Pachnoda marginata, Popillia femoralis.*

Lepidoptera Notodontidae: *Anaphe* spp. **Saturniidae:** *Bunaea alcinoe, Microgone herilla.*

Hymenoptera Apidae: *Apis adansoni, A. mellifica* var. *adansoni.* **Formicidae:** *Anomma nigricans, Camponotus* spp., *Oecophylla* spp., *Pachycondyla tarsata.* **Polistidae:** *Euchromia lethe.*

Cape Coast

Coleoptera Dynastidae: *Oryctes owariensis.*

Central African Republic (CAR)

Orthoptera Acrididae: *Zonocerus variegates.*

Isoptera Termitidae: *Apicotermes* sp., *Bellicositermes natalensis, Macrotermes bellicosus, Protermes* sp.

Coleoptera Cerambycidae: *Plocaederus frenatus, Stenodontes* (=*Mallodon*) *downesi.* **Elateridae:** *Tetralobus flabellicornis.* **Scarabaeidae:** *Goliathus cacicus, G. cameronensis, G. goliathus, G. regius.*

Lepidoptera Lasiocampidae: *Mimopacha* aff. *knoblauchi.* **Notodontidae:** *Anaphe aranus, A. caenus, A. venata, Cerurina marshalli.* **Nymphalidae:** *Cymothoe aramis.* **Papilionidae:** *Papilio* sp. **Saturniidae:** *Imbrasia epimethea, I. obscura, I. truncata, Pseudantheraea discrepans.* **Sphingidae:** *Coelonia fulvinotata, Lophostethus demolini.*

Hymenoptera Formicidae: *Pachycondyla tarsata.*

Democratic Republic of Congo (DRC)

Odonata Libellulidae: *Trithemis arteriosa, Cyrtacanthacris septemfasciata.*

Orthoptera Acrididae: *Homoxyrrhepes punctipennis.* **Gryllidae:** *Brachytrupes membranaceus.* **Titigoniidae:** *Ruspolia differens.*

Isoptera Termitidae: *Macrotermes bellicosus, M. muelleri, M. natalensis, M.* spp., *Megagnathotermes katangensis, Odontotermes* spp., *Pseudacanthotermes militalis, P. spiniger, Termes gabonensis, Trinervitermes* sp.

Hemiptera-Homoptera Cicadidae: *Afzeliada afzelii, A. duplex, Ioba horizontalis, I. leopardina, Munza furva, Sadaka radiate, Ugada limbalis, U. limbimaculata.*

Coleoptera Cerambycidae: *Pycnopsis brachyptera, Sternotomis itzingeri katangensis, Zographus aulicus.* **Curculionidae:** *Rhynchophorus phoenicis,* **Dynastidae:** *Augosoma centaurus, Oryctes boas, O. owariensis.* **Ditiscidae:** *Cybister distinctus.* **Scarabaeidae:** *Gnathocera* sp., *Goliathus* sp., *Platygenia barbata.*

Lepidoptera Hesperiidae: *Coeliades libeon.* **Lasiocampidae:** *Gonometa* sp. **Lymantriidae:** *Rhypopteryx poecilanthes.* **Noctuidae:** *Nyodes prasinodes, Prodenia* sp. **Notodontidae:** *Acherontia atropus, Anaphe gribodoi, A. panda, A.* spp., *Antheua insignata, A.* sp., *Drapetides uniformis, Elaphrodes lacteal, Rhenea mediata.* **Nymphalidae:** *Cymothoe caenis.* **Psychidae:** *Eumeta cervina, Eumeta rougeoti.* **Saturniidae:** *Argemia* sp., *Athletes gigas, A. semialba, Bunaea* *alcinoe, Bunaeopsis aurantiaca, Bunaeopsis* sp., *Cinabra hyperbius, Cirina forda, Gonimbrasia hecate, G. richelmanni, G. zambesina, Goodia kunzei, Gynanisa maja, Imbrasia alopia, I. anthina, I. dione, I. eblis, I. epimethea, I. macrothyris, I. melanops, I. nictitans, I. obscura, I. oyemensis, I. petiveri, I. rectilineata, I. rhodina, I. rubra, I. truncata, I. tyrrhea, I. whalbergii, Lobobunaea goodie, L. phaedusa, L. saturnus, Melanocera parva, Micragone ansorgei, Microgone cana, M. herilla, Nudaurelia authina, Nyodes prasinodes, Pseudantheraea arnobia, P. discrepans, Saturnia* sp., *Tagoropsis flavinata, T. natalensis, Urota sinope, Usta Terpsichore.* **Sphingidae:** *Acherontia Atropos, Hippotion eson, Platysphinx* sp.

Hymenoptera Apidae: *Apis adansoni, A. mellifera, A. mellifica* var. *adansoni, Apotrigona* ref. *ferruginea, A.* ref. *komiensis, Axestotrigona simpsoni, Dactylurina staudingeri, Hypotrigona araujor, Trigona* (=*Hypotrigona*) *braunsi, T.* (=*Axestotrigona*) *erythra interposita, T. gribodoi, T. lendliana, T. occidentalis, T. richardsi, T.* (*Hypotrigona*) spp. **Eumenidae:** *Synagris* sp. **Formicidae:** *Carebara vidua, Oecophylla smaragdina longinoda, Sternotornis* sp. **Sphecidae:** *Sceliphron* sp. **Vespidae:** *Synagris* sp.

People's Republic of the Congo (PRC)

Orthoptera Acrididae: *Acanthacris ruficornis, Afroxyrrhepes procera, Cantatops spissus, Chirista compta, Cyrtacanthacris septemfasciata, Gastrimargus africanus, Heteracris guineensis, Locusta migratoria, L. migratoria migratorioides, Ornithacris turbida, Oxycantatops congoensis, O. spissus, Phymateus viridipes, Schistocerca gregaria.* **Gryllidae:** *Ampe* sp., *Brachytrupes membranaceus.*

Isoptera Termitidae: *Macrotermes bellicosus.*

Hemiptera-Homoptera Cicadidae: *Afzeliada* sp., *Platypleura adouma, Ugada giovanninae, U. limbata, U. limbimaculata.*

Hemiptera-Heteroptera Belostomatidae: *Lethocerus* spp.

Coleoptera Curculionidae: *Rhynchophorus phoenicis,* **Dynastidae:** *Augosoma centaurus, Oryctes boas, O. owariensis.*

Lepidoptera Hesperiidae: *Coeliades libeon.* **Notodontidae:** *Anaphe* spp. **Saturniidae:** *Cirina forda, Imbrasia epimethea, I. ertli, I. obscura, I. oyemensis, I. truncata, Lobobunaea phaedusa, Pseudantheraea discrepans.*

Hymenoptera Formicidae: *Oecophylla* spp.

Côte d'Ivoire

Isoptera Termitidae: *Bellicositermes* sp.

Gabon

Coleoptera Cerambycidae: *Ancylonotus tribulus, Petrognatha gigas.* **Ditiscidae:** *Cybister insignis.*

Lepidoptera Saturniidae: *Anthocera teffraria, A.* spp., *Bunaea alcinoe, Melanocera menippe, Saturnia marchii, Urota Sinope.*

Ghana

Coleoptera Curculionidae: *Rhynchophorus phoenicis.* **Dynastidae:** *Oryctes owariensis.*

Guinea

Coleoptera Dynastidae: *Oryctes* spp.

Kenya

Isoptera Termitidae: *Microtermes* spp., *Odontotermes* spp.

Diptera Chaobridae: *Chaoborus edulis.*

Lepidoptera Cossidae: *Xyleutes capensis.*

Apidae: *A. mellifera.* **Megachilidae:** *Chalicodema* sp., *Halictus* sp., *Osmia* sp.

Madagascar

Orthoptera Acrididae: *Cyrtacanthacris septemfasciata,* Locusta cernensis, L. migratoria capito.*

Hemiptera-Homoptera Flatidae: *Phromnia rubra, Flugoridae: Zana tenebrosa.*

Hemiptera-Heteroptera Nepidae: *Nepa* spp.

Coleoptera Carabidae: *Scarites* sp. **Cerambycidae:** *Batocera rufomaculata, Megopis mutica.* **Cicindellidae:** *Proagsternus* sp. **Curculionidae:** *Aphiocephalus limbatus, Eugnoristus monachus, Rhyna* sp., *Rhynchophorus* spp. **Dynastidae:** *Oryctes nasicornis.* **Ditiscidae:** *Cybister hova, C. operosus, Rhantus consimilis.* **Lucanidae:** *Cladognathus serricornis, Prosopocoilus serricornis.* **Scarabeidae:** *Proagosternus* sp., *Tricholespis* sp.

Lepidoptera Lasiocampidae: *Borocera madagascariensis, Cnethocampa diegoi, Livethra cajani, Rombyx radama.* **Psychidae:** *Deborrea malagassa.* **Saturniidae:** *Antherina suraka, Tagoropsis* sp.

Hymenoptera Apidae: *Trigona madecassa.* **Polistidae:** *Polistes hebraeus.*

Malawi

Hymenoptera Formicidae: *Carebara vidua.*

Ephemeroptera Caenidae: *Caenis kungu, Povilia adusta.*

Orthoptera Acrididae: *Acanthacris ruficornis, Cyrtacanthacris aeruginosa.* **Tettiginiidae:** *Ruspolia differens.*

Diptera Chaobridae: *Chaoborus edulis.*

Hemiptera-Homoptera Cicadidae: *Ioba* sp., *Monomotapa* sp., *Orapa* sp., *Platypleura* sp., *Pycna* sp.

Hemiptera-Heteroptera Nepidae: *Sphaerocoris* sp. **Pentatomidae:** *Nezara robusta.*

Scutelleridae: *Sphaerocoris* sp.

Lepidoptera Saturniidae: *Gonimbrasia belina, Gynanisa maja.*

Mali

Lepidoptera Saturniidae: *Cirina forda butyrospermi.*

Mauritius

Hymenoptera Apidae: *Apis unicolor unicolor.*

Mozambique

Orthoptera Acrididae: *Zonocerus elegans.*

Coleoptera Cerambycidae: *Stenodontes (=Mallodon) downesi.* **Curculionidae:** *Rhynchophorus phoenicis.*

Lepidoptera Saturniidae: *Bunaea alcinoe.*

Hymenoptera Formicidae: *Pheiodole* sp.

Namibia

Lepidoptera Saturniidae: *Cirina forda, Gonimbrasia belina, Gynanisa maja, Heniocha apollonia, H. dyops, H. marnois, Imbrasia tyrrhea, Pseudobunaea irius, Rohaniella pygmaea, Usta wallengrenii.*

Niger

Orthoptera Acrididae: *Acridoderes strenuus, Acrotylus blondelli, A. longipes, Anacridium burri, Anacridium melanorhodon, A. wernerellum, Catantops axillaris, C. haemorrhoidalis, C. stylifer, Diabolocatantops axillaris, Gastrimargus africanus, G. proceus, Humbe tenuicornis, Orthacanthacris humilicrus.*

Nigeria

Orthoptera Acrididae: *Cyrtacanthacris aeruginosus unicolor, Zonocerus variegates.* **Gryllide:** *Brachytrupes membranaceus, B.* sp.

Isoptera Termitidae: *Macrotermes bellicosus, M. natalensis.*

Coleoptera Cerambycidae: *Analeptes trifasciata.* **Curculionidae:** *Rhynchophorus phoenicis.* **Dynastidae:** *Oryctes boas.*

Lepidoptera Notodontidae: *Anaphe panda, A. reticulata, A. venata.* **Saturniidae:** *Cirina forda.*

Hymenoptera Apidae: *Apis mellifera.*

Rwanda

Orthoptera Acrididae: *Acanthacris ruficornis.*

Sahel (Sub-Saharan Africa)

Orthoptera Acrididae: *Acanthacris ruficornis, Cataloipus fuscocoeruleipes.*

Sao Tome and Principe

Coleoptera Cerambycidae: *Macrotoma edulis.*

Senegal

Coleoptera Cerambycidae: *Ancylonotus tribulus, Petrognatha gigas.* **Ditiscidae:** *Cybister distinctus.* **Hydrophilidae:** *Hydrophilus marginatus, H. senegalensis.*

Hymenoptera Apidae: *Apis mellifera, Trigona ferruginea gambiensis, T. occidentalis, T. ruspolii, T. senegalensis.*

Sierra Leone

Coleoptera Curculionidae: *Rhynchophorus phoenicis.* **Dynastidae:** *Oryctes boas.* **Ditiscidae:** *Cybister distinctus.*

Republic of South Africa (RSA)

Orthotera Acrididae: *Cyrtacanthacris septemfasciata, Locusta migratria, L. pardalina, Locustana pardalina, Schistocerca* spp., *Zonocerus elegans.*

Isoptera Termitidae: *Macrotermes natalensis, M. swaziae, Odontotermes badius, Termes capencis, T. fatale.* **Hodotermitidae:** *Hodotermes* sp., *Microhodotermes viator.*

Hemiptera-Heteroptera Pentatomidae: *Euchosternum delegorguei.*

Coleoptera Buprestidae: *Sternocera orissa.* **Carabidae:** *Thrincopyge alacris.* **Cerambycidae:** *Plocaederus frenatus, Stenodontes (Mallodon) downesi.* **Curculionidae:** *Polyclaeis equestris, P. plumbeus.* **Dynastidae:** *Oryctes boas, O. monoceros, O. owariensis.*

Lepidoptera Lasiocampidae: *Bombycomorpha pallida, Gonometa postica.* **Psychidae:** *Eumeta cervina.* **Saturniidae:** *Argema mimosae, Bunaea alcinoe, Cirina forda, C. similis, Gonimbrasia belina, G. zambesina, Gynanisa maja, Microgone cana, Urota sinope.* **Sphingidae:** *Herse convolbuli.*

Hymenoptera Apidae: *Trigona togoensis* var. *junodi.* **Formicidae:** *Carebara vidua.*

Sudan

Hemiptera-Homoptera Cicadidae: *Agonoscelis versicolor.*

Hemiptera-Heteroptera Pentatomidae: *Agonoscelis pubescens, A. versicolor.*

Tanzania

Orthoptera Acrididae: *Cyrtacanthacris septemfasciata, Locusta migratoria migratorioides, Schistocerca gregaria.* **Gryllidae:** *Brachytrupes membranaceus.* **Tettigoniidae:** *Ruspolia differens.*

Isoptera Termitidae: *Acanthotermes acanthothorax, Macrotermes* spp., *Pseudacanthotermes militalis, P.* spp., *Termes* spp.

Coleoptera Curculionidae: *Sipalinus aloysii-sabaudiae.*

Diptera Chaobridae: *Chaoborus edulis.*

Lepidoptera Cossidae: *Xyleutes capensis.* **Notodontidae:** *Anaphe panda.* **Saturniidae:** *Bunaea alcinoe, B. aslauga.*

Hymenoptera Apidae: *Apis adansoni, Apis mellifera, A. mellifica* var. *adansoni, Trigona erythra togoensis.* **Formicidae:** *Oecophylla smargdina longinoda.* **Megachilidae:** *Chalicodema* sp., *Halictus* sp., *Osmia* sp.

Uganda

Orthoptera Acrididae: *Locusta migratoria migratorioides, Schistocerca gregaria, Cyrtacanthacris septemfasciata.* **Gryllidae:** *Brachytrupes membranaceus, Gryllus bimaculatus.* **Gryllotalpidae:** *Gryllotalpa africana.* **Tettigoniidae:** *Ruspolia differens, R. nidula, R. viridulus.*

Isoptera Termitidae: *Acanthotermes acanthothorax, Macrotermes falciger, Macrotermes subhyalinus, Pseudacanthotermes militalis, P. spiniger.*

Coleoptera Curculionidae: *Rhynchophorus phoenicis, Rhynchophorus* spp.

Diptera Chaoboridae: *Chaoborus (=Neochaoborus) anomalus, C. edulis, C. pallidipes, Procladius umbrosus, Tanypus guttatipennis.*

Zambia

Orthoptera Acrididae: *Acanthacris ruficornis, Acorypha nigrovariegata, Acrida sulphuripennis, Afroxyrrhepes* sp., *Cardeniopsis guttatus, Caredeniopsis nigropunctatus, Catantops ornatus, C.* sp., *C. tatarica, Cyathosternum* spp., *Cyrtacanthacris aeruginosa, C. septemfasciata, C. tatarica, Locusta migratoria, I. migratoria migratorioides, L. pardalina, Oedaleus nigrofasciatus, Poecilocerastis tricolor, Schistocerca gregaria flaviventris.* **Gryllidae:** *Acheta* spp., *Brachytrupes membranaceus, Gryllus bimaculatus, Gymnogryllus* spp. **Tettigoniidae:** *Ruspolia differens.*

Isoptera Termitidae: *Macrotermes falciger, M. mossambicus, M. subhyalinus, M. vitrialatus, Odontotermes badius, Pseudacanthotermes spiniger.*

Hemiptera-Homoptera Cicadidae: *Ioba leopardina, Platypleura stridula, Ugada limbalis.*

Coleoptera Cerambycidae: *Acanthophorus capensis, A. confinis, A. maculatus.* **Scarabaeidae:** *Pachylomera fermoralis.*

Lepidoptera Lasiocampidae: *Bombycomorpha pallida, Catalebeda jamesoni, Gonometa postica, Mimopacha* aff. *Knoblauchi.* **Limacodidae:** *Hadraphe ethiopica.* **Noctuidae:** *Busseola fusca, Euxoa segetis, Helicoverpa armigera, Heliothis obsoleta, Sphingomorpha chlorea, Spodoptera exempta, S. exigua.* **Notodontidae:** *Anaphe panda, A. venata, Desmeocraea* sp., *Elaphrodes lacteal.* **Saturniidae:** *Anthocera zambezina, Bunaea alcinoe, Bunaeopsis aurantiaca,*

Bunaeopsis sp., *Cinabra hyperbius, Cirina forda, Gonimbrasia belina, G. zambesina, Gynanisa maja, Holocerina agomensis, Imbrasia cytherea, I. dione, I. epimethea, I. rubra, Lobobunaea christyi, L. saturnus, Melanocera parva, Micragone ansorgei, Pseudantheraea discrepans*. **Sphingidae:** *Herse convolbuli, Nephele comma*.

Hymenoptera Apidae: *Apis adansoni, A. mellifera capensis, Bombinae* spp. **Formicidae:** *Carebara vidua, Pheiodole* sp.

Zimbabwe

Orthoptera Acrididae: *Acanthacris ruficornis, Acrida bicolor, Cyathosternum* spp., *Cyrtacanthacris septemfasciata, Locusta migratoria, L. migratoria migratorioides, Ornithacris cyanea (magnifica), O.* sp., *Truxaloides constrictus*. **Gryllidae:** *Acheta* spp., *Brachytrupes membranaceus*. **Gryllotalpidae:** *Gryllotalpa africana*. **Tettigoniidae:** *Ruspolia differens*.

Isoptera Termitidae: *Macrotermes falciger, M. mossambicus, M. natalensis, M. subhyalinus, Macrotermes* spp., *Odontotermes* spp., *Platypleura quadraticollis*.

Hemiptera-Homoptera Cicadidae: *Ioba leopardina*. **Psyllidae:** *Arytaira mopane*.

Hemiptera-Heteroptera Coreidae: *Pe(n)tascelis remipes, P. wahlbergi*. **Pentatomidae:** *Euchosternum delegorguei, E. pallidus*.

Coleoptera Buprestidae: *Sternocera funebris, Sternocera orissa*. **Scarabaeidae:** *Euchloropus laetus, Eulepida anatine, E. anatine, E. nitidicollis*.

Lepidoptera Lasiocampidae: *Bombycomorpha pallida, Brachiostegia* sp., *Gonometa postica, G. robusta*. **Notodontidae:** *Anaphe panda*. **Saturniidae:** *Anthocera zambezina, Athletes gigas, A. semialba, Bunaea alcinoe, B.* sp., *Bunaeopsis aurantiaca, Cirina forda, Gonimbrasia belina, Goodia kunzei, Gynanisa maja, Heniocha dyops, Imbrasia epimethea, I. ertli, Lobobunaea saturnus, Pseudobunaea irius*. **Sphingidae:** *Herse convolbuli*.

Hymenoptera Formicidae: *Carebara vidua, Pheiodole* sp.

7. The Middle and Near East

Algeria
Orthoptera Acrididae: *Acridium peregrinum*.

Egypt
Orthoptera Acrididae: *Gryllus aegyptius, G. lineola, G. locust, G. tataricus*.

Hemiptera-Heteroptera Corixidae: *Corixa esculenta*.

Coleoptera Scarabaeidae: *Ateuches sacer, Scarabaeus sacer*. **Tenebrionidae:** *Blaps sulcate*.

Hymenoptera Apidae: *Apis unicolor fasciata*.

Arab

Hymenoptera Apidae: *Apis unicolor fasciata.*

Iran (Persia)

Orthoptera Acrididae: *Locusta persarum.*

Hemiptera-Homoptera Pseudococcidae: *Najacoccus sepentinus, Trabutina* sp.

Coleoptera Curculionidae: *Larinus mellificus, L. onopordi, L. syriacus.*

Iraq

Orthoptera Acrididae: *Schistocerca gregaria.*

Hemiptera-Homoptera Cicadellidae: *Euscelis decorates, Opsius jucundus.* **Pseudococcidae:** *Najacoccus serpentinus, Trabutina mannipara.*

Coleoptera Curculionidae: *Larinus mellificus, L. onopordi, L. syriacus.*

Israel

Hemiptera-Homoptera Pseudococcidae: *Trabutina mannipara.*

Coleoptera Curculionidae: *Larinus rudicollis.*

Kuwait

Orthoptera Acrididae: *Cyrtacanthacris septemfasciata.*

Levant

Hymenoptera Cynipidae: *Aulacides levantina, Cynips rosae.*

Morocco

Hemiptera-Homoptera Dactylopiidae: *Dactylopius coccus.*

Coleoptera Cerambycidae: *Cerambyx cerdo, Cyrtognathus forficatus, Dorysthenus forficatus.* **Meloidae:** *Alosimus tenuicornis.*

Orthoptera Acrididae: *Acridium peregrinum, Schistocerca gregaria.*

Saudi Arabia

Orthoptera Acrididae: *Cyrtacanthacris septemfasciata, Locusta gregaria.*

Coleoptera Tenebrionidae: *Blaps sulcata, Tenebrio* sp., *Ulomoides dermestoides.*

Syria

Coleoptera Cuculionidae: *Larinus rudicollis.*

Tunisia

Coleoptera Tenebrionidae: *Blaps sulcate.*

Turkey

Coleoptera Ditiscidae: *Dytiscus marginalis.* **Hydrophilidae:** *Hydrophilus piceus.* **Tenebrionidae:** *Blaps sulcata, B.* sp., *Pimelia* sp., *Tenebrio* sp. **Hymenoptera Cynipidae:** *Aulacides levantina.*

8. Oceania

Australia

Orthoptera Acrididae: *Chortoicetes terminifera.* **Gryllidae:** *Teleogryllus commodus.*

Blattaria Blattellidae: *Cosmozosteria* sp.

Anoplura Pediculidae: *Pediculus humanus.*

Hemiptera-Homoptera Pseudococcidae: *Apiomorpha pomiformis, Cystococcus echiniformis, C. pomiformis, C.* sp. **Psyllidae:** *Austrotachardia acacia, Eucalyptolyma* sp., *Glycaspis* spp., *Spondyliaspis eucalypti.*

Coleoptera Carabidae: *Euryscaphus* sp. **Cerambycidae:** *Agrianome spinicollis, Appectrogastra flavipilis, Bardistus cibarius, Cnemoplites edulis, C. flavipilis, Eurynassa australis, E. odewahni, Mnemopulis edulis, Paroplites australis.* **Scarabaeidae:** *Anoplognathus viridiaeneus.*

Lepidoptera Cossidae: *Catoxophylla cyanauges, Cossus* sp., *Endoxyla n.* sp., *Xyleutes amphiplecta, X. biarpiti, X. boisduvali, X. eucalypti, Xyleutes leuchomochla, Zeuzera citurata, Zeuzera eucalypti.* **Danaidae:** *Euploea hamata.* **Eupterotidae:** *Ochrogaster lunifer, Panacela* sp. **Hepialidae:** *Abantiades marcidus, Oxycanus* sp., *Strigops grandis, Trictena argentata, T. argyrosticha, T. atripalpis, Zelotypia stacyi.* **Noctuidae:** *Agrotis infusa.* **Sphingidae:** *Coenotes eremophilae, Hyles lineata livornicoides.*

Hymenoptera Apidae: *Trigona carbonaria, T. cassiae, T. (Hypotrigona)* spp. **Formicidae:** *Camponotus consobrinus, C. inflatus, Colobopsis grasseri, Leptomyrmex varians, Melophorus bagoti, M. cowleyi, M. midas, Myrmecia pyriformis, Myrmecia sanguinea, Oecophylla smaragdina, O. smaragdina virescens, Plagiolepis* sp. **Perilampidae:** *Trachilogastir* sp.

Papua New Guinea (PNG)

Orthoptera Acrididae: *Locusta migratoria, Valanga irregularis, Valanga* sp. **Gryllidae:** *Metioche* sp., *Teleogryllus comma, T.* spp. **Gryllotalpidae:** *Gryllotalpa africana, G.* spp. **Tettigoniidae:** *Caedicina* sp.

Phasmida Phasmatidae: *Eurycantha horrida, Extatosoma tiaratum.*

Mantodea Mantidae: *Hierodula sternosticta, H.* spp., *Tenodera* sp.

Anoplura Pediculidae: *Pediculus humanus.*

Hemiptera-Homopera Cicadidae: *Baeturia* sp., *Diceropyga* sp. **Pseudococcidae:** *Apiomorpha pomiformis.*

Hemiptera-Heteroptera Coreidae: *Mictis* sp.

Coleoptera Cerambycidae: *Batocera* spp., *Dihamnus* spp., *Haplocerambyx severus.* **Curculionidae:** *Rhnychophorus ferrugineus, R. ferrugineus papuanus, R. schach, R.* spp. **Dynastidae:** *Oryctes centaurus, O. rhinoceros, O.* spp., *Scapanes* sp., *Xylotrupes gideon.* **Scarabaeidae:** *Lepidiota vogeli.*

Lepidoptera Papilionidae: *Papilio lagleizei.*

Hymenoptera Formicidae: *Oecophylla smaragdina, O. virescens.*

New Caledonia
Coleoptera Cerambycidae: *Aplagiognathus costata.*

New Zealand
Coleoptera Cerambycidae: *Prionoplus reticularis.* **Scarabaeidae:** *Pyronota festiva.*

Lepidoptera Hepialidae: *Aenetus virescens.*

Hymenoptera Vespidae: *Vespula germanica.*

Solomon Islands
Coleoptera Dynastidae: *Oryctes rhinoceros.*

9. Asia

Cambodia
Coleoptera Ditiscidae: *Cybister* sp. **Hydrophilidae:** *Hydrophilus (Dytiscus) hastatus.*

China
Thysanura Lepismatidae: *Ctenolepisma villosa, Lepisma saccharina.*

Ephemeroptera Ephemerellidae: *Ephemerella jianghongensis.*

Odonata Aeschnidae: *Anax parthenope Julius;* **Gomphidae:** *Gomphus cuneatus;* **Lestidae:** *Lestes praemorsa, Crocothemis servilia;* **Libellulidae:** *Orthetrum albistylum, O. triangulare melania, Pantala flavescens, Sympetrum uniforme.*

Orthoptera Acrididae: *Acrida chinensis, Myrmecia sanguinea, A. oxycephala, A. turrita, Arcyptera fusca, Atractomorpha psittacina, A. sinensis, Bryodema gebleri, Calliptamus abbreviates, C. barbarous cephalotes, C. italicus, Ceraeri kiangsu, Chondacris rosea, Dociostaurus kraussi nigrogeniculatus, Euprepocnemis shirakii, Gomphocerus sibiricus, Locusta migratoria, L. migratoria manilensis, Oedaleus decorus, Oxya chinensis, O. intricata, O. japonica japonica, Pararcyptera microptera, Patanga japonica, Rammeacris kiangsu, Sphingonotus* spp., *Stauroderus scalaris.*

Gryllidae: *Brachytrupes portentosus, Gryllus bimaculatus, G. chinensis, G. testaceus, Loxoblemmus doenitzi, Scapsipedus aspersus, Teleogryllus derelictus, T. taiwanemma, Velarifictorus asperses.* **Gryllotalpidae:** *Gryllotalpa africana, G. formosana, G. gryllotalpa, G. orientalis, G. unispina, G.* spp. **Tettigoniidae: Damalacantha vacca sinica,** *Mecopoda elongate.*

Isoptera Rhinotermitidae: *Coptotermes formosanus.* **Termitidae:** *Macrotermes acrocephalus, M. annandalei, M. barneyi, M. denticulatus, M. jinhongensis, M. menglongensis, M. yunnanensis, Odontotermes angustignathus, O. annulicornis, O. conignathus, O. faveafrons, O. formosanus, O. gravelyi, O. hainanensis, O. yunnanensis.*

Blattaria Blattellidae: *Blatella germanica.* **Blattidae:** *Blatta orientalis, Periplaneta americana, Periplaneta australasiae, P. fuliginosa.* **Corydiidae:** *Eupolyphaga sinensis, Polyphga plancyi.* **Phyllodromiidae:** *Opisthoplatia orientalis.*

Mantodea Mantidae: *Hierodula patellifera, H. saussurei, Mantis religiosa, Paratempdera angustipennis, P. sinensis, Statilia maculata, Tenodera aridifolia, T. capitata, T. sinensis.*

Hemiptera-Homoptera Asterolecaniidae: *Asterolecanium bambusae.* **Cicadidae:** *Cicada flammata, Cryptotympana atrata, C. facialis, Dundubia intemerata, Graptopsaltria tienta, Huechys sanguinea, H. thoracica, Oncotympana maculaticollis, Platypleura kaempferi.* **Coccidae:** *Ericerus pela.* **Eriosomatida:** *Melaphis chinensis, Nurudae rosea, N. shiraii, N. sinica.* **Flatidae:** *Flata limbata, Lawana imitata.* **Fulgoridae:** *Lycorma delicatula.* **Lacciferidae:** *Lccifer lacca.* **Membracidae:** *Darthula hardwicki.* **Pseudococcidae:** *Phenacoccus prunicola.*

Hemiptera-Heteroptera Belostomatidae: *Kirkaldgia degrollei, Lethocerus indicus, Sphaerodema rustica.* **Coreidae:** *Dicranocephalus wallichi bowringi, Mictis tenebrosa.* **Corixidae:** *Micromecta quadriseta, Sigara substriata.* **Naucoridae:** *Sphaerodema rustica.* **Notonectidae:** *Anisops fieberi, Enithares sinica, Notonecta chinensis.* **Pentatomidae:** *Aspongopus chinensis, Coridicus chinensis, Cyclopelta parva, Erthesina fullo, Eurostus validus, Eusthenes curpreus, E. saevus, Nezara antennata, N. viridula, Tessaratoma papillosa.*

Coleoptera Anobiidae: *Lasioderma serricorne.* **Bruchidae:** *Bruchus pisorum, B. rufimanus.* **Buprestidae:** *Buprestis* sp., *Chalcophora yunnana, Coraebus sidae, C. sauteri, Sphenoptera kozlovi, Sternocera aequisignata.* **Cerambycidae:** *Anoplophora chinensis, A. malasiaca, A. nobilis, Apriona germari japonica, Aromia bungii, Batocera horsfieldi, Psacothea hilaris, Purpricenus temminekii, Stromatium longicorne, Xylorhiza* sp. **Chrysomelidae:** *Sagra femorata purpurea.* **Curculionidae:** *Cyrtotrachelus buqueti borealis, C. longimanus, C. rufopectinipes birmanicus, Macrochirus longipes, Otidognathus davidis, Rhynchophorus (=Calandra) chinensis.* **Dynastidae:** *Allomyrina dichotoma, Oryctes rhinoceros.* **Ditiscidae:** *Acilius* sp., *Cybister bengalensis, C. flavocinctus, C. frimbiolatus, C. guerini, C. japonicus, C. lewisianus, C. occidentalis,*

C. singulatus, C. sticticus, C. sugillatus, C. tripunctatus, C. tripunctatus asiaticus, C. tripunctatus orientalis, C. ventralis, Dytiscus habilis, D. marginalis, D. validus, D. sp., *Eretes sticticus, Gaurodytes fulvipennis, Platynectes guttula.* **Hydrophilidae:** *Hydrophilus acuminatus, H. bilineatus, H. bilineatus cashimirensis, H. (Stethoxus) cavisternus, H. (Dytiscus) hastatus.* **Ipidae:** *Sphaerotrypes yunnanensis, Xyleborus emarginatus.* **Meloidae:** *Epicauta gorhami, Lytta caraganae, Myrabris cichorii, M. phalerata.* **Scarabaeidae:** *Anomala corpulenta, A. cuprea, A. exoleta, Catharsius molossus, Dicranocephalus wallichi, Gymnopleurus mopsus, Heliocopris bucephalus, Holotrichia diomphalia, H. diomphilia, H. lata, H. morosa, H. oblita, H. ovata, H. parallela, H. sinensis, H. srobiculata, H. szechuanensis, H. titanis, Liocola brevitarsis, Oxycetonia jucunda, Polyphylla laticollis, Protaetia aerate, Scarabaeus molossus, S. perigrinus, Trematodes tenebrioides.* **Scolitidae:** *Spaerotrypes yunnanensis, Tomicus piniperda, Xyleborus emarginatus.* **Tenebrionidae:** *Blaps rhynchoptera, Palembus dermestoides, Tenebrio molitor, T. obscurus, Tribolium confusum.*

Neuroptera Corydalidae: *Acanthacorydalis orientalis.*

Diptera Calliphoridae: *Calliphora megacephala.* **Hyppoboscidae:** *Hyppobosca longipennis.* **Muscidae:** *Musca domestica.* **Tabanidae:** *Atylotus horvathi, Tabanus amaenus, T. budda, T. kiansuensis, T. kinoshitai, T. mandarinus, T. rubidus, T. yao.* **Tipulidae:** *Tipula paludosa.*

Lepidoptera Bombycidae: *Andraca bipunctata, Bombyx mandarina, Bombyx mori, Rondotia menciana.* **Carposinidae:** *Carposina niponensis.* **Cossidae:** *Cossus chinensis, C. cossus, C. hunanensis, Holcocerus vicarius.* **Gelechiidae:** *Pectinophora gossypiella.* **Geometridae:** *Biston marginata,* **Hepialidae:** *Endoclyta sinensis, Hepialus albipictus, H. altaicola, H. armoricanus, H. baimaensis, H. cingulatus, H. deqinensis, H. deudi, H. dongyuensis, H. ferrugineus, H. ganna, H. gonggaensis, H. jinshaensis, H. kangdingensis, H. kangdingroides, H. lijiangensis, H. litangensis, H. luquensis, Hepialus macilentus, H. markamensis, H. meiliensis, H. menyuanicus, H. nebulosus, H. oblifurcus, H. pratensis, H. renzhiensis, H. sichuanus, H. varians, H. xunhuaensis, H. yeriensis, H. yuloangensis, H. yunlongensis, H. yunnanensis, H. yushuensis, H. zhangmoensis, H. zhayuensis, H. zhongzhiensis, Napialus hunanensis.* **Hesperiidae:** *Erionata torus, Parnara guttata guttata.* **Lasiocampidae:** *Dendrolimus houi, D. kikuchii, D. punctatus, D. punctatus wenshanensis, D. superans.* **Limacodidae:** *Cania bilineata, Cnidocampa flavescens, Thosea sinensis.* **Noctuidae:** *Agrotis ypsilon, Anomis flava, Hydrillodes morosa, Naranga aenescens, Sesamia inferens, Spodoptera exigua, S. litura.* **Notodontidae:** *Leucodonta bicoloria, Notodonta dembowskii, Phalera assimilis, P. bucephala, Semidonta biloba.* **Papilionidae:** *Papilio machaon, P. polytes, Papilio xuthus.* **Pieridae:** *Pieris rapae.* **Pyralidae:** *Aglossa dimidiata, Chilo luteellus, C.* sp., *Cnaphalocrocis medinalis, Dichocrocis punctiferalis, Omphisa fuscidentalis, Ostrinia fumalis, O. furnacalis, Plodia interpunctlla, Proceras venosatum, Proceras venosatum, Sylepta derogate, Tryporyza incertulas.* **Saturniidae:** *Antheraea pernyi,*

Saturnia pyretorum. **Satyridae:** *Mycalesis gotoma.* **Sesiidae:** *Paranthrene regalis, P. tabaniformis.* **Sphingidae:** *Clanis bilineata, C. bilineata tsingtauica, C. deucalion, Herse convolbuli, Smerinthus planus, Sphinx* spp., *Theretra oldenlandiae.* **Tortricidae:** *Leguminivora glycinivorella.* **Xyloryctidae:** *Linoclostis gonatias.*

Hymenoptera Agaonidae: *Blastoophaga pumilae.* **Apidae:** *Apis cerana, Apis dorsata, Apis florea, A. mellifera.* **Bombidae:** *Bombus speciosus, B.* spp. **Formicidae:** *Camponotus japonicas, Carebara lignata.* **Formicidae:** *Formica aquilonia, F. beijingensis, F. fusca, F. japonica, F. rufa, F. sanguinea, F. uralensis, F. yessensis, Lasius flavus, Oecophylla smaragdina, Polyrhachis dives, P. illaudata, P. lamellidens, P. mayri, P. vicina, Tetramorium caespitum.* **Megachilidae:** *Megachile* spp. **Polistidae:** *Polistes chinensis, Polistes gigas, P. hebraeus, P. mandarinus, P. sagittarius, P. sulcatus.* **Sphecidae:** *Ammophila infesta.* **Vespidae:** *Dolichovespula flora, Provespa barthelemyi, Vespa analis, V. auraria, V. basalis, V. bicolor bicolor, V. crabro, V. ducalis, V. magnifica, V. mandarinia, V. soror, V. tropica, V. variabilis, V. velutina, V. velutina auraria, V.* spp. **Xylocopidae:** *Xylocopa violacea.*

India

Ephemeroptera Baetidae: *Cloeon kimminsi;* **Ephemeridae:** *Ephemera danica.*

Odonata Aeschnidae: *Aeschna mixta, A.* spp.; **Coenagrionidae:** *Enallagma* sp.; **Cordulegasteridae:** *Cordulegaster* sp.; **Gomphidae:** *Ictinogomphus rapax, Stylurus* sp.; **Libellulidae:** *Acisoma parnorpaides, Brachythemis contaminata, Diplacodes trivialis, D.* sp., *Libellula carolona, Neurothemis fluctuans, Pachydiplax* sp., *Sympetrum* sp., *Urothemis* sp.

Plecoptera Nemouridae: *Nemoura* sp. **Pteronarcidae:** *Pteronarcys dorsata.*

Orthoptera Acrididae: *Acrida exaltata, A. gigantea, Acridium melanocorne, A. perigrinum, Catantops annexus, Ceracris nigricornis nigricornis, Chondacris rosea, Choroedocus robustus, Cyrtacanthacris tatarica, Diabolocanthops innotabilis, Hieroglyphodes assamensis, Hieroglyphus concolour, H. oryzivorus, H.* sp., *Leptysma marginicollis, Leptisma* sp., *Locusta mahrattarum, L. (=Gryllus?) onos, L.* sp., *Oxya* sp., *Patanga succincta, Phlaeoba antannata, Schistocerca gregaria, S.* spp. **Gryllidae:** *Brachytrupes portentosus, B.* sp., *Gryllodes melanocephalus, Gryllus assimilis, G. bimaculatus, G. campestris, G.* sp., *Tarbinskiellus orientalis.* **Gryllotalpidae:** *Gryllotalpa africana, G. gryllotalpa, G.* spp. **Raphidophoridae:** *Diestrammena* sp. **Schizodactylidae:** *Schizodactylus monstrosus, S. tuberculatus.* **Tettigoniidae:** *Arachnacris* sp., *Chloracris brullei, Conocephalus* spp., *Eucocephalus pallidus, Holochlora albida, H. indica, Lima corded, Mecopoda elongate, Microcentrum rhombifolium, M.* sp., *Neoconocephalus* sp., *Tettigonia* sp., *Thylotropides ditymus.*

Isoptera Rhinotermitidae: *Reticulitermes flavipes.* **Termitidae:** *Macrotermes* spp., *Microtermes obesus, Odontotermes feae, O. formosanus, O. obesus, O.* spp.

Blataria Blaberidae: *Gomphadorhina* sp. **Cryptocercidae:** *Cryptocercus* sp.

Mantodea Mantidae: *Hierodula coarctata, H. westwoodi, Mantis religiosa.*

Hemiptera-Homoptera Cicadidae: *Cicada pruinosa, C. verides, C.* spp., *Cryptotympana aquilla, Cyclochila virens, Dundubia spiculata, Euterphosia crowfooti, Pomponia linearis, Pycna repandar.*

Hemiptera-Heteroptera Alydidae: *Leptocorisa acuta.* **Belostomatidae:** *Lethocerus indicus, L.* spp., *Lohita grandis.* **Coreidae:** *Dalader acuticosta, Mictis tenebrosa.* **Gerridae:** *Gerris spinole, G.* sp. **Nepidae:** *Laccotrephes* sp., *Ranatra* sp. **Notonectidae:** *Anisops* spp. **Pentatomidae:** *Alcaerrhynchus grndis, Aspongopus nepalensis, Bagrada picta, Cyclopelta subhimalayensis, Dolycoris indicus, Erthesina fullo, Halyomorpha picus, Nezara viridula, Ochrophora montana, Tessaratoma quadrata.* **Pyrrhocoridae:** *Antilochus coqueberti.*

Coleoptera Anobiidae: *Anthrenus* sp., *Xestobium* sp. **Buprestidae:** *Sternocera sternicornis, Sternocera* sp. **Cerambycidae:** *Acanthopus serraticornis, Anoplophora* sp., *Aplosonyx chalybseus, Aristobia* sp., *Batocera parryi, B. roylei, B. rubra, B. rufomaculata, Coelosterma scabrator, C.* sp., *Diastocera wallichi, Dihamnus cervinus, Dorysthenes buqueti, Dorysthenes montanus, Glenea (Stiroglenis) obese, Hoplocerambyx spinicornis, Lamia 8-maculata, L. rubus, Macrotoma crenata, Monochamus versteegi, Neocerambyx paris, Nupserha fricator, Oplatocera* sp., *Stromatium barbatum, Xylotrechus quadripes, X. smei, Xysterocera globosa, Xystrocera* sp.

Chrysomeridae: *Aplosonyx chalybaeus.* **Curculionidae:** *Balaninus album, Cyrtotrachelus buqueti borealis, Rhynchophorus (=Calandra) chinensis, Rhnychophorus ferrugineus, R. phoenicis.* **Dynastidae:** *Allomyrina dichotoma, Megasoma elephas, Oryctes rhinoceros, O.* spp., *Xylotrupes gideon.* **Ditiscidae:** *Cybister* sp., *Dytiscus* sp., *Eretes sticticus, Laccophilus* sp. **Elateridae:** *Cardiophorus aequabilis.* **Hydrophilidae:** *Hydrophilus acuminatus, H. olivaceus, H.* sp. **Lucanidae:** *Calodes cuvera, C. siva, Cyclommatus albersi, Cyclommatus pehengenesis chiangmainesis, Dorcus* spp., *Eurytrachelus titanus, Hexathrius forsteri, Lucanus cantor, L. elaphus, L. laminifer, Odontolabris cuvera, O. gazelle, Prosopocoilus* sp. **Lyctidae:** *Meligethes aeneus.* **Meloidae:** *Myrabris cichorii, M. himalayaensis.* **Nitidulidae:** *Epuraea* sp. **Passalidae:** *Aceraius helferi, Aulacocyclus bicuspid, Odontotaenius* sp., *Passalus interruptus.* **Scarabaeidae:** *Agestrata* sp., *Anomala* spp., *Cassolus humeralis, Catharsius* sp., *Copris corpulentus, C. furciceps, C. puntatus, C. (Microcopris) vittalisi, C. (Paracopris)* sp. *Heliocopris bucephalus, Lepidiota stigma, L.* sp., *Onitis castaneus, O. feae, O. subopacus, Paraphytus hindu, Polyphylla* sp., *Propomacrus* sp. **Trictenotomidae:** *Trictenotoma* sp.

Lepidoptera Arctiidae: *Diacrisia oblique.* **Bombycidae:** *Bombyx mori.* **Gracillaridae:** *Stomphosistis thraustica.* **Lasiocampidae:** *Chatra grisea, Malacosoma* spp. **Saturniidae:** *Antheraea assamensis, A. mylitta, A. paphia, Samia Cynthia, Samia ricini.*

Hymenoptera Apidae: *Apis cerana indica, A. dorsata, A. florea, A. mellifera, A.* spp. **Eumenidae:** *Eumenes* sp. **Formicidae:** *Bothroponera rufipes, Crematogaster dohni, Myrmica rubra, Oecophylla smaragdina, Polistes stigmata, P.* spp. **Vespidae:** *Icaria artifex, Vespa bicolor bicolor, V. mandarinia, V. orientalis, V. tropica, V.* spp., *Vespula vulgaris.* **Xylocopidae:** *Xylocopa* sp.

Indonesia

Odonata Aeschnidae: *Anax* sp.; **Libellulide:** *Cratilla lineate assidua, Crocothemis servilia, C.* sp., *Neurothemis fluctuans, N. ramburii, N.* sp., *Orthecum glaucum, Orthetrum Sabina, Pantala flavescens, Potamarcha obscura, Trithemis aurora.*

Orthoptera Acrididae: *Acridium aerigonosum, Valanga nigricornis.* **Gryllidae:** *Brachytrupes portentosus.* **Gryllotalpidae:** *Gryllotalpa africana, G.* spp.

Phasmida Phasmatidae: *Extatosoma* spp.

Isoptera Termitidae: *Macrotermes* spp., *Termes atrox, T. destructor, T. fatale, T. mordax, T. sumatranum.*

Blattaria Panesthiidae: *Panesthia angustipennis.* **Blattidae:** *Polyzosteria* sp.

Anoplura Pediculidae: *Pediculus humanus.*

Hemiptera-Homoptera Cicadidae: *Cosmopsaltria waine, Pomponia merula.*

Hemiptera-Heteroptera Alydidae: *Leptocorisa acuta, L. oratorius.* **Coreidae:** *Stenocoris varicornis.* **Pentatomidae:** *Nezara viridula.* **Rhopalidae:** *Leptocoris acuta.*

Coleoptera Cerambycidae: *Batocera albofasciata, B. rubus, B. wallecei, B.* spp., *Dihamnus* spp., *Macrotoma* sp., *Osphryon* sp., *Rosenbergia mandibularis, Xixuthrus* sp. **Curculionidae:** *Behrensiellus glabratus, Rhnychophorus ferrugineus, R. ferrugineus papuanus.* **Dynastidae:** *Chalcosoma atlas, C. moellenkampi, Xylotrupes Gideon.* **Ditiscidae:** *Cybister guerini, C. tripunctatus.* **Lucanidae:** *Odontolabis* spp., **Scarabaeidae:** *Cotilis* sp., *Lepidiota hypoleuca, L. stigma, L.* sp., *Leucopholis rorida, Melolontha hypoleuca.*

Lepidoptera Hyblaeidae: *Hyblaea puera.* **Noctuidae:** *Hyblea puera.* **Saturniidae:** *Coscinocera anteus, Syntherata apicalis, S. weyneri.* **Sphingidae:** *Acherontia achesis, Daphnis hypothous, Oxyamblyx dohertyi.* **Uraniidae:** *Nyctalemon patroclus goldiei.*

Hymenoptera Apidae: *Apis cerana, Apis cerana indica, Apis dorsata, Melipona favosa, M. minuta, M. vidua.* **Formicidae:** *Crematogaster* sp., *Oecophylla smaragdina.* **Vespidae:** *Provespa anomala, P.* sp., *Ropalidia* spp., *Vespa analis, V. ducalis, V. tropica, V.* spp., *V. velutina.* **Xylocopidae:** *Xylocopa latipes.*

Japan

Thysanura Lepismatidae: *Lepisma saccharina.*

Ephemeroptera Baetidae: *Cloeon dipterum;* **Ehemeridae:** *Ephemera strigata.*

Odonata Aeschnidae: *Anax parthenope Julius.* **Cordulegasteridae:** *Anotogaster sieboldii.* **Gomphidae:** *Dabidius nanus.* **Libellulidae:** *Crocothemis servilia, Orthetrum japonicum, O. triangulare melania, Sympetrum darwinianum, S. eroticum eroticum, S. frequens, S. infuscatum, S. pedemontanum elatum.*

Plecoptera Perlidae: *Kamimuria tibialis, Oyamia gibba, Paragnetia tinctipennis, Perlodes frisonana.* **Pteronarcidae:** *Pteronarcys reticulata.*

Orthoptera Acrididae: *Acrida cinerea, A. peregrinum, Atractomorpha bedeli, Locusta migratoria, Oedaleus infernalis, Oxya chinensis, O. japonica japonica, O. ninpoensis, O. sinuosa, O. yezoensis, Tryxalis nasuta.* **Gryllidae:** *Gryllus ritzemae, Homoeogryllus japonicas, Loxoblemmus arietulus, L. doenitzi, Modicogryllus nipponensis, Scapsipedus parvus, Teleogryllus emma, Velarifictorus micado.* **Gryllotalpidae:** *Gryllotalpa africana.* **Tettigoniidae:** *Conocephalus gladiatus, C. maculatus, Euconocephalus thunbergii, Gampsocleis buergeri, Hexacentrus japonicus japonicas, Homorocoryphus ineosus, Phaneroptera falcate, Pseudorhynchus japonicas, Tettigonia orientalis orientalis.*

Blattaria Blattidae: *Periplaneta Americana.* **Phyllodromiidae:** *Opisthoplatia orientalis.*

Mantodea Mantidae: *Hierodula bipapilla, H. patellifera, Paratenodera (=Tempdera) aridifolia, P. superstitiosa, Statilia maculata, Tenodera aridifolia, T. capitata.*

Anoplura Pediculidae: *Pediculus humanus, P. humanus corporis.*

Hemiptera-Homoptera Aphidide: *Schlechtendalis chinensis.* **Cicadidae:** *Cryptotympana facialis, Graptopsaltria nigrofuscata, Meimuna opalifera, Oncotympana maculaticollis, Platypleura kaempferi, Tanna japonensis japonensis, Terpnosia vacua.* **Coccidae:** *Ericerus pela.* **Eriosomatidae:** *Melaphis chinensis.*

Hemiptera-Heteroptera Aphelochiridae: *Apherocheirus vittatus.* **Belostomatidae:** *Diplonychus japonicas, Lethocerus deyrolei.* **Corididae:** *Hesperocorixa distanti distanti.* **Nepidae:** *Laccotrephes flavovenosa, Ranatra chinensis.* **Pentatomidae:** *Nezara viridula.*

Coleoptera Buprestidae: *Chalcophora japonica.* **Cerambycidae:** *Anoplophora malasiaca, Apriona germari japonica, A. japonica (=A. rugicolis), Apriona rugicollis, Aromia moschota ambrosiaca, Batocera lineolata, Chloridolum thaliodes, Eupogonius tenuicornis, Malambyx japonicus, Oberea japonica, Prionus insularis, Thyestilla (=Thyestes) gebleri, Xylotrechus chinensis.* **Coccinellidae:** *Harmonia axyridis (=Ptychanatis axyridis).* **Cicindellidae:** *Cicindela chinensis japonica.* **Curculionidae:** *Curculio sikkimensis (=Balanius dentipes?).* **Ditiscidae:** *Cybister brevis, C. japonicus, C. tripunctatus orientalis, Dytiscus validus, D. sp., Rhantus pulverosus.* **Gyrinidae:** *Dineutus marginatus, D. orientalis, Gyrinus curtus, G. japonicas.* **Hydrophilidae:** *Hydrophilus acuminatus, H. affinis, Sternolophus rufipes.* **Lampyridae:** *Luciola cruciata, L. lateralis.* **Lucanidae:** *Lucanus maculifemoratus, Prismognathus angularis angularis, Prosopocoilus inclinatus*

inclinatus. **Meloidae:** *Epicauta gorhami, Epicauta megalocephala.* **Mordellidae:** *Metoeus stanus.* **Psephenidae:** *Psephemus* sp. **Scarabaeidae:** *Mimela splendens.* **Tenebrionidae:** *Palembus dermestoide.*

Neuroptera Corydalidae: *Parachauliodes japonicas, Protohermes grandis.* **Myrmeleontidae:** *Haenomyia micans.*

Diptera Calliphoridae: *Calliphora lata.* **Cecidomyiidae:** *Pseudasphondylia matatabi.* **Dryomyzidae:** *Dryomyza formosa.* **Muscidae:** *Fannia canicularis, Musca domestica, Musca stabulans.* **Syrphidae:** *Eristalis tenx, Eristalis* sp. **Tabanidae:** *Tananus chrysurus, T. mandarinus, T. rubidus, T. rufidens, T. trigonus.* **Tachinidae:** *Blepharipa zebina, Saturnia sericariae.* **Tipulidae:** *Antocha (Proantocha) spinifera.*

Trichoptera Hydropsychidae: *Cheumatopsyche brevilineatus, Hydropsyche orientalis, H. ulmeri, H.* sp., *Macronema radiatum.* **Leptoceridae:** *Stenopsyche sauteri, Stenopsyche marmorata* . **Limnephilidae:** *Allophylax* sp. **Phryganeidae:** *Neuronia melaleuca, N. regina, Phryganea japonica.* **Rhyacophilidae:** *Rhyacophila nigrocephala, R.* sp.

Lepidoptera Bombycidae: *Bombyx mandarina, B. mori.* **Brahmaeidae:** *Brahmaea japonica.* **Cossidae:** *Cossus insularis, C. jezoensis, Zeuzera multistrigata.* **Geometridae:** *Phthonandria atrilineata.* **Hepialidae:** *Endoclyta excrescens, E. sinensis.* **Hesperiidae:** *Parnara guttata guttata.* **Lasiocampidae:** *Dendrolimus spectabilis.* **Limacodidae:** *Cnidocampa flavescens.* **Noctuidae:** *Agrotis ypsilon, Mamestra brassicae.* **Notodontidae:** *Craniophora ligustri.* **Psychidae:** *Eumeta minuscula.* **Pyralidae:** *Chilo suppressalis, Conogethes punctiferalis, Glyphodes pyloalis, Numonia pyrivorella, Pyrausta polygoni, Scirpophaga incertulas.* **Saturniidae:** *Antheraea pernyi, A. yamamai yamamai, Caligula japonica japonica, Rhodinia fugox.* **Sesiidae:** *Paranthrene regalis.* **Sphingidae:** *Acherontia lechesis, A. styx crathis, Herse convolbuli, Langia zenzeroides nawai, Macroglossum stellatarum, Marumba gaschkewitschii echephron, Pachilia ficus, Pergesa elpenor lewisi, Psilogramma increta, Smerinthus planus, Theretra japonica, T. nessus, T. oldenlandiae.*

Hymenoptera Apidae: *Apis cerana japonica, Apis mellifera.* **Bombidae:** *Bombus diversus tersatus, B. ignitus, B. kainowskyi, B. speciosus.* **Braconidae:** *Euurobracon penetrator (=E. yokohamae).* **Cynipidae:** *Trichagalma serratae.* **Formicidae:** *Formica sanguinea, F. yessensis, Lasius fliginosus, Monomorium nipponensis.* **Polistidae:** *Polistes fadwigae, P. hebraeus, P. jadwigae jadwigae, P. mandarinus, P. rothney, P. snelleni.* **Sphecidae:** *Ammophila infesta.* **Vespidae:** *Vespa affinis, V. analis, V. crabro, V. mandarinia, V. simillima, Vespula flavipes lewisii, V. germanica, V. koreensis, V. rufa, V. shidai.* **Xylocopidae:** *Xylocopa appendiculata circumvolans, X. dissimilis, X. pictifrons, X. violacea.*

Laos

Odonata Agrionidae: *Ceriagrion* sp.; **Libellulidae:** *Crocothemis* sp.

Orthoptera Acrididae: *A. willemsei, Caelifera* sp., *Chondracris rosea brunneri, Ducetia japonica, Patanga succincta.* **Gryllidae:** *Acheta domesticus, Brachytrupes portentosus, Gryllus bimaculatus, Teleogryllus mitratus.* **Gryllotalpidae:** *Gryllotalpa africana.* **Tettigoniidae:** *Euconocephalus* spp., *Mecopoda elongate, Pseudophyllus titan.*

Hemiptera-Homoptera Cicadidae: *Dundubia intemerata, Meimuna opalifera, Orientopsaltria* spp.

Hemiptera-Heteroptera Belostomatidae: *Lethocerus indicus.* **Nepidae:** *Ranatra chinensis.* **Notonectidae:** *Anisops* spp. **Pentatomidae:** *Tessaratoma papillosa, T. quadrata.*

Coleoptera Cerambycidae: *Batocera rubus.* **Chrysomelidae:** *Sagra femorata.* **Dynastidae:** *Xylotrupes gideon.* **Ditiscidae:** *Cybister lewisianus, C. rugosus, C.* sp., *Eretes sticticus.* **Hydrophilidae:** *Hydrophilus affinis, H. bilineatus cashimirensis, H. (Dytiscus) hastatus.* **Scarabaeidae:** *Anomala cupripes, Heliocopris bucephalus, H. dominus, Holotrichia* spp.

Lepidoptera Bombycidae: *Bombyx mori.* **Hesperiidae:** *Erionata thrax thrax, Erionata torus.* **Pyralidae:** *Omphisa fuscidentalis.*

Hymenoptera Apidae: *Apis florea, A.* spp. **Formicidae:** *Oecophylla smaragdina, Polistes stigmata.* **Vespidae:** *Vespa affinis, V. soror, V. tropica, V.* spp.

Malaysia

Odonata Libellulidae: *Orthetrum* sp.

Orthoptera Acrididae: *Acrida* sp., *Aiolopus* sp., *Atractomorpha psittacina, Locusta migratoria manilensis, Oxya japonica japonica, Valanga nigricornis.* **Gryllidae:** *Nisitrus vittatus.* **Gryllotalpidae:** *Gryllotalpa longipennis.* **Tettigoniidae:** *Hexacentrus unicolor, Macrolyristes imperator, Mecopoda elongate, Onomarchus* sp.

Phasmida Phasmatidae: *Eurycnema versifasciata, Haaniella echinata, H. grayi grayi.* **Necrosciidae:** *Platycrana viridana.*

Isoptera Termitidae: *Macrotermes gilvus, Microcerotermes dubius.*

Blattaria Panesthiidae: *Panesthia* sp. **Blattidae:** *Periplaneta americana, Tenodera* sp.

Hemiptera-Homoptera Cicadidae: *Dundubia jacoona, D.* spp., *Graptopsaltria nigrofuscata, Orientopsaltria* spp., *Platylomia* spp., *Pomponia imperatorial.*

Hemiptera-Heteroptera Alydidae: *Leptocorisa acuta, L. oratorius.* **Pentatomidae:** *Nezara viridula.*

Coleoptera Buprestidae: *Chrysochroa* spp. **Cerambycidae:** *Batocera* spp., *Hoplocerambyx spinicornis, Macrotoma* sp., *Rhaphipodus* sp. **Chrysomeridae:** *Aplosonyx albicornis.* **Curculionidae:** *Rhnychophorus ferrugineus.* **Dynastidae:** *Chalcosoma atlas, C. moellenkampi, Oryctes rhinoceros, Xylotrupes gideon.*

Ditistidae: *Cybister* sp. **Lucanidae:** *Dorcus* spp., *Odontolabis* spp. **Passalidae:** *Aceraius* spp. **Scarabaeidae:** *Anomala concha, A. coxalis, A. lasiocnemis, A. latefemorata, Exopholis hypoleuca, Leucopholis staudingeri.*

Diptera Stratiomyidae: *Helmetia illucens.*

Lepidoptera Cossidae: *Zeuzera* sp. **Hesperiidae:** *Anchistroides nigrita, Erionata thrax thrax.* **Sphingidae:** *Hippotion celerio.*

Hymenoptera Apidae: *Apis cerana, A. cerana indica, A. dorsata, A. nigrocincta. Trigona apicalis.* **Formicidae:** *Camponotus gigas, Oecophylla smaragdina.* **Vespidae:** *Ropalidia* spp., *Vespa affinis, V. tropica, V.* spp. **Xylocopidae:** *Xylocopa latipes, X.* sp.

Mongolia

Orthoptera Acrididae: *Locusta* (=*Gryllus*?) *onos.*

Anoplura Pediculidae: *Pediculus humanus, P. humanus corporis.*

Myanmar

Odonata Agrionidae: *Ceriagrion* sp., *Crocothemis* sp.

Orthoptera Acrididae: *Acrida willemsei, Chondracris rosea pbrunner, Ducetia japonica, Patanga succincta.* **Gryllidae:** *Gryllus bimaculatus, Teleogryllus mitratus.* **Tettigoniidae:** *Euconocephalus* spp., *Pseudophyllus titan.*

Hemiptera-Homoptera Cicadidae: *Meimuna opalifera, Platypleura insignis.*

Hemiptera-Heteroptera Belostomatidae: *Lethocerus indicus.* **Notonectidae:** *Anisops* spp., *Notonecta* sp.

Coleoptera Chrysomelidae: *Sagra femorata.* **Curculionidae:** *Rhnychophorus ferrugineus.* **Dynastidae:** *Oryctes rhinoceros, Xylotrupes gideon.* **Ditiscidae:** *Cybister lewisianus, C. rugosus, C.* sp., *Eretes sticticus.* **Hydrophilidae:** *Hydrophilus bilineatus cashimirensis, H. (Dytiscus) hastatus.* **Scarabaeidae:** *Anomala cupripes, Heliocopris bucephalus, H. dominus, Holotrichia* spp.

Lepidoptera Bombycidae: *Bombyx mori.* **Hesperiidae:** *Erionata thrax thrax.*

Hymenoptera Apidae: *Apis florea.* **Formicidae:** *Oecophylla smaragdina.* **Polistidae:** *Polistes stigmata.* **Vespide:** *Vespa auraria, V. velutina auraria.*

Nepal

Hymenoptera Apidae: *Apis dorsata, A. laboriosa, A. mellifera.*

North Korea

Orthoptera Acrididae: *Acrida turrita, Atractomorpha lata, Oxya japonica japonica, O. sinuosa, Teleogryllus emma.* **Gryllotalpidae:** *Gryllotalpa africana.* **Tettigoniidae:** *Gampsocleis mikado.*

Mantodea Mantidae: *Hierodula patellifera, Tenodera aridifolia, T. capitata.*

Hemiptera-Homoptera Cicadidae: *Cryptotympana facialis.*

Coleoptera Cerambycidae: *Apriona germari japonica.* **Cicindellidae:** *Cicindela chinensis.* **Ditiscidae:** *Cybister japonicus, C. tripunctatus orientalis.* **Hydrophilidae:** *Hydrophilus acuminatus.* **Meloidae:** *Epicauta megalocephala, Meloe violaceus.* **Scarabaeidae:** *Aphodius* sp., *Gymnopleurus sinuatus, Mimela luciaula.*

Diptera Tabanidae: *Tabanus tropicus, T.* sp.

Lepidoptera Bombycidae: *Bombyx mori.* **Noctuidae:** *Mamestra brassicae.* **Pyralidae:** *Glyphodes pyloalis.*

Hymenoptera Apidae: *Apis cerana japonica, A. indica* var. *japonica.* **Formicidae:** *Formica rufa.* **Sphecidae:** *Ammophila infesta.*

Philippines

Orthoptera Acrididae: *Acridium manilense, A. ranunculum, A. rubescens, Curtilla africana, Locusta migratoria, Oedipoda subfasciata?* **Gryllotalpidae:** *Gryllotalpa africana, G.* spp.

Isoptera Termitidae: *Macrotermes gilvus.*

Coleoptera Cerambycidae: *Batocera albofasciata, B. numitor.* **Curculionidae:** *Pachyrrhynchus moniliforis, Rhynchophorus* (=*Calandra*) *chinensis, Rhnychophorus ferrugineus.* **Dynastidae:** *Oryctes rhinoceros.* **Hydrophilidae:** *Hydrophilus picicornis.* **Scarabaeidae:** *Lepidiota punctum, Leucopholis irrorata, L. pulverulenta.* **Tenebrionidae:** *Palembus dermestoides.*

Hymenoptera Apidae: *Apis cerana, Apis cerana indica, Apis dorsata, Apis* (=*Megapis*) *zonata, Trigona biroi.* **Formicidae:** *Camponotus* spp., *Oecophylla smaragdina.*

Singapore

Hemiptera-Heteroptera Belostomatidae: *Lethocerus indicus.*

South Korea

Thysanura Lepismatidae: *Ctenolepisma villosa.*

Odonata Libellulidae: *Crocothemis servilia.*

Orthoptera Acrididae: *Acrida cinerea, Atractomorpha lata, Oxya japonica japonica, O. sinuosa.* **Gryllidae:** *Homoeogryllus japonicas, Teleogryllus emma.* **Gryllotalpidae:** *Gryllotalpa africana.* **Tettigoniidae:** *Gampsocleis mikado.*

Mantodea Mantidae: *Hierodula patellifera, Tenodera aridifolia, T. capitata, T. sinensis.*

Hemiptera-Heteroptera Nepidae: *Laccotrephes flavovenosa.*

Coleoptera Cerambycidae: *Apriona germari japonica, Thyestilla* (=*Thyestes*) *gebleri.* **Coccinellidae:** *Harmonia axyridis* (=*Ptychanatis axyridis*). **Cicindellidae:** *Cicindela chinensis.* **Ditiscidae:** *Cybister japonicus, C. tripunctatus orientalis.*

Hydrophilidae: *Hydrophilus acuminatus.* **Meloidae:** *Epicauta megalocephala,* **Scarabaeidae:** *Gymnopleurus sinuatus, Mimela luciaula.*

Diptera Tabanidae: *Tabanus* sp.

Lepidoptera Bombycidae: *Bombyx mori.* **Cossidae:** *Cossus jezoensis.* **Geometridae:** *Phthonandria atrilineata.*

Hymenoptera Apidae: *Apis cerana japonica, A. indica* var. *japonica.* **Vespidae:** *Vespa manchurica, Vespula koreensis, V.* spp.

Sri Lanka

Orthoptera Acrididae: *Acrida* sp. **Gryllotalidae:** *Gryllotalpa Africana.* **Tettigoniidae:** *Holochlora albida, Mecopoda elongate.*

Isoptera Termitidae: *Odontotermes feae.*

Hemiptera-Heteroptera Belostomatidae: *Lethocerus indicus.* **Pentatomidae:** *Bagrada picta.*

Coleoptera Buprestidae: *Chrysobothris* sp. **Cerambycidae:** *Batocera albofasciata, B. rubus, Neocerambyx paris, Xysterocera globosa.* **Curculionidae:** *Hypomeces squamosus, Rhnychophorus ferrugineus.* **Dynastidae:** *Oryctes rhinoceros, Xylotrupes gideon.* **Ditiscidae:** *Cybister lewisianus, C. tripunctatus, Eretes sticticus.* **Hydrophilidae:** *Hydrophilus olivaceus.* **Passalidae:** *Passalus interruptus.* **Scarabaeidae:** *Adoretus compressus, Lepidiota stigma.* **Simuliidae:** *Simulium* sp.

Lepidoptera Arctiidae: *Diacrisia obliqua.* **Bombycidae:** *Bombyx mori.* **Saturniidae:** *Antheraea assamensis, A. paphia, Samia cynthia, Samia ricini.*

Hymenoptera Apidae: *Apis cerana, A. cerana indica, A. dorsata, A. florea, A. laboriosa.* **Xylocopidae:** *Xylocopa* sp.

Taiwan

Hymenoptera Vespidae: *Vespa affinis, V. analis, V.basalis ,V. mandarinia, V.velutina.*

Thailand

Odonata Aeschnidae: *Aeschna* spp., *Anax guttatus.* **Agrionidae:** *Ceriagrion* sp. **Corduliidae:** *Epophtalmia vittigera bellicose, Macromia* sp. **Libellulidae:** *Crocothemis* sp., *Libellula pulchella, Rhyothemis* sp. **Macromiidae:** *Macromia* sp.

Orthoptera Acrididae: *Acrida willemsei, A.* sp., *Aeolopus tamulusus, Atractomorpha* sp., *Aularches miliaris, Chondacris rosea, C. rosea brunneri, C. rosea pbrunner, C.* sp., *Chorthippus* sp., *Cyrtacanthacris tatarica,C.* sp., *Ducetia japonica, Euprepocnemis shirakii, Hieroglyphus banian, Locusta migratoria, Oxya japonica japonica, O.* sp., *Parapleurus* sp., *Patanga avis, P. japonica, P. succincta, Pseudocephalus litan, Trilophidia annulata.* **Gryllidae:** *Acheta confirmata, A. domesticus,*

A. testacea, Brachytrupes portentosus, Gryllus bimaculatus, G. testaceus, G.
sp., *Gymnogryllus* spp., *Homeoxipha* sp., *Madasumma* sp., *Pteronemobius* sp.,
Teleogryllus mitratus, T. spp., *Velarifictorus* sp. **Gryllotalpidae:** *Gryllotalpa*
africana, Gryllotalpa africana microphtalma. **Tettigoniidae:** *Conocephalus*
maculatus, C. spp., *Ducetia japonica, Euconocephalus incertus, E.* spp., *M.* sp.,
Neoconocephalus incertus, Onomarchus sp., *Pseudophyllus titan, P.* sp., *Scudderia*
sp. **Tetrigidae:** *Euparatettix* sp.

Phasmida Phasmatidae: *Eurycnema versirubra, E.* sp.

Isoptera Kalotermitidae: *Kalotermes flavicollis,* **Rhinotermitidae:** *Coptotermes*
havilandi, Reticulitermes flavipes. **Termitidae:** *Macrotermes gilvus, Odontotermes*
spp., *Termes flavicole.*

Blattaria Blattidae: *Blatta orientalis.* **Blattidae:** *Stylophyga rhombifolia.*

Mantodea Mantidae: *Hierodula* spp., *Mantis religiosa, Sinensisa* sp., *Tenodera*
sinensis.

Hemiptera-Homoptera Cicadidae: *Catharsius molasses, Chremistica* sp.,
Cosmopsaltria waine, Dundubia emanatura, D. intemerata, D. mannifera, D . spp.,
Meimuna opalifera, Orientopsaltria spp., *Platylomia radha, P.* spp., *Pomponia*
sp., *Rihana* sp. **Lacciferidae:** *Lccifer lacca.*

Hemiptera-Heteroptera Belostomatidae: *Diplonychus rusticus, D.* sp.,
Lethocerus europdeum, L. indicus, Sphaerodema rustica. **Coreidae:** *Anoplocnemis*
phasianus, Homoeocerus sp., *Onchorochira nigrorufa, Prionolomia* sp. **Gerridae:**
Cylindrostethus scrutator. **Naucoridae:** *Sphaerodema molestum, S. rustica.*
Nepidae: *Laccotrephes griseus, L. ruber, Nepa* spp., *Ranatra chinensis,*
R. longipes thai, R. variipes. **Notonectidae:** *Anisops barbatus, A. bouvieri,*
A. spp., *Notonecta undulata.* **Pentatomidae:** *Amissus testaceus, Proxys* sp.,
Tessaratoma javanica, T. papillosa.

Coleoptera Buprestidae: *Buprestis* sp., *Catoxamtha* sp., *Chrysobothris femorata,*
Sternocera aequisignata, S. ruficornis. **Cerambycidae:** *Aeolesthes* sp., *Apriona*
germari japonica, Aristobia approximator, Batocera rubus, Dorysthenes buqueti,
D. granulosus, D. walkeri, D. sp., *Hoplocerambyx spinicornis, Macrotoma fisheri,*
Pachyterla dimidiata, Plocaederus obesus, P. rufinornis, P. sp., *Stromatium eascymans,*
Threnetica lacrymans, Xysterocera globosa. **Chrysomelidae:** *Sagra femorata, S.* sp.
Curculionidae: *Aristobla approximator, Arrhines hirtus, Arrhines* spp., *Astycus*
gestvoi, Cnaphoscapus decorates. **Curculionidae:** *Cyrtotrachelus buqueti borealis,*
C. dochrous, C. longimanus, Episomus aurivillius, E. sp., *Hypodisa talaca,*
Hypomeces squamosus, Onaphoscapus decorates, Pollendera atomaria,
Rhnychophorus ferrugineus, R. schach, R. spp., *Tanymeces* sp., *Xanthochellus*
sp. **Dynastidae:** *Chalcosoma atlas, Eupatorus gracilicornis, Oryctes rhinoceros,*
Xylotrupes gideon. **Ditiscidae:** *Copelatus* sp., *Cybister lewisianus, C. rugosus,*
C. tripunctatus, C. tripunctatus asiaticus, C. sp., *Cypris* sp., *Eretes sticticus,*
Hydaticus rhantoides, Laccophilus pulicarius, Rhantaticus congestus.
Hydrophilidae: *Hydrobiomorpha spinicollis, Hydrochara* sp., *Hydrophilus*

bilineatus, H. bilineatus cashimirensis, H. (Stethoxus) cavisternus, H. (Dytiscus)
hastatus, H. olivaceus, H. sp., *Sternolophus rufipes.* **Scarabaeidae:** *Adoretus*
compressus, A. convexus, A. cribatus, A. pachysomatus, A. spp., *Agestrata*
orichalca, Anomala anguliceps, A. antiqua, A. bilunulata, A. blaisei, A. cantori,
A. chalcites, A. chlrochelys, A. cupripes, A. fusikibia, A. laotica, A. lignea,
A. pallida, A. parallera, A. puctulicollis, A. rugosa, A. scherei, A. shanica,
A. vuilletae, A. spp., *Aphodius (Pharaphodius) crenatus, A. (Ph.) marginellus,*
A. (Ph.) putearius, A. (Ph.) sp., *Apogonia* sp., *Brahmina Mikado, B. parvula,*
Catharsius birmanicus, C. molossus, Chaetadoretus cribratus, Chantonhumectus
hidalguensis, Copris (s.str.) *carinicus, C. corpulentus, C. furciceps, C.* (s.str.)
nevinsoni, C. (Paracopris) puctulatus, C. (Microcopris) reflexus, C. sinicus,
Cyclocephala fasciolata, Empectida tonkinensis, Exolontha castanea, Exopotus sp.,
Gymnopleurus aethiops, G. melanarius, Heliocopris bucephalus, H. dominus, H.
sp., *Heteronychus lioderes, Holotrichia cephalotes, H. hainanensis, H. nigricollis,*
H. pruinosella, H. siamensis, H. spp., *Lepidiota bimaculata, L. discidens,*
L. hauseri, L. stigma, L. stigma var. *alba, L.* sp., *Leucopholis* sp., *Liatongus affinis,*
L. (Paraliatongus) rhadamistus, L. tridentatus, L. venator, L. (Paraliatongus)
rhadamistus, Maladera sp., *Megistophylla andrewesi, Melolontha malaccensis,*
Microtrichia sp., *Mimela ferreroi, M. ignistriata, M. linping, M. schneideri,*
M. schulzei, M. sp., *Miridiba tuberculipennis obscura, Oniticellus cinctus,*
Onitis falcatus, O. niger, O. subopacus, O. virens, O. sp., *Onthophagus*
avocetta, O. bonasus, O. khonmiinitnoi, O. luridipennis, O. mouhoti, O.
orientalis, O. papulatus, O. proletarius, O. ragoides, O. rectecornutus, O.
sagittarius, O. seniculus, O. taurinus, O. tragoides, O. tragus, O. tricornis, O.
trituber, O. sp., *Pachnessa* sp., *Polyphylla tonkinensis, Protaetia fusca, P.* sp.,
Psilophosis sp., *Sophrops abscessus, S. bituberculatus, S. excises, S. foveatus,*
S. opacidorsalis, S. paucisetosa, S. rotundicollis, S. simplex, S. spancisetosa, S. sp.
Tenebrionidae: *Tenebrio molitor.*

Diptera Calliphoridae: *Calliphora megacephala.* **Simuliidae:** *Simulium*
aureohirtum, S. sp.

Lepidoptera Bombycide: *Bombyx mori.* **Cossidae:** *Xyleutes (=Duomitus)*
leuconotus, Zeuzera coffeae. **Hesperiidae:** *Erionata thrax thrax.* **Lycaenidae:**
Lyphyra brassolis. **Pieridae:** *Catopsilia pomona.* **Pyralidae:** *Omphisa*
fuscidentalis.

Hymenoptera Apidae: *Apis andreniformis, A. cerana, A. cerana indica,*
A. dorsata, A. florea. **Eumenidae:** *Eumenes petiolata.* **Formicidae:** *Carebara*
castanea, C. lignata, Crematogaster sp., *Oecophylla smaragdina.* **Halictidae:**
Nomia spp., *Trichona* spp. **Polistidae:** *Polistes stigmata, P.* spp. **Vespidae:**
Vespa affinis indosinensis, V. auraria, V. cincta, V. magnifica, V. soror, V. tropica,
V. velutina, V. spp. **Xylocopidae:** *Xylocopa confusa, Xylocopa latipes.*

Tibet

Coleoptera Hydrophilidae: *Hydrophilus acuminatus.*

Timor
Hymenoptera Apidae: *Apis dorsata.*

Viet Nam
Odonata Agrionidae: *Ceriagrion* sp.; **Libellulidae:** *Crocothemis* sp.

Orthoptera Acrididae: *Acrida willemsei, Chondracris rosea pbrunner, Ducetia japonica, Oxya japonica japonica, Patanga succincta.* **Gryllidae:** *Brachytrupes portentosus, Gryllus bimaculatus, Teleogryllus mitratus.* **Tettigoniidae:** *Euconocephalus* spp., *Pseudophyllus titan.*

Hemiptera-Homoptera Cicadidae: *Meimuna opalifera.*

Hemiptera-Heteroptera Belostomatidae: *Lethocerus indicus.* **Notonectidae:** *Anisops* spp.

Coleoptera Cerambycidae: *Apriona germari japonica.* **Chrysomelidae:** *Sagra femorata.* **Curculionidae:** *Rhynchophorus* spp. **Dynastidae:** *Xylotrupes gideon, X. gideon siamensis.* **Ditiscidae:** *Cybister lewisianus, C. rugosus, C.* sp., *Eretes sticticus.* **Hydrophilidae:** *Hydrophilus bilineatus, H. bilineatus cashimirensis, H. (Stethoxus) cavisternus, H. (Dytiscus) hastatus.* **Scarabaeidae:** *Anomala cupripes, Heliocopris dominus, Holotrichia* spp.

Lepidoptera Bombycidae: *Bombyx mori.* **Hesperiidae:** *Erionata thrax thrax.* **Pyralidae:** *Brihaspa astrostigmella.*

Hymenoptera Apidae: *Apis florea, A. mellifera.* **Polistidae:** *Polistes stigmata.*

B. Districts

1. Europe
Europe general
Orthoptera Acrididae: *Locusta viridissima.*
Blattaria Blattellide: *Ectobios lapponicus.*
Anoplura Haematopinidae: *Haematopinus suis.*
Coleoptera Meloidae: *Lytta vesicatoria.*
Hymenoptera Formicidae: *Formica major, F. minor.*

South Europe
Orthoptera Acrididae: *Gryllus aegyptius, G. lineola, G. locust, G. tataricus.*

Lapland
Hemitera-Homoptera Psyllidae: *Chermes* sp.

2. North America

North America general
Coleoptera Bruchidae: *Algarobius* spp., *Mimosestes* spp.

3. West Indies

West Indies general
Orthoptera Acrididae: *Oedipoda corallipes, Xanthipus corallipes zapotecus.*
Coleopter Cerambycide: *Macrodontia cervicornis, Stenodontes damicornis.*
Curculioide: *Rhynchophorus palmarum.*
Orthoptera Acrididae: *Oedipoda (=Xanthippus) corallipes.*

4. South America

South America general
Isoptera Termitidae: *Termes arborum.*

Amazonia
Isoptera Termitidae: *Termes flavicole.*
Anoplura Pediculidae: *Pediculus humanus.*
Coleoptera Bruchidae: *Caryoborus serripes, Pachymerus cardo.* **Chrysomelidae:** *Speciomerus giganteus.* **Curculionidae:** *Rhynchophorus cruentatus, R. palmarum.* **Dynastidae:** *Dynastes hercules, Megaceras* sp., *Megasoma actaeon.*
Hymenoptera Formicidae: *Atta cephalotes.* **Polybiidae:** *Polybia* spp.

Bolivia-Brazil
Hymenoptera Apidae: *Trigona jati.*
Bolivia-Peru-Brazil-Colombia
Coleoptera Curculionidae: *Rhynchophorus palmarum.*
Isoptera Termitidae: *Nasutitermes corniger.*
Coleoptera Curculionidae: *Rhynchophorus palmarum.*

Brazil-Colombia
Hymenoptera Formicidae: *Atta sexdens.*
Anoplura Pediculidae: *Pediculus humanus.*

Brazil-Guyana
Coleoptera Curculionidae: *Rhynchophorus palmarum.*

Brazil-Venezuela

Diptera Simuliidae: *Simulium rubithorax.*

Hymenoptera Formicidae: *Atta sexdens.*

Colombia-Peru

Coleoptea Crculionidae: *Rhynchophorus palmarum.*

Colombia-Venezuela

Coleoptera Curculionidae: *Anthonomus* spp., *Rhynchophorus palmarum.* **Dynastidae:** *Podischnus agenor.* **Lariidae:** *Caryobruchus scheelaea.*

Neuroptera Corydalidae: *Corydalus* spp.

Trichoptera Hydropsychidae: *Leptnema* spp.

Hymenoptera Polybiidae: *Polybia ignobilis.* **Polistidae:** *Mischocyttarus* spp.

5. Africa

Africa General

Orthoptera Acrididae: *Acorypha clara, Aiolopus thalassinus, Amblyphymus* sp., *Anacridium burri, A. moestum, Caloptenopsis nigrovariegata, Cataloipus congnatus, Eyprepocnemis plorans, Gastrimargus determinatus, Heteracris coerulescens, H.* sp., *H. tenuicornis, Humbe tenuicornis, Lamarckiana bolivariana, L. punctosa, Mesopsis abbreviates, Oedaleus carvalhoi, O. flavus, O. senegalensis, Orthochtha magnifica, O. turbid, Pycnodictya flavipes, Pyrgomorpha vignaudii, Roduniella insipid, Sherifura hanoingtoni, Stenohippus mundus, Tristria conops, T. discoidalis, Truxalis burtti, Tylotropidius didymus.* **Catantopidae:** *Catantops melanostichus, C. quadrates, C. stramineus, Diabolocantatops axillaris, Exoporpacris modica, Harpezocantatops stylifer, Oxycantatops congoensis, O. spissus, Parapropacris notate, Phaeocantatops decorates.* **Gryllacrididae:** *Gryllacris africana.* **Gryllidae:** *Acheta smeathmanni, Gymnogryllus leucostictus.* **Hemiacrididae:** *Acanthoxia gladiator, Hieroglyphus daganensis, Mazaea granulosa.* **Tettigoniidae:** *Anabrus simplex, Anoedopoda erosa, Conocephalus* spp., *Lanista* sp., *Leprocristus* sp., *Pseudorhynchus lanceolatus.*

Isoptera Termitidae: *Cubitermes* spp., *Termes smeathmanni.* **Hodotermitidae:** *Microhodotermes* spp.

Mantodea Hymenopodidae: *Pseudoharpax virescens.* **Mantidae:** *Epitenodera houyi, Mantis religiosa, Miomantis paykullii, Psedoharpax virescens, Sphodromantis centralis.*

Anoplura Pediculidae: *Pediculus humanus corporis.*

Hemiptera-Homoptera Cicadidae: *Andropogon gayanus.*

Coleoptera Curculionidae: *Rhynchophorus quadranglulus.* **Ditiscidae:** *Cybister binotatus, Dytiscus (Macrodytes) circumflex.* **Scarabaeidae:** *Platygenia* spp.

Lepidoptera Brahmaeidae: *Dactylocerus Lucina.* **Notodontidae:** *Busseola fusca, Ipanaphe carteri.* **Psychidae:** *Eumeta* sp. **Saturniidae:** *Epiphora bauhiniae, Gonimbrasia rhodina, Nudaurelia anthinoides, N. macrothyris, N. richelmanni.*

Hymenptera Apidae: *Meliponula bocandei, Trigona clypeata, T. schmidti, T. (Hypotrigona)* spp. **Eumenidae:** *Synagris cornuta flavofasciata.* **Formicidae:** *Camponotus fulvopilosus, Carebara junodi, Messor aegyptiacus, Oecophylla smaragdina, Pachycondyla* sp.

Central area of Africa

Lepidoptera Saturniidae: *Cirina forda, Gonimbrasia alopia, G. anthina, G. belina, G. macrothyris, Imbrasia dione, I. melanops, I. obscura, I. truncata, Melanocera nereis, Pseudantheraea discrepans.* **Sphingidae:** *Acherontia atropos, Platysphinx* spp.

Congo Basin

Isoptera Termitidae: *Macrotermes nobilis.*

Eastern area of Africa

Orthoptera Acrididae: *Locusta migratoria.*

Isoptera Termitidae: *Macrotermes subhyalinus.*

Equatorial area of Africa

Lepidoptera Notodontidae: *Anaphe* spp. **Psychidae:** *Eumeta cervina.* **Saturniidae:** *Microgone* sp.

Northern area of Africa

Orthoptera Acrididae: *Schistocerca gregaria gregaria.*

Lepidoptera Saturniidae: *Imbrasia melanops, I. obscura, I. truncata.*

Northern central area of Africa

Lepidoptera Sturniidae: *Pseudantheraea discrepans.*

Southern area of Africa

Orthoptera Acrididae: *Locusta devastator, L. migratoria, Schistocerca gregaria flaviventris.*

Isoptera Termitidae: *Termes fatale.*

Hemiptera-Homoptera Psyllidae: *Psylla* sp.

Coleoptera Cerambycidae: *Ceroplesis burgeoni.*

Lepidoptera Eupterotidae: *Striphnopteryx edulis.* **Saturniidae:** *Cirina forda, C. similis, Gonimbrasia belina, G. cytherea, Goodia kunzei, Gynanisa maja, Heniocha apollonia, H. dyops, H. marnois, Imbrasia dione, I. ertli, I. tyrrhea, Melanocera menippe, Microgone cana, Rohaniella pygmaea, Usta wallengrenii.* **Sphingidae:** *Herse convolbuli, Hippotion eson.*

Hymenoptera Formicidae: *Carebara lignata.*

Southern central area of Africa

Lepidoptera Saturniidae: *Athletes gigas, A. semialba, Bunaeopsis aurantiaca, Cinabra hyperbius, Cirina forda, Gonimbrasia anthina, G. hecate, G. macrothyris, G. rectilineata, Imbrasia ertli, I. tyrrhea, Lobobunaea saturnus, Melanocera parva, Micragone ansorgei, M. cana, Tagoropsis natalensis, Urota sinope.* **Sphingidae:** *Herse convolbuli, Hippotion eson.*

Southern Sahara

Hymenoptera Apidae: *Apis unicolor adansoni.*

Tropical area of Africa

Lepidoptera Sphingidae: *Platysphinx stigmatica.*

Western area of Africa

Coleoptera Cerambycidae: *Ancylonotus tribulus.*

Lepidoptera Saturniidae: *Cirina forda butyrospermi.* **Sphingidae:** *Lophostethus demolini.*

6. The Middle and Near East

Barbary

Orthoptera Acrididae: *Gryllus aegyptius, G. lineola, G. locust, G. tataricus.*

Crimea

Orthoptera Acrididae: *Locusta migratoria, L. tartarica.*

Kalahari Desert

Orthoptera Acrididae: *Locusta tartarica.*

Near East

Orthoptera Acrididae: *Dociostaurus moroccanus.*

PART III

References

Acuña, A.M., L. Caso, M.M. Aliphat and C.H. Vergara. 2011. Edible insects as part of the traditional food system of the Popoloca Town of Los Reyes Metzontla, Mexico. J. Ethnobiol. 31(1): 150–169.

Adalla, C.B. and C.R. Cervancia. 2010. Philippine edible insects: a new opportunity to bridge the protein gap of resource-poor families and to manage pests. pp. 151–160. *In*: P.B. Durst, D.V. Johnson, R.N. Leslie and K. Shono [eds.]. Forest Insects as Food: Humans Bite Back. FAO Regional Office for Asia and the Pacific, Bangkok, Thailand.

Adriaens, E.L. 1951. Recherches sur l'alimentation des populations au Kwango. Bull. Agric. Congo Belge. 62: 473–550 [cited from DeFoliart 2002; van Huis 2005].

Agea, J.G., D. Biryomumaisho, M. Buyinza and G.N. Nabanoga. 2008. Commercialization of *Ruspolia nitidula* (nsenene grasshoppers) in central Uganda. African J. of Food Agriculture and Development 8(3): 319–332.

Akre, R.D. 1992. Yellow jackets anyone? A new, easily collected taste treat. The Food Insect Newsletter 5(2): 5.

Albisetti, C. and A.J. Venturell. 1962. Enciclopédia bororo: Vocabulário e etnografia. Campo Grande : Instituto de Pesquisas Etnográficas.

Aldasoro Maya, E.M. 2003. Étude ethnoentomologique chez les Hñähñu de "El Dexthi" (vallée du Mezquital, État de Hidalgo, Mexique). pp. 63–72. *In*: E. Motte-Florac and J.M.C. Thomas [eds.]. Les "insectes" dans la tradition orale, Peeters, Leuven, Belgium.

Aldrich, J.M. 1912. Flies of the leptid genus *Atherix* used as food by California Indians. Entomol. News 23: 159–161.

Aldrich, J.M. 1921. *Coloradia pandora* Blake, a moth of which the caterpillar is used as food by Mono Lake Indians. Ann. Entomol. Soc. Am. 14: 36–38.

Ambro, R.D. 1967. Dietary-technological-ecological aspects of Lovelock Cave coprolites. Berkeley: Univ. California Archololl. Survey Reports No. 70: 37–47 [cited from Sutton 1988].

Ancona, L. 1933. Les jumiles de cuautla *Euschistes zopilotensis*. An. Inst. Biol., Mexico Vol. 4: 103–108 [cited from Brues 1946].

Anonymous. 1893. The edible qualities of ants. Insect Life 5: 268.

Araujo, Y. and P. Beserra. 2007. Diversidad de invertebrados consumidos por las etnias Yanomami y Yekuana del Alto Orinoco, Venezuela. Interciencia 32(5): 318–323.

Arnett, R.H. 1985. American Insects: A Handbook of the Insects of America North of Mexico.: Vasn Nostrand Reinhold. Florence, Kentucky, USA [cited from DeFoliart 2002].

Aruga, H. 1983. Life of a wasp, *Vespula schrenckii*. Insects and Nature 18(11): 9–14 (in Japanese).

Ashiru, M.O. 1988. The food value of the larvae of *Anaphe venata* Butler (Lepidoptera: Notodontidae). Ecol. Food Nutrition 22: 313–320.

Bahuchet, S. 1985. Les pygmées Aka et la forêt Centrafricaine. Selaf, Paris, France [cited from Malaisse 2005].

Balée, W. 2000. Antiquity of traditional ethnobiological knowledge in Amazonia: The Tupí-Guarani Family and time. Ethnohistory 47(2): 399–422.

Balick, M.J. 1988. Jessenia and Oenocarpus neotropical oil palms worthy of domestication. FAO Plant Production and Protection Paper 88.

Balinga, M.P., P.M. Mapunzu, J.-B. Moussa and G. N'gasse. 2004. Contribution des insectes de la foret a la securite alimentaire. L'example des chenilles d'Afrique Centrale. pp. 3–36. FAO Departement des Forets: Contribution des insectes de la foret a la securite alimentaire. Produits forestiers non ligneux, Document de Travail No. 1, FAO, Rome, Italy.

Bancroft, E. 1769. Essay on the natural history of Guiana in S. America. Becket and De Hondt, London, England [cited from Bodenheimer 1951; DeFoliart 1990].

Bani, G. 1995. Some aspects of entomophagy in the Congo. Food Insects Newsletter 8(3): 4–5.

Banjo, A.D., O.A. Lawal and E.A. Songonuga. 2006. The nutritional value of fourteen species of edible insects in southwestern Nigeria. African J. Biotechnol. 5(3): 298–301.

Barrère, P. 1741. [Yasumatsu 1948 (in Japanese)].

Barreteau, D. 1999. Les Mofu-Gudur et leaurs croquets. L'homme et l'animal dans le basin du Lac Tchad. Actes du Colloque du réseau Mega-Tchad, Editions IRD, collections Cooloqie et Séminaires, n°00/354, Univ. Nanterre, Nanterre, France, pp. 133–169 [cited from Malaisse 2005].

Barrett, S.A. 1936. The army worm: a food of the Pomo Indians. *In*: R.H. Lowie [ed.]. Essays in Anthropology Presented to A.L. Kroeber. Berkeley: Univ. Calif. Press. USA [cited from DeFoliart 2002].

Bean, L.J. and D. Theodoratus. 1978. Western Pomo and Northeastern Pomo. pp. 289–305. *In*: R.F. Heizer [ed.]. Handbook of North American Indians, Vol. 8 California. Smithonian Institution, Washington DC, USA.

Beccari, O. 1904. Wanderings in the Great Forests of Borneo. Archibald Constable & Co. Ltd., London, England.

Beckerman, S. 1977. The use of palms by the Bari Indians of the Maracaibo Basin. Principes 22(4): 143–154.

Bejsak, V.R. 1992. Recipe—A golden oldie for Europeans! The Food Insect Newsletter 5(3): 6.

Bell, W.H. and E.F. Castetter. 1937. American Insects: A handbook of the Insects of America North of Mexico. Van Nostrand Reinhold, Florence, Kenntucky, USA [cited from DeFoliart 1991].

Benhalima, S., M. Dakki and M. Mouna. 2003. Les insectes dans le coran et dans la société islamique (Maroc). pp. 533–540. *In*: E. Motte-Florac and J.M.C. Thomas [eds.]. Les "insectes" dans la tradition orale, Peeters, Leuven, Belgium.

Bennett, G. 1834. Wanderings in New South Wales, Batavia, Pedir Coast, Singapore, and China, being a journal of a naturalist. Vol. I, Richard Bentley, London, England.

Bequaert, J. 1921. The predaceous enemies of ants. Bull. Am. Mus. Nat. Hist. 45: 271–331 [cited from DeFoliart 2002].

Berensberg, H.P. 1907. The use of insects as food delicacies, medicines or in manufactures. Natal Agric. J. and Mining Rec. 10: 757–762 [cited from Bodenheimer 1951].

Bergeron, D., R.J. Bushway, F.L. Roberts, I. Kornfield, J. Okedi and A.A. Bushway. 1988. The nutrient composition of an insect flour sample from Lake Victoria. Uganda. J. Food Comp. Anal. 1: 371–377.

Bergier, E. 1941. Peuples entomophages et insectes comestibles: étude sur les moeurs de l'homme et de l'insecte. Imprimerie Rullière Frères, Avignon, Vancluse, France [cited from van Huis 2003].

Bernatzik, H.A. 1936. Owa Raha. Wien-Leipzig-Olten [cited from Bodenheimer 1951].

Bodenheimer, F.S. 1951. Insects as human food, a chapter of the ecology of man. Dr. W. Junk, Publishers, Hague, Holland.

Bonwick, J. 1898. Daily Life and Origin of the Tasmanians. Sampson Low, Marston and Co., Ltd., London, England.

Boulidam, S. 2008. Gathering non-timber forest products in a market economy: A case study of Sahakone Dan Xang fresh food market, Xaithany District, Vientiane Capitao, Lao PDR.

FAO Workshop "Edible Forest Insects, Humans Bite Back!!", Chaing, Mai, Thailand, February 19–21, 200.

Bouvier, G. 1945. Quelques questions d'entomologie vétérinaire et lutte contre certains arthropods en Afrique tropicale. Acta Trop. 2: 42–59.

Bragg, P. 1990. Phasmida and coleoptera as food. Amateur Entomol. Bull. 49: 157–158.

Bréhion, A. 1913. Utilisation des insectes en Indochine. Préjugéges et moyens de defense contre quelques-uns d'entre eux. Bull. Mus. Nt. D'Hist. Nat. Paris 19: 277–281 [cited from Bodenheimer 1951].

Brilhante, N.A. and P.C. Mitoso. 2005. Manejo de abelhas nativas como componentes agroflorestais por populações tradicionais do estado do Acre. http://www.ufac.br/ orgaosup/pz/arboreto/publicacoes.htm [cited from Costa-Neto and Ramos-Elorduy 2006].

Bristowe, W.S. 1932. Insects and other invertebrates for human consumption in Siam. Trans. Entomol. Soc. Lond. 8: 387–404.

Brygoo, E. 1946. Essai de bromatologie entomologique: Les insects comestibles. Thése de Doctorat. Bergerac [cited from Bodenheimer 1951].

Bryk, F. 1927. Termitenfang am Fusse des Mount Elgon. Entomol. Rundschau 44: 1–3 [cited from DeFoliart 2002].

Burgett, M. 1990. Bakuti—A Nepalese culinary preparation of giant honey bee brood. The Food Insect Newsletter 3(3): 1.

Callen, E.O. 1969. Diet as revealed by coprolites. pp. 235–243. *In*: D.R. Brothwell and E.S. Higgs [eds.]. Science in Archaeology. Thames & Hudson, London, England and New York, N.Y., USA.

Callewaert, R. 1922. Witte mieren in Kasai. Congo 2(3): 366–380 [cited from Malaisse 2005].

Campbell, T.G. 1926. Insect foods of the aborigines. Australian Mus. Mag. 2: 407–410.

Capiomont and Leprieur. 1874. Ann. Soc. Entomol. Fr. Ser. 5, 4: 65 [cited from DeFoliart 2002].

Carr, L.G.K. 1951. Interesting animal foods, medicines, and omens of the eastern Indians, with comparisons to ancient European practices. J. Washington Acad. Sci. 41: 229–235.

Carrera, M. 1992. Entomofagia humana. Revista Brasileira de Entomologia 36(4): 889–894 [cited from Costa-Neto and Ramos-Elorduy 2006].

Carvalho, H.C.M. 1951. Relações entre os indios do Alto Xingu e a fauna reginal. Publicações Avulsas do Museu Nacional [cited from Costa-Neto and Ramos-Elorduy 2006].

Cerda, H. and F. Torres. 1999. Collection [cited from Paoletti and Dufour 2005].

Césard, N. 2006. Des libellules dans l'assiette : les insectes consommés à Bali. Insectes 1(140): 3–6.

Chagnon, N.A. 1968. Yąnomamö, The fierce people. Holt, Rinehart and Winston, Inc., New York, N.Y., USA.

Chakravorty, J., S. Ghosh and V.B. Meyer-Rochow. 2011. Practices of entomophagy and entomotherapy by members of the Nyshi and Galo tribes, two ethnic groups of the state of Arunachal Pradesh (North-East India). J. Ethnobiol. Ethnomed. 7(5): 1–14.

Chamberlain, R.V. 1911. The ethnobotany of the Gosiute Indians. Am. Anthropol. Assoc. Memoirs, No. 2 [cited from Heizer 1945].

Chavanduka, D.M. 1975. Insects as a source of protein to the African. The Rhodesia Sci. News 9: 217–220.

Chen, P.P. and S. Wongsiri. 1995. A survey of insects as human food in north and north-east of Thailand. An Open-file Report, Faculty of Agr. Naresuan Univ., Phitsanulok, Thailand (in Thai) [cited from Chen et al. 1998].

Chen, P.P., S. Wongsiri, T. Jamyanya, T.E. Rinderer, S. Vongsamanode, M. Matsuka et al. 1998. Honey bees and other edible insects used as human food in Thailand. Am. Entomol. 44: 24–29.

Chen, X. and Y. Feng. 1999. The Edible Insects of China. Chinese Science and Technology Pub. Co., China (in Chinese).

Chen, X., Y. Fen and Z. Chen. 2009. Common edible insects and their utilization in China. Entomol. Res. 39(5): 299–303.

Chen, Z. 1982. Illustrated Encyclopedia on Chinese Medicinal Materials. Vol. 1. Kōdansha, Tokyo, Japan (in Japanese).

Chin, T. and C.G. Chin. 2006. Culture and Utilization of Meal Worm. Jin Dun Publication Company, Beijing, China (in Chinese).

Chinn, M. 1945. Notes pour l'étude de l'alimentation des indigènes de la province de Coquilhatville. Ann. Soc. Belg. Méd. Trop. 25: 57–149 [cited from Malaisse 1997 and DeFoliart 2002].

Choo, J. 2008. Potential ecological implication of human entomophagy by subsistence groups of the Neotropics. Terrestrial Arthropod Reviews 1: 81–93.

Chowdhury, S.N. 1982. Eri silk industry. Directorate of sericulture and weaving. Govt. of Assam, Guahati, Assam, India [cited from DeFoliar 2002].

Chung, A.Y.C. 2008. An overview of ethnoentomological practices in Borneo. FAO Workshop "Edible Forest Insects, Humans Bite Back!!", Chiang Mai, Thailand, February 19–21, 2008.

Chung, A.Y.C. 2010. Edible insects and entomophagy in Borneo. pp. 141–150. *In*: P.B. Durst, D.V. Johnson, R.N. Leslie and K. Shono [eds.]. Forest Insects as Food: Humans Bite Back. FAO Regional Office for Asia and the Pacific, Bangkok, Thailand.

Chung, A.Y.C., V.K. Chey, S. Unchi and M. Binti. 2002. Edible insects and entomophagy in Sabah, Malaysia. Malayan Nature J. 56(2): 131–144.

Clastres, P. 1972. The Guayaki. pp. 138–74. *In*: M.G. Bicchieri [ed.]. Hunters and Gatherers Today. Holt, Rhinhart and Winston, New York, N.Y., USA [cited from DeFoliart 1990].

Clausen, L.W. 1954. Insect Fact and Folklore. Macmillan Company, New York, N.Y., USA.

Clavigero, F.J. 1937. The history of lower California [cited from Heizer 1945].

Cleland, J.B. 1966. The ecology of the aboriginal in south and central Australia. pp. 111–158. *In*: B.C. Cotton [ed.]. Aboriginal Man in South and Central Australia. Part I. Government Printer, Adelaide [cited from DeFoliart 2002; Yen 2005].

Coimbra, C.E.A., Jr. 1984. Estúdios de ecologia humana entre os Surui do Parque Indígena Aripuana, Rondónia. 1. O uso de larvas de Coleopteros (Bruchidae e Curcurionidae) na alimentação. Revista Brasileira de Zoologia 2: 35–47 [cited from Paoletti and Dufour 2005].

Common, I.F.B. 1954. A study of the ecology of the adult bogong moth, *Agrotis infusa* (Boisd.) (Lepidoptera: Noctuidae), with special reference to its behaviour during migration and aestivation. Aust. J. Zool. 2: 223–263.

Conway, J.R. 1986. The biology of honey ants. Am. Biol. Teacher 48: 335–343.

Conway, J.R. 1991. A honey of an ant. Biology Digest 18(4): 11–15.

Conzemius, E. 1932. Ethnographical survey of the Miskito and Sumu Indians of Honduras and Nicaragua. Smithonian Inst. Bur. Am. Ethnol. Bull. 106, U.S. Govt. Print Off., Washington, D.C., USA [cited from DeFoliart 2002].

Costa-Neto, E.M. 1994. Etnoictiologia alagoana, com énfase na utização medicinal de insetos. Maceió: Universidade Federal de Algagoas [cited from Costa-Neto and Ramos-Elorduy 2006].

Costa-Neto, E.M. 1996. Ethnotaxonomy and use of bees in northeastern Brazil. The Food Insect Newsletter 9(3): 1–3.

Costa-Neto, E.M. 1998. Folk taxonomy and significance of "abeia" (Insecta, Hymenoptera) to the Pankararé, Northeastern Bahia State, Brazil. J. Ethnobiol. 18(1): 1–13.

Costa-Neto, E.M. 1999. Healing with animals in Feira de Santana City, Bahia, Brazil. J. Ethnopharmacol. 65: 225–230.

Costa-Neto, E.M. 2000. Introduão à etnoentomologia: considerações metodológicas e estudo de casos. Feira de Santana: UEFS [cited from Costa-Neto and Ramos-Elorduy 2006].

Costa-Neto, E.M. 2003a. Consideration on the man/insect relationship in the State of Bahia, Brazil. pp. 95–104. *In*: E. Motte-Florac and J.M.C. Thomas [eds.]. Insects in Oral Literature and Traditions. Peeters, Leuven, Belgium.

Costa-Neto, E.M. 2003b. Etnoentomologia no povoado de Pedra Branca, município de Santa Terezinha, Bahia, um estudo de caso das interações seres humanos/insectos. Tese (Doutorado em Ecologiae e Recurosos Naturais). São Carlos: Universidade Federal de São Carlos [cited from Costa-Neto and Ramos-Elorduy 2006].

Costa-Neto, E.M. and M.N. de Melo. 1998. Entomotherapy in the county of Matinha dos Pretos, State of Bahia, Northern Brazil. The Food Insect Newsletter 11(2): 1–3.

Costa-Neto, E.M. and J.M. Pacheco. 2005. Utilização medicinal de insetos no povoado de Pedra Branca, Santa Terezinha, Bahia, Brazil. Biotemas 18(1): 113–133.

Costa-Neto, E.M. and J. Ramos-Elorduy. 2006. Los insectos comestibles de Brasil: Etnicidad, diversidad e importancia en la alimentación. Boletin Sociedad Entomologica Aragonesa, No. 38: 423–442.

Cowan, F. 1865. Curious Facts in the History of Insects; Including Spiders and Scorpions. Lippincott, Philadelphia, USA [cited from DeFoliart 2002].

Crane, E. 1967. The past in the present. For those interested in history. Bee World 48: 3+36–37.

Crane, E. 1992. The past and present status of beekeeping with stingless bees. Bee World 73(1): 29–42 [cited from Costa-Neto and Ramos-Elorduy 2006].

Curran, C.H. 1939. On eating insects—You eat them unknowingly almost every day. If the idea repels you consider that they have nourished mankind for countless centuries without ill effect and are still openly relished in many parts of the world. Nat. Hist. February-1939: 84–89.

Cutright, P.R. 1940. The Great Naturalists Explore South America. Macmillan Company, New York, N.Y., USA.

Cuvier, G.R.C.F. 1827/1835. Animal Kingdom. Vol. 16 [cited from Animal Kingdom. Henry G. Bohn, London, England. 1863].

Dalziel, J.M. 1937. Useful plants of West Tropical Africa. Crown Agents for Colonies, London, England [cited from Irvine 1957].

Davis, E.L. 1963. The Desert culture of the Western Great Basin: A lifeway of seasonal transhumance. Am. Antiquity 29: 202–212.

De Colombel, V. 2003. Les insects chez dix populations de langue Tchadique (Cameroun). pp. 45–62. *In*: E. Motte-Florac and J.M.C. Thomas [eds.]. Les insectes dans la tradition orale. Peeters, Leuven, Belgium.

De Lisle, M. 1944. Note sur la faune coléoptérologique du Cameroun. Bull. Soc. Etud. Cameroun, No. 5: 55–71 [cited from Bodenheimer 1951].

Decary, R. 1937. L'entomophagie chez les indigenes de Madagascar. Bull. Soc. Entomolol. Fr. 42: 168–171.

DeFoliart, G.R. 1989. The human use of insects as food and as animal feed. Bull. Entomol. Soc. Am. 35: 22–35.

DeFoliart, G.R. 1990. Hypothesizing about palm weevil and palm Rhinoceros beetle larvae as traditional cuisine, tropical waste recycling, and pest and disease control on coconut and other palms—Can they be integrated? The Food Insect Newsletter 3(2): 1, 3, 4 and 6.

DeFoliart, G.R. 1991. Toward a recipe file and manuals on how to collect edible wild insects in North America. The Food Insect Newsletter 4(3): 1, 3, 4 and 9.

DeFoliart, G.R. 1999. Insects as food: why the western attitude is important. Ann. Rev. Entomol. 44: 21–50.

DeFoliart, G.R. 2002. The Human Use of Insects as a Food Resource: A Bibliographic Account in Progress. http://www.food-insects.com/

DeFoliart, G.R. 2005. Overview of role of edible insects in preserving biodiversity. pp. 123–140. *In*: M.G. Paoletti [ed.]. Ecological Implications of Minilivestock, Potential of Insects, Rodents, Frogs and Snails. Science Publishers, Inc., Enfield, USA.

Dei, G.J.S. 1989. Hunting and gathering in a Ghanaian rain forest community. Ecol. Food Nutr. 22: 225–243.

Delmet, C. 1975. Extraction d'huile comestible d'Agonoscelis versicolor Fab. (Heteroptera, Pentatomidae) au Djebel Gouli, Soudan. L'homme et l'animal. Premier Colloque d'Ethnozoologie. Paris, Inst. Int. Ethnosciences, pp. 255–258 [cited from Ramos-Elorduy 2003].

Denevan, W.M. 1971. Campa subsistence in the Gran Pajonal, eastern Peru. Geogr. Rev. 61: 496–518.

Dias, C. de S. 2003. Notas preliminaries sobre a criação de abelhas sem ferrão no Vale do Paraguaçu. Eymba acuay 21: 2–3 [cited from Costa-Neto and Ramos-Elorduy 2006].

Dickson, H.R.P. 1949. The Arab of the desert. A Glimpse into Badawin life in Kuwait and Saudi Arabia. George Allen & Unwin Ltd., London, England [cited from DeFoliart 2002].

Distant, W.L. 1889–1892. A monograph of oriental Cicadidae. Taylor & Francis, London, England [cited from DeFoliart 2002].

Distant, W.L. 1902. Fauna of British India, including Ceylon and Burma: Rhynchota I (Heteroptera). Taylor and Francis, London, England.

Ditlhogo, M. 1996. The economy of *Imbrasia belina* (Westwood) in north-eastern Botswana. pp. 46–68. *In*: B.A. Gashe and S.F. Mpuchane [eds.]. "Phane", Proc. 1st Multidisciplinary Symp. on Phane, June 1996 Botswana. Dep. of Bio. Sci. & The Kalahari Conservation Society. NORAD, Botswana.

Donovan, E. 1842. Natural History of the Insects of China. Henry G. Bohn, London, England.

Dornan, S.S. 1925. Pygmies and bushmen of the Kalahari. Seeley & Co. Ltd., London. England [cited from Bodenheimer 1951].

Doughty, C.M. 1923. Travels in Arabia desert. Boni & Liveright Inc., 2 vols. [cited from DeFoliart 2002].

Drucker, P. 1937. Culture element distributions: V. Southern California. Univ. Calif. Anthropol. Records 1(1). Berkeley, California, USA [cited from Heizer 1945].

Dube, S., N.R. Dlamini, A. Mafunga, M. Mukai and Z. Dhlamini. 2013. A survey on entomophagy in Zimbabwe. ajfand (African J. Food, Agricul. Nutr. Deveop.) 13(1): 7242–7253.

Dufour, D.L. 1987. Insects as food: a case study from the northwest Amazon. Am. Anthropol. 89: 383–397.

Ealand, C.A. 1915. Insects and Man. Grant Richards, Ltd., London, England.

Eastwood, R. 2010. Collecting and eating *Lyphyra brassolis* (Lepidoptera: Lycaenidae) in southern Thailand. J. Res. on the Lepidoptera 43: 19–22.

Ebeling, W. 1986. Handbook of Indian Foods and Fibers of Arid America. Univ. Calif. Press, Berkeley, California, USA.

Egan, H. 1917. Pioneering the West, 1846 to 1878. Richmond, Utah: Privately Published [cited from Sutton 1988].

Embrapa. 2000. Os biótopos acridianos e a relação homemgafanhoto.http://www. cnpm,embrapa.br/projects/grshop_us/36.html [cited from Costa-Neto and Ramos-Elorduy 2006].

Ene, J.C. 1963. Insects and man in West Africa. Ibadan Univ. Press [cited from DeFoliart 2002].

Esaki, T. 1958. A story on eating insects. *In*: T. Esaki and S. Esaki [eds.]. Collection of essays by Teizo Esaki. Hokuryukan, Tokyo, Japan pp. 345 (in Japanese).

Esparza-Frausto, G., F.J. Macías-Rodoríguez, M. M.artínez-Salvador, M.A. Jiménez-Guevara and J. Méndez-Gallegos. 2008. Insectos comestibles asociados a las magueyeras en el Ejido Tolosa, Pinos, Zacatecas, México. Agrociencia 42(2): 243–252.

Essig, E.O. 1934. The value of insects to the California Indians. Scientist Month. 38: 181–186.

Essig, E.O. 1947. College Entomology. Macmillan, New York, N.Y., USA.

Essig, E.O. 1949. Man's six-legged competitors. Sci. Month. 69: 15–19.

Essig, E.O. 1965. A History of Entomology. Hafner Publishing Co., N.Y., USA.

Fabre, J.-H. 1922. Souvenirs Entomologiques 5. XV. La cigale - La transformation. Librairie Delagrave, Paris, France.

Fabre, J.-H. 1924. Souvenirs Entomologiques 10. VI. L'ergate - cossus. Librairie Delagrave, Paris, France.

Fagan, M.M. 1918. The use of insect galls. Am. Nat. 52: 155–176.

Fasoranti, J.O. and D.O. Ajiboye. 1993. Some edible insects of Kwara State, Nigeria. Am. Entomol. 39: 113–116.

Faure, J.C. 1944. Pentatomid bugs as human food. J. Entomol. Soc. S. Africa 7: 110–112.

Felger, R.S. and M.R. Moser. 1985. People of the Desert and Sea. Ethnobotany of the Seri Indians. Univ. Arizona Press, Tucson. USA.

Fladung, E.B. 1924. Insects as food. Maryland Acad. Sci. Bull. October: 5–8 [cited from Bodenheimer 1951].

Flood, J. 1996. Moth Hunters of the Australian Capital Territory. Canberra, Australia.

Forbes, J. 1813. Oriental memoirs Vol. 1. White, Cochrane, and Co., London, England.

Fowler, C.S. 1986. Subsistence. pp. 64–97. *In*: W.L. d'Azevedo [ed.]. Handbook of North American Indians, Vol. 11, Great Basin. Smithonian Institution, Washington, D.C., USA.

Fowler, C.S. and D. Fowler. 1981. The Southern Paiute, A.D. 1400–1776. pp. 129–162. *In*: D.R. Wilcox and W.B. Masse [eds.]. The prehistoric period in the North American Southwest, A.D. 1450–1700. Arizona State University Anthropological Research papers 24, Tempe, Arizona State Univ. [cited from Tarre 2003].

Frison, G.C. 1971. Shoshonean antelope procurement in the upper Green River Basin, Wyoming. Plains Anthropogist 16: 258–284 [cited from Sutton 1988].

Fukuhara, N. 1986. Cooked rice grasshopper. Story of rediscovery of *Oxia ninpoensis*. Insectarium 23: 262–269 (in Japanese).

Fuller, C. 1918. Notes on white ants. On the behaviour of true ants towards white ants. Bull. South Afr. Biol. Soc. 1(1): 16–20 [cited from Malaisse 2005].

Gade, D.W. 1985. Savanna woodland, fire, protein and silk in highland Madagascar. J. Ethnobiol. 5: 109–122.

Gast, M. 2000. Moissonsdu désert. Ibis, Paris, France [cited from Malaisse 2005].

Gelfand, M. 1971. Diet and tradition in an African culture. E. & S. Livingstone, London, England [cited from DeFoliart 2002].

Gessain, M. and T. Kinzler. 1975. Miel et insectes à mile chez les Bassari et d'autres populations du Sénégal Oriental. pp. 163–171. *In*: R. Pujol [ed.]. L'Homme et l'animal, Premier Colloque d'ethnozoologie, Paris, France [cited from van Huis 2005].

Ghaly, A.E. 2009. The use of insects as human food in Zambia. OnLine J. Biol. Sci. 9(4): 93–104.

Ghesquiére, G. 1947. Les insectes palmicoles comestibles. pp. 17–31. *In*: P. Lepesme [ed.]. Les insectes des palmiers, Paris, France [cited from Bodenheimer 1951; DeFoliart 2002].

Ghosh, C.C. 1924. A few insects used as food in Burm. Rept. Proceed. 5th Entomol. Meeting, Posa, 1923. Calcutta, India [cited from Bodenheimer 1951].

Gibbs, H.D., F. Agcaoili and G.R. Shilling. 1912. Some Filipino foods. Philippine J. Sci. Sect. A-7: 383–401.

Gilmore, R.M. 1963. Fauna and ethnozoology of South America. pp. 345–464. *In*: J.H. Steward [ed.]. Handbook of South American Indians. Vol. 6. Cooper Square Publ. Inc. New York, N.Y., USA.

Goldman, I. 1963. Tribes of Uapes—Caqueta region. pp. 770. *In*: J.H. Steward [ed.]. Handbook of South American Indians. Vol. 3. Cooper Square Publ. Inc. New York, N.Y., USA.

Gope, B. and B. Prasad. 1983. Preliminary observations on the nutrition value of some edible insects of Manipur. J. Adv. Zool. 4: 55–61 [cited from Malaisse 1997].

Gordon, D.G. 1996. The compleat cockroach. Univ. Illinois Press, Illinois, USA [cited from D.G. Gordon [author] and S. Matuura [translator] "Gokiburi Daizen". Seido-sha, Tokyo, Japan. 199 (in Japanese)].

Gourou, P. 1948. Les Pays Tropicaux. Principes d'une geographie humaine et economique. 2nd ed. Paris, France [cited from DeFoliart 2002].

Green, S.V. 1998. The bushman as an entomologist. Antenna-London 22(1): 4–8 [cited from Malaisse 2005].

Grimaldi, J. and A. Bikia. 1985. Le grand livre de la cuisine Camerounaise [cited from DeFoliart 1990].

Grivetti, L.E. 1979. Kalahari agro-pastoral-hunter-gatherers: the Tswana example. Ecol. Food Nutr. 7: 235–256.

Günther, K. 1931. A Naturalist in Brazil. The Flora and Fauna and the People of Brazil. George Allen and Unwin Ltd. London, England.

Hall, H.J. 1977. A paleoscatalogical study of diet and disease at Dirty Shame Rocksheleter, southeast Oregon. Tebiwa: Misc. Papers Idaho State Univ. Mus. Nat. Hist., no. 8 [cited from DeFoliart 1991].

Hanboonsong, Y. 2010. Edible insects and associated food habits in Thailand. pp. 173–182. *In*: P.B. Durst, D.V. Johnson, R.N. Leslie and K. Shono [eds.]. Forest Insects as Food: Humans Bite Back. FAO Regional Office for Asia and the Pacific, Bangkok, Thailand.

Hanboonsong, Y., A. Rattanapan, Y. Utsunomiya and K. Masumoto. 2000. Edible insects and insect-eating habit in northeast Thailand. Elytra 28: 355–364.

Hanboonsong, Y., A. Rattanapan, Y. Waikakul and A. Liwvanich. 2001. Edible insects survey in northeastern Thailand. Khon Kaen Agriculture Journal 29(1): 35–45 (in Thai with English summary).

Hanboonsong, Y., T. Jamjanya and P.B. Durst. 2013. Six-legged livestock: edible insect farming, collecting and marketing in Thailand. FAO of United Nations, Regional Office for Asia and the Pacific, Bankok, Thailand.

Harris, W.V. 1940. Some notes on insects as food. Tanganyika Notes and Records 9: 45–48.

Hawkes, K., K. Hill and J.F. O'connell. 1982. Why hunters gather: optimal foraging and the Achè of the eastern Paraguay. Am. Ethnol. 9: 379–398.

Hayashi, T. 1903. Insects usable as medicine. Insect World 7(72): 340–341 (in Japanese).

Hayashi, T. 2005. Ant Power for Pain. Toransu-shuppan, Tokyo, Japan (in Japanese).

Hearn, L. 1890. Two Years in the French West Indies. Harper & Bros. Publ., New York, N.Y., USA.

Heimpel, W. 1996. Moroccan locusts in Quattunan. Rev. d'Assyriologie 90: 101–120 [cited from Lanfranchi 2005].

Heizer, R.F. and L.K. Napton. 1969. Biological and cultural evidence from prehistoric human coprolites. Science 165: 563–568.

Heymans, J.C. and A. Evrard. 1970. Contribution à l' étude de la composition alimentaire des insectes comestibles de la Province du Katanga. Probl. Soc. Congolais (Bull. Trimest. C.E.P.S.I.) [cited from Malaisse 1997].

Hill, K., K. Hawkes, H. Kaplan and A.M. Hurtado. 1984. Seasonal variance in the diet of Ache hunter-gatherers in eastern Paraguay. Human Ecol. 12(2): 101–135.

Hirase, S. and K. Shitomi. 1799. Illustrated specialty of marine and terrestrial materials in Japan. pp. 3–80. *In*: T. Miyamoto, T. Haraguchi and K. Tanigawa [eds.]. Collection of Common People's Life Histories in Japan. Vol. 10 Life of Farmers and Fishermen. San-ichi-shobo, Tokyo, Japan] (in Japanese).

History of Nagano Publishing Body. 1988. History of Nagano Prefecture, Forklore Vol. 2, Southern area of Shinshu (1) Daily life. Nagano Prefecture, Nagano (in Japanese).

Hitchcock, S.W. 1962. Insects and indians of the Americas. Bull. Entomol. Soc. Am. 8: 181–187.

Hoare, A.L. 2007. The use of non-timber forest products in the Congo Basin: Constrains and opportunities. The Rainforest Foundation, London, England. ISBN : 978-1-906131-03-6.

Hockings, H.J. 1884. Notes on two Australian species of Trigona. Trans. Ent. Soc. Lond.: 149–157 [cited from Bodenheimer 1951].

Hoffmann, W.E. 1947. Insects as human food. Proc. Entomol. Soc. Washington 49: 233–237.

Honda, K. 1981. Canada Eskimo. Asahi News Paper Co., Tokyo, Japan (in Japanese).

Hope, F.W. 1842. Observations respecting various insects which at different times have offered food to Man. Trns. Entomol. Soc. Lond. 1842: 129–150.

Hori, S. 2006. Aqueous bugs. Bombardier beetles harbored in water. Nasu Culture Research Association [ed.]. Archives of Nasu Culture, Nature, History, Folklore (in Japanese).

Howard, L.O. 1886. The edibility of the periodical cicada. Proc. Entomol. Soc. Washington 1: 29.

Howard, L.O. 1915. The edibility of insects. J. Econ. Entomol. 8: 549.

Howard, L.O. 1916. *Lachnosterna* larvae as a possible food supply. J. Econ. Entomol. 9: 389–392.

Hrdlicka, A. 1908. Physiological and medical observations among the Indians of southwestern United States and northern Mexico. Smithsonian Inst. Bur. Am. Ethnol. Bull. 34: 25, 264–265 [cited from DeForiart 2002].

Hu, P. and L.-S. Zha. 2009. Records of edible insects from China. Agric. Sci. Technol. 10(6): 114–118.

Hugh-Jones, S. 1979. The Palm and the Pleiades. Initiation and Cosmology in Northwest Amazonia. Cambridge Univ. Press, London, England.

Hunter, J.M. 1984. Insect clay geophagy in Sierra Leone. J. Cult. Geo. 4: 2–13.

Huntingford, G.W.B. 1955. The economic life of the Dorobo. Anthropos 50: 602–634.
Hurtado, A.M., K. Hawkes and H. Kaplanm. 1985. Female subsistence strategies among Achè hunter-gatherers in Eastern Paraguay. Human Ecol. 13: 1–28.
Huyghe, P. 1992. An acquired taste. The Science November/December: 8–11.
Ichikawa, M. 1982. Life of Hunters in Forests, Mbti-pygmee. Jinbun-shoin, Kyoto, Japan (in Japanese).
Ikeda, J., S. Dugan, N. Feldman and R. Mitchell. 1993. Native Americans in California surveyed on diets, nutrition needs. Calif. Agri. 47(3): 8–10.
Inagaki, K. 1984. Pharmacognostical studies on the traditional Chinese crude drugs of insects. Private Publication (in Japanese).
İncekara, Ü. and H. Türkez. 2009. The genotoxic effects of some edible insects on human whole blood cultures. Mun. Ent. Zool. 4(2): 531–532, 755–757.
Irvine, F.R. 1957. Indigenous African methods of beekeeping. Bee World 38: 113–128.
Isaacs, J. 1987. Bush food: Aboriginal food and herbal medicine. Weldon Pty Ltd., Sydney, NSW, Australia [cited from Yen 2005].
Ishimori, N. 1944. Agriculture, Insects and Silkworms. Kyoiku-Kagakusha, Tokyo, Japan (in Japanese).
Iwano, H. 2000. Ultimate gourmet "The taste of the banana skipper". Tamamushi No. 38: 33–35 (in Japanese).
Jackson, A.P. 1954. Ample food without ploughing. NADA, S. Rhodesia Native Affairs Dept. Ann. 31: 64–66 [cited from DeFoliart 2002].
James, E. 1823. Account of an expedition from Pittsburgh to the Rocky Mountains. 3 Vols. London, England [cited from Sutton 1988].
James, K.W. 1983. Analysis of indigenous Australian foods. Food Technol. Australia 35: 342–343.
Jamjanya, T., S.N. Nakorn, P. Whangsomnuk, P. Taapsiipear and Y. Ponsaen. 2001. The variety of edible insect in Khonkaen Province. Khon Kaen Agr. J. 29(1): 1–9.
Jolivet, P. 1971. A propos des insectes "a boissons" et des insectes "a sauce". L'Entomologiste 27: 3–9 [cited from DeFoliar 2002].
Jones, V.H. 1945. The use of honey-dew as food by Indians. The masterkey 19: 145–149.
Jones, V.H. 1948. Prehistoric plant materials from Castle Park. pp. 94–99. *In*: R.F. Burgh and C.R. Scoggin [eds.]. The Archaeology of Castle Park, Dinosaur National Monument. Appendix III. Boulder. Univ. Colo. Ser. Anthropol. No. 2 [cited from DeFoliart 2002].
Jonjuapsong, L. 1996. In Northeast Thailand, insects are supplementary food with a lot of value. The Food Insect Newsletter 9(2): 4–7.
Junod, H.A. 1927. The Life of a South African Tribe. Macmillan, London, England.
Junod, H.A. 1962. The life of a South African Tribe. Vol. 1 and Vol. 2. Univ. Books, Inc., New York, N.Y., USA.
Katagiri, M. and R. Awatsuhara. 1996. Investigation of edible insects in "Ina area" and comparison among the contents of fatty-acids in imago of locusts by various extracting methods. Bull. Iida Women's Junior College No. 1 3: 82–89 (in Japanese).
Kato, D. and G.V. Gopi. 2009. Ethnozoology of Galo tribe with special reference to edible insects in Arunachal Pradesh. Indian J. Traditional Knowledge 8(1): 81–83.
Kelly, I.T. 1932. The ethnography of the Surprise Valley Paiute. Univ. Calif. Publs. Am. Archaeol. Ethnol. 31(3): 67–210 [cited from DeFoliart 2002].
Kerr, A.J. 1931. An edible larva (*Zeuzera coffeae*). Siam Soc. Nat. Hist. Suppl. VIII-No. 3: 217–218.
Kevan, D.K. McE. 1991. The eating of stick insects by humans. Food Insect Newsletter 4(2): 7.
Khen, C.V. and S. Unchi. 1998. Edible sago grubs. Malaysian Nat. 52: 29–30.
Khun, P. 2008. Insects as diet foods for humans. Siam, Insect Zoo. http://www.malaeng.com/blog/index.php.
Knox, R. 1817. An historical relation of the island of Ceylon. London [cited from Bodenheimer 1951].
Koizumi, K. 1935. Medicinal animals in Japan. Maruyama-shoten, Tokyo, Japan (in Japanese).
Kolben, P. 1738. The present state of Cape of Good Hope. Vol. II, London, England [cited from Bodenheimer 1951].

Komuro, S. 1968. Dobsonflies around Sai River, Miyagi Prefecture. Insects and Nature 3(7): 24–25 (in Japanese).

Kropf, A.A. 1899. A. Kaffir-English Dictionary. Lovedale, U.K. [cited from Malaisse 2005].

Künckel d'Herculais, M.J. 1891. La note sur les criquets pélerins de l'extrême sud de l'Algérie et sur les populations acridiophages. Bull. Soc. Entomol. Fr. Séance, Feb. 11: 24–26.

Kurumi, S. 1918. Edible wild plants and insects. *Dozoku to Densetsu* (Folk Habits and Legend) 1(2): 16–18 (in Japanese).

Kuwabara, M. 1997a. Present status of the custom to eat insects in Thailand. Iden 51: 67–72 (in Japanese).

Kuwabara, M. 1997b. Retrospect and present status of the custom to eat insects in Thailand. pp. 120–146. *In*: J. Mitsuhashi [ed.]. People Eating Insects. Heibonsha, Tokyo, Japan (in Japanese).

Kuwana, I. 1930. Insects in general. pp. 460–545. *In*: S. Uchida [ed.]. A Illustrated Book of Economic Animals. Hokuryukan, Tokyo, Japan (in Japanese).

Labat, J.B. 1722. Nouveau voyage aux Iles de l'Amérique [cited from Bodenheimer 1951].

Laredo, G. 2004. Pitéu crocante. Globo Rural 230: 96–97 [cited from Costa-Neto and Ramos-Elorduy 2006].

Latham, P. 1999. Edible caterpillars of the Bas Congo region of the Democratic Republic of the Congo. Antenna of Bull. Roy. Entomol. Soc. 23: 134–139.

Latham, P. 2002. Edible caterpillars in Bas Congo. Trop. Agr. Assoc. Newsletter 22(1): 14–17 (URL: http://www.taa.org.uk/TAAScotland/Edible Caterpillars2.htm).

Latham, P. 2003. Edible caterpillars and their food plants in Bas-Congo. Mystole Publications, Canterbury, England.

Latz, P. 1995. Bushfires and bushtucker: Aboriginal plant use in Central australia. IAD Press, Alice Springs, N. Terr., Australia [cited from Yen 2005].

Leibowitz, J. 1943. A new source of trehalose. Nature 152: 414.

Leksawasdi, P. 2008. Compendium of research on selected edible insects in northern Thailand. Abstract of a workshop on Asia-Pacific resources and their potential for development, Forest insects as food: humans bite back, FAO Regional Office for Asia and the Pacific, Bangkok [org.]. 19–21 February. 2008. Chiang Mai, Thailand.

Leksawasdi, P. 2010. Compendium of research on selected edible insects in northern Thailand. pp. 183–188. *In*: P.B. Durst, D.V. Johnson, R.N. Leslie and K. Shono [eds.]. Forest Insects as Food: Humans Bite Back. FAO Regional Office for Asia and the Pacific, Bangkok, Thailand.

Leksawasdi, P. and P. Jirada. 1983. Malang mun (Formicidae: Hymenoptera) Part 1. 9th conference on Sciences and Technology for Development of North Eastern Region, Faculty of Sciences, Khon Kaen University, 27–29 October [cited from Leksawasdi 2010].

Lenko, K. and N. Papavero. 1979. Insetos no Folclore. Conselho Estadual de Artes e Ciencias Humanas, Sao Paulo, Brazil [cited from Mill 1982].

Lévy-Luxereau, A. 1980. Note sur quelques criquets de la région de Maradi (Niger), et leur noms Hausa. J. d'Agric. Trad. et de Bota. Appl. 27(3-4): 263–272.

Li, Shizhen. 1596. Compendium of Materia Medica Chapters 39–42 (Insects) [Japanese version, Shunyo-Do Shoten, Tokyo, Vol. 10, pp. 586, 1930].

Lima, D.C. de O. 2000. Conhecimento e práticas populares envolvendo insetos na região em tornoda Usina Hidroelétrica de Xingó (Sergipe e Alagoas). Monografia (Bacharelado em Ciências Biológicas). Recife: Universidade Federal de Pernambuco [cited from Costa-Neto and Ramos-Elorduy 2006].

Linné, C. 1811. Lachesis Lapponica (translated from original MS. and published by J.E. Smith) London, England [cited from Fagan 1918].

Liú, Y. 2006. Mealworms. Technics for the Production and Application. Chinese Agricultural Publication Co., Beijing, China (in Chinese).

Livingstone, D. 1857. Missionary travels and Researches in South Africa. John Murrey, London, England.

Lizot, J. 1977. Population, resources and warfare among the Yanomami. Man(N.S.) 12: 497–517.

Logan, J.W.M. 1992. Termites (Isoptera), a pest or resource for small farmers in Africa. Trop. Sci. 32: 71–79.

Long, A.M. 1901. Red ants as an article of food. J. Bombay Nat. His. Soc. 13: 536.

Low, T. 1989. Bush tucker. Australia's wild food harvest. Angus and Robertson. North Ryde, Australia.

Lukiwati, D.R. 2010. Teak caterpillars and other edible insects in Java. pp. 99–103. *In*: P.B. Durst, D.V. Johnson, R.N. Leslie and K. Shono [eds.]. Forest Insects as Food: Humans Bite Back. FAO Regional Office for Asia and the Pacific, Bangkok, Thailand.

Lumholtz, C. 1890. Among cannibals [cited from Bodenheimer 1951].

Lumsa-ad, C. 2001. A study on the species and the nutrition values of edible insects in upper Southern Thailand. Khon Kaen Agriculture Journal 29(1): 45–49 (in Thai).

Luo, Ke. 1990. Ke Processed insects in China. The Food Insect Newsletter 3(1): 6.

Luttrell, D. 1992. Collecting termites in the Pacific Northwest. The Food Insect Newsletter 5(1): 5.

MacDonald, W.W. 1956. Observation on the biology of Chaoborids and Chironomids in Lake Victoria and on the feeding habits of the 'elephant-snout fish' (*Mornyrus kannume* Forsk.). J. Animal Ecol. 25: 36–53.

Madsen, D.B. and J.E. Kirkman. 1988. Hunting hoppers. Am. Antiquity 53: 593–604.

Mainichi News Paper Co. 1979. Villains in Human Life. Mainichi News Paper Co., Tokyo, Japan (in Japanese).

Makita, Y. 2002. Sole culture in the world. - Characteristic thing of winter in Ina district. http://www.valley.ne.jp/~zaza/zazamushi.htm (in Japanese).

Malaisse, F. 1997. Se nourir en forêt claire africaine. Approche écologique et nutritionnelle. Les Presses Agronomiques de Gembloux, Gembloux, Belgium.

Malaisse, F. 2005. Human consumption of Lepidoptera, termites, Orthoptera, and ants in Africa. pp. 175–230. *In*: M.G. Paoletti [ed.]. Ecological Implications of Minilivestock. Potential of Insects, Rodents, Frogs and Snails. Science Publishers Inc., Enfield, USA.

Malaisse, F. and G. Parent. 1980. Les chenilles comestibles du Shaba méridional (Zaïre). Nat. Belges 61: 2–24.

Malaisse, F. and G. Parent. 1997a. Minor wild edible products of the Miombo area. Geo-Eco-Trop. 20 [cited from Malaisse and Parent 1997b].

Malaisse, F. and G. Parent. 1997b. Chemical composition and energetic value of some edible products provided by hunting or gathering in the open forest (Miombo). Geo-Eco-Trop. 21: 65–71.

Malaisse, F. and G. Lognay. 2003. Les chenilles comestibles d'Afrique tropicale. pp. 279–304. *In*: E. Motte-Florac and J.M.C. Thomas [eds.]. "Insects" in Oral Literature and Traditions. Peters, Leuven, Belgium.

Malouf, C. 1951. Gosiute Indians. pp. 25–172. *In*: D.A. Horr [ed.]. "Shoshone Indians" (1974). Garland Publ. Inc., New York, N.Y., USA [cited from DeFoliart 2002].

Mao, K. 1997. Traditional entomophagy. pp. 69–89. *In*: J. Mitsuhashi [ed.]. People Who Eat Insects. Heibon-sha, Tokyo, Japan (in Japanese).

Mapunzu, M. 2002. Contribution de l'exploitation des chenilles et autres larves comestibles dans la lutte contre l'insécurité alimentaire et la pauvreté en République Démocratique du Congo. *In*: N'Gasse [ed.]. Contribution des insectes de la forêt à la sécurité alimentaire. L'exemple des chenilles d'Afrique Centrale. FAO, Rome, Italy. URL: http://www.fao.org/docrep/007/j3463f/j3463f00.htm.

Mapunzu, M. 2004. Contribution de l'exploitation des chenilles et autres larves comestibles dans la lutte contre l'insécurité alimentaire et la pauvreté en République démocratique du Congo. pp. 66–86. *In*: FAO Departement des Forets: Contribution des insectes de la foret a la securite alimentaire. Produits forestiers non ligneux. Document de Travail No. 1, FAO, Rome, Italy.

Marais, E. 1996. Omaungu in Namibia: *Imbrasia belina* (Saturniidae: Lepidoptera) as a commercial resource. pp. 23–31. *In*: B.A. Gashe and S.F. Mpuchane [eds.]. "Phane", Proc.

1st Multidisciplinary Symp. on Phane, June 1996 Botswana, Dep. of Biol. Sci. & The Kalahari Conserv. Soc. (org.). Botswana.

Marconi, S., P. Manzi, L. Pizzoferrato, E. Buscardo, H. Cerda, D.L. Hernandez and M.G. Paoletti. 2002. Nutritional evaluation of terrestrial invertebrates as traditional food in Amazonia. Biotropica 34(2): 273–280.

Marques, J.G.W. and E.M. Costa-Neto. 1997. Insects as folk medicines in the state of Alagoas, Brazil. The Food Insect Newsletter 10(1): 7–10.

Masumoto, K. and Y. Utsunomiya. 1997. Beetles as food material observed in northern Thailand. Elytra 25: 424.

Matsumura, G. 1918. View on Insects 8, Medicinal insects. Insect World 22: 343–344 (in Japanese).

Matsuura, M. 1998a. Dishes of oriental bee larvae in eastern Java Island, Indonesia. Bee Sci. 19(4): 149–154 (in Japanese).

Matsuura, M. 1998b. Why hornets sting? Publishing Department, Hokkaido Univ., Sapporo, Japan (in Japanese).

Matsuura, M. 2002. Eating honets. Search for entomophagy culture. Publishing Department of Hokkaido Univ., Sapporo, Japan (in Japanese).

Matsuura, S. 1999. See Gordon 1996.

Maxwell-Lefroy, H. 1906. Indian insect pest. Office of the Superintendent of Goberment Printing, India (reprinted in 1971 by Today and Tomorrow Publishers) [cited from DeFoliart 2002].

May, R.J. 1984. KAIKAI ANIANI (A guide to bush foods markets and culinary arts of Papua New Guinea). Robert Brown and Assoc. Bathurst, Australia.

Mayr, G. 1855. [cited from Yasumatsu 1948 (in Japanese)].

Mbata, K.J. 1995. Traditional use of arthropods in Zambia: I. The food insects. Food Insects Newsletter 8(3): 1, 5–7.

Mbata, K.J. and E.N. Chidumayo. 2003. Traditional values of caterpillars (Insecta: Lepidoptera) among the Bisa people of Zambia. Insect Sci. Applic. 23: 341–354.

Mbata, K.J., E.N. Hidumayo and C.M. Lwatula. 2002. Traditional regulation of edible caterpillar exploitation in Kopa area of Mpika district in northern Zambia. J. Insect Conserv. 6: 115–130.

McFadden, M.W. 1966. Discovery of fossils of *Hermetia illucens* (Linnaeus) in Mexico (Diptera: Atratiomyidae). Proc. Entomol. Soc. Wash. 68: 56.

Mekloy, P. 2002. Catching cicadas. As night falls, a forest hot spring becomes the eerie scene of an unusual hunt. Bangkok Post, Monday 22 April, 2002.

Mendes dos Santos, G. 1995. Agricultura e coleta enawene-nawe: relações sociais e representações simbólicas. Estudo das potencialidades e do manejo dos recursos naturais na Área Indigena Enawene-Nawe. OPAN/GERA-UFMT [cited from Costa-Neto and Ramos-Elorduy 2006].

Menzel, P. and F. D'Aluisio. 1998. Man Eating Bugs: The Art and Science of Eating Insects. Ten Speed Press, Berkeley, California, USA.

Mercer, C.W.L. 1993. Insects as food in Papua New Guinea. Proc. Seminar 'Invertebrates (minilivestock) farming', EEC-DGXII/CTA/IFS/DMMMSU/ITM, Philippines, November 1992, Tropical Animal Production Unit., Inst. Trop. Med., Antwerpen, Belgium.

Métraux, A. 1963. Tribes of the middle and upper Amazon River. pp. 687–712. *In*: J.H. Steward [ed.]. Handbook of South American Indians, Vol. 3. Cooper Square Publ. New York, N.Y., USA.

Meyer-Rochow, V.B. 2005. Traditional food insects and spiders in several ethnic groups of northeast India, Papua New Guinea, Australia and New Zealand. pp. 389–413. *In*: M.G. Paoletti [ed.]. Ecological Implications of Minilivestock, Potential of Insects, Rodents, Frogs and Snails. Science Publishers, Inc., Enfield, USA.

Meyer-Rochow, V.B. 1973. Edible insects in three different ethnic groups of Papua and New Guinea. Am. J. Clin. Nutr. 26: 673–677.

Meyer-Rochow, V.B. and S. Changkija. 1997. Uses of insects as human food in Papua New Guinea, Australia, and North-east India: Cross-cultural considerations and cautious conclusions. Ecol. Food Ntr. 36: 159–185.

Mignot, J.-M. 2003. La classification des arthropods selon les Masa Bugudum (Nord-Cameroum): Premier apercu. pp. 105–121. *In*: E. Motte-Florac and J.M.C. Thomas [eds.]. Les "insectes" dans la tradition orale. Peeters, Leuven, Belgium.

Mill, A.E. 1982. Amazon termite myths: Legends and folk-lore of the Indians and caboclos. Bull. Roy. Entomol. Soc., London 6(2): 214–217.

Milton, K. 1984. Protein and carbohydrate resources of the Maku Indians of northwestern Amazonia. Am. Anthropol. 86: 7–27.

Milton, K. 1997. Real men don't eat deer. Discover, June [cited from Costa-Neto and Ramos-Elorduy 2006].

Mitamura, T. 1995. Observation of insects in Laos. - Edible insects. Insects in Fukushima No. 13: 32–33 (in Japanese).

Mitsuhashi, J. 1997. People Who Eat Insects. Heibon-sha, Tokyo Japan (in Japanese).

Mitsuhashi, J. 2003. Cochineal and kermes. pp. 998–999. *In*: J. Mitsuhashi [ed.]. Encyclopedia of Entomology. Asakura-shoten, Tokyo, Japan (in Japanese).

Mitsuhashi, J. 2005. Edible insects in Japan. pp. 251–262. *In*: M.G. Paoletti [ed.]. Ecological Implications of Minilivestock. Potential of Insects, Rodents, Frogs and Snails. Science Publishers Inc., Enfield, USA.

Mitsuhashi, J. 2008. A Complete World Entomophagy. Yasaka-shobo, Tokyo, Japan (in Japanese).

Mitsuhashi, J. 2010. World Entomophagy Past and Present. Kōgyō Chōsa-kai, Tokyo, Japan (in Japanese).

Mitsuhashi, J. 2012. Encyclopedia of Entomophagy Culture. Yasaka-shobo, Tokyo, Japan (in Japanese).

Miyake, T. 1919. Researches on edible and medicinal insects. Special Report from Agricultural Experiment Station. No. 31. Tokyo, Japan (in Japanese).

Mizuno, A. and T. Ibaraki. 1983. *Akabachi* at the foot of Mt. Hakusan. Hakusan 10(3): 12–15 (in Japanese).

Mojeremane, W. and A.U. Lumbile. 2005. The characteristics and economic values of *Colophospermum mopane* (Kirk ex Benth.) J. Léonard in Botswana. Pakistan J. Biol. Sci. 8(5): 781–784.

Motschoulsky, M.V. 1856. Etudes entomologiques. 5 [cited from Bodenheimer 1951].

Moussa, J.-B. 2004. Les chenilles comestibles de la République de Congo : Intérêt alimentaire et circuits de commercialization. Cas de Brazzaville. FAO Departement des Forets: Contribution des insectes de la foret a la securite alimentaire. Produits forestiers non ligneux, Document de Travail No. 1, FAO, Rome, Italy.

Muir, J. 1911. My first summer in the Sierra. Houghton Mifflin Co., Boston [cited from Nature Writings, Lib. Am., New York, N.Y., USA. 1997].

Mukaiyama, M. 1987. Eating habit - Insects and other small animals. Collection of M. Mukaiyama's Publications "Career in mountainous regions - Ina Valley in Shinshu. 1. Food, clothing and shelter in mountainous regions." Shinyo-sha, Nagano, Japan (in Japanese).

Mungkorndin, S. 1981. Forest as a source of food to rural communities in Thailand. FAO Regular Programme No. RAPA 52, FAO, Bankok, Thailand.

Munthali, S.M. and D.E.C. Mughogho. 1992. Economic incentives for conservation: beekeeping and Saturniidae caterpillar utilization by rural communities. Biodiversity and Conservation 1: 143–154.

Mushambanyi, T.M.B. 2000. Etude préliminaire orientée vers la production des chenilles consommables par l'élevage des papillons (*Anaphe infracta*: Thaumetopoeidae) à Lwiro, Sud-Kivu. République Démocratique de Congo. Tropicultura 18: 208–211.

Nadchatram, M. 1963. The winged stick insect, *Eurynema versifasciata* Servile (Phasmida, Phasmatidae), with special reference to its life-history. Malayan Nature J. 17: 33–40.

Nagano Fishery Experiment Station, Suwa. 1985. Species composition of aquatic insects in Tenryu River [cited from Makita 2002] (in Japanese).

Nakai, I. 1988. Zaza mushi, a special product from Ina district, Nagano Prefecture, and its specific composition. Bulletin of The Ikeda High School affiliated to Osaka Educational University No. 20: 141–46 (in Japanese).

Nakao, S. 1964. Custom of Thailand related to entomophagy. Bulletin of Kurume University 13: 81–85 (in Japanese).

Nanba, T. 1980. Color illustration of Japanese and Chinese Medicinal Materials (Vol. II). Hoikusha, Osaka, Japan (in Japanese).

Nandasena, M.R.M.P., D.M.S.K. Disanayak and L. Weeratunga. 2010. Sri Lanka as a potential gene pool of edible insects. pp. 161–164. *In*: P.B. Durst, D.V. Johnson, R.N. Leslie and K. Shono [eds.]. Forest Insects as Food: Humans Bite Back. FAO Regional Office for Asia and the Pacific. Bangkok, Thailand.

Naumann, I. 1993. CSIRO handbook of Australian insect names. 6th ed. CSIRO. East Melbourne Australia [cited from Yen 2010].

Netolitzky, F. 1918/1920. Kaefer als Nahrung und Heilmittel. Koleopterologische Rundschau Vols. 7 and 8.

Nishikawa, K. 1974. An eight years underground travel in trackless western Asia (supplement) - Walking in Tibet. Fuyo-shobo, Tokyo, Japan (in Japanese).

Noetling, F. 1910. The food of the Tasmanian aborigines. Papers and Proc., Roy. Soc. Of Tasmania, Australia [cited from Bodenheimer 1951].

Noice, H.H. 1939. Bach of beyond. G.P. Putman's Sons, New York, N.Y., USA [cited from DeFoliart 2002].

Nonaka, K. 1991. Entomophagy in South Korea. Behavior and Culture No. 18: 25–44 (in Japanese).

Nonaka, K. 1996. Ethnoentomology of the Central Kalahari San. African Study Monographs, Suppl. 22: 29–46.

Nonaka, K. 1998. Tasting nature - Way to taste *hachinoko*, and its application to activation of villages. Research Bulletin of Mie University (Human Science) No. 15: 141–154 (in Japanese).

Nonaka, K. 1999a. Changes in agriculture at continental area of Southeast Asia. - On the relation of natural circumstance and changes in social condition. I.F. Report 26: 54–61 (in Japanese).

Nonaka, K. 1999b. Entomophagy of Sago weevil (*Rhynchophorus ferrugineus*) in Slawesi and Maluku district in eastern Indonesia. Sago Palm 7: 8–14 (in Japanese).

Nonaka, K., S. Sivilay and S. Boulidam. 2008. The biodiversity of edible insects in Vientiane. Res. Inst. for Humanity and Nature, Japan and Nat. Agr. and Forestry Res. Inst., Laos.

Nunome, J. 1979. Origin of sericulture and ancient silk. Yūzankaku, Tokyo, Japan (in Japanese).

Oberprieler, R. 1995. The Emperor Moths of Namibia. Ekogilde, Hartbeespoort, R.S.A. [cited from Malaisse 2005].

Ohtsuka, R., T. Kawabe, T. Inaoka, T. Suzuki, T. Hongo, T. Akimichi et al. 1984. Composition of local and purchased foods consumed by the Gidra in lowland Papua. Eco., Food Nutr. 15: 159–169.

Okamoto, H. and S. Muramatsu. 1922. Researches on edible and medicinal insects. Korea Govermental Experimental Station Research Reports No. 7 (in Japanese).

Oliveira, J.F.S., J.P. de Carvalho, R.F.X.B. de Sousa and M.M. Simão. 1976. The nutritional value of four species of insects consumed in Angola. Ecol. Food Nutr. 5: 91–97.

Olivier, A.G. 1813. Travels in the Ottoman Empire, Egypt and Persia. Engl. Transl., London. 2 Vols., I [cited from Bodenheimer 1951].

Olivier, G.A. 1801–1807. Voyage dans l'empire Othoman. Vol. 2 [cited from Fagan 1918].

Olmsted, D.L. and O.C. Stewart. 1978. Achumawi. pp. 225–235. *In*: R.F. Heizer [ed.]. Handbook of North American Indians Vol. 8 California. Smithonian Institution, Washington, D.C., USA.

Ono, R. 1844. Enlightenment of medicinal materials [cited from T. Sugimoto [Editor-Writer]. Enlightenment of medicinal materials by Ranzan Ono. Publishing Department of Waseda University, Tokyo, Japan] (in Japanese).

Onore, G. 1997. A brief note on edible insects in Ecuador. Ecol. Food Nutr. 36: 277–285 [cited from Paoletti and Dufour 2005].

Onore, G. 2005. Edible insects in Ecuador. pp. 343–352. *In*: M.G. Paoletti [ed.]. Ecological Implications of Minilivestock, Potential of Insects, Rodents, Frogs and Snails. Science Publishers, Inc., Enfield, USA.

Orcutt, C.R. 1887. A lemonade and sugar tree. West Am. Sci. 3: 45–47 [cited from Heizer 1945].

Ott, J. 1998. The delphic bee: bees and toxic honeys as pointers to psychoactive and other medicinal plants. Economic Botany 52(3): 260–266.

Overstreet, R.M. 2003: Flavor buds and other delights. J. Parasitol. 89: 1093–1107.

Owen, D.F. 1973. Man's Environmental Predicament. An Introduction to Human Ecology in Tropical Africa. Oxford University Press, London, England.

Pagezy, H. 1975. Les interrelations homme-faune de la forêt du Zaïre. pp. 63–88. *In*: R. Pujol [ed.]. L'homme et l'animal. Premier colloque d'ethonozoologie. Inst. Int. Ethnosci. Paris, France [cited from Malaisse 2005].

Panagiotakopulu, E. and P.C. Buckland. 1991. Insect pests of stored products from late Bronze Age Santorini, Greece. J. stored Prod. Res. 27: 179–184.

Paoletti, M.G. and A.L. Dreon. 2005. Minilivestock, environment, sustainability, and local knowledge disappearance. pp. 1–18. *In*: M.G Paoletti [ed.]. Ecological Implications of Minilivestock, Potential of Insects, Rodents, Frogs and Snails. Science Publishers, Inc., Enfield, USA.

Paoletti, M.G. and D.L. Dufour. 2005. Edible invertebrates among Amazonian Indians: A critical review of disappearing knowledge. pp. 293–342. *In*: M.G. Paoletti [ed.]. Ecological Implications of Minilivestock, Potential of Insects, Rodents, Frogs and Snails. Science Publishers, Inc., Enfield, USA.

Paoletti, M.G., D.L. Dufour, H. Cerda, F. Torres, L. Pizzoferrato and D. Pimentel. 2000. The importance of leaf- and litter-feeding invertebrates as sources of animal protein for the Amazonian Amerindians. Proc. R. Soc. Lond. B-267: 2247–2252.

Paoletti, M.G., E. Buscardo, D.J. Vanderjagt, A. Pastuszyn, L. Pizzoferrato, Y.-S. Huang et al. 2003. Nutrient content of termites (*Syntermes* soldiers) consumed by Makiritare Amerindians of the Alto Orinoco of Venezuela. Ecol. Food Nutr. 42: 173–187.

Parent, G., F. Malaisse and C. Verstraeten. 1978. Les miels dans la forêt Claire du Shaba méridional. Bull. Rech. Agron. Gembloux 13: 161–176.

Parsons, M. 1999. The butterflies of Papua New Guinea.: their systematics and biology. Academic Press, London, England [cited from Ramandey and van Mastrigt 2010].

Pathak, K.A. and K.R. Rao. 2000. Insects as human food of tribals in north eastern region of India. Indian J. Entomol. 62: 97–100.

Patterson, J.E. 1929. The pandra moth, a periodic pest of western pine forests. USDA Tech. Bull. No. 137.

Paulian, R. 1943. Les Coléopteres. Paris [cited from Bodenheimer 1951].

Peigler, R.S. 1993. Wild silks of the world. Am. Entmol. 39(3): 151–161.

Peigler, R.S. 1994. Non-sericultural uses of moth cocoons in diverse cultures. Proc. Denver Mus. Nat. Hist. Ser. 3, No. 5.

Pemberton, R.W. 1994. The revival of rice-field grasshoppers as human food in South Korea. Pan-Pacific Entomol. 70: 323–327.

Pemberton, R.W. 1995. Catching and eating dragonflies in Bali and elsewhere in Asia. Am. Entomol. 41: 97–99.

Pemberton, R.W. 2005. Contemporary use of insects and other arthropods in traditional Korean medicine (Hanbang) in south Korea and elsewhere. pp. 459–473. *In*: M.G. Paoletti [ed.]. Ecological Implications of Minilivestock, Potential of Insects, Rodents, Frogs and Snails. Science Publishers, Inc., Enfield, USA.

Pennino, M., E.S. Dierenfeld and J.L. Behler. 1991. Retinol, α-tocopherol and promimate nutrient composition of invertebrates used as feed. Int. Zoo Yb. 30: 143–149.

Pereira, N. 1954. Os indios Maués. Rio de Janeiro, Organização Simoes [cited from Mill 1982; Posey 2003].

Pierce, W.D. 1915. The use of certain weevils and weevil products in food and medicine. Proc. Entomol. Soc. Washington 17: 151–154.

Platt, B.S. 1980. Table of representative values of foods commonly used in tropical countries. HMSO Medical Research Council, Special Report Series 302, London, England [cited from Malaisse 1997].

Plinius, Gaius Secundus Major. AD77-79. Naturlis Historia Vols. 11, 21 and 29. S. Nakano, S. Nakano and M. Nakano. [translated]. Natural Histories by Plinius. Yuzankaku Publishing Company, Tokyo, Japan. 1986.] (in Japanese).

Pomeroy, D.E. 1976. Studies on a population of large termite mounds in Uganda. Ecol. Entomol. 1: 49–61.

Posey, D.A. 1978. Ethnoentomological survey of Amerind groups in lowland Latin America. Florida Entomol. 61: 225–229.

Posey, D.A. 1979. Ethnoentomology of the Gorotire Kayapo of Central Brazil. PhD. Univ. Georgia, Athens, USA [cited from Paoletti and Dufour 2005].

Posey, D.A. 1983a. Folk apiculture of the Kayapó Indians of Brazil. Biotropika 15: 154–158.

Posey, D.A. 1983b. Indigenous knowledge and development: an ideological bridge to the future. Ciência e Cultura 35: 877–894.

Posey, D.A. 1987. Enthnoentomological survey of Brazilian Indians. Entomologia. Generalis 12: 191–202.

Posey, D.A. and J.M.F. de Camargo. 1985. Additional notes on beekeeping of Meliponinae by the kayapo Indians of Brazil. Ann. Carnegie Mus. Nat. Hist. 54(8): 247–274 [cited from DeFoliart 2002].

Powers, S. 1877. Tribes of California. Contributions to North American Ethnology, Vol. III. U.S. Geograph. & Geol. Surv. Of Rocky Mtn. Region, Dept. Interior, USA [cited from DeFoliart 1991].

Provancher, L. 1882. A propos de fourmis. Le Naturaliste Canadien 13(145): 30–31 [cited from DeFoliart 2002].

Provancher, L. 1890. Des insects comme aliment. Naturaliste Canadien 20: 114–127 [cited from The Food Insect Newsletter 6(2): 3].

Quin, P.J. 1959. Foods and feeding habits of the Pedi with special reference to identification, classification, preparation and nutritive value of the respective foods. Witwatersrand University Press, Johannesburg, Republic of South Africa.

Ramandey, E. and H. van Mastrigt. 2010. Edible insects in Papua, Indonesia: from delicious snack to basic need. pp. 105–114. In: P.B. Durst, D.V. Johnson, R.N. Leslie and K. Shono [eds.]. Forest Insects as Food: Humans Bite Back. FAO Regional Office for Asia and Pacific. Bangkok, Thailand.

Ramos-Elorduy, J. 1997. Insects: A sustainable source of food? Ecol. Food Nutr. 36: 247–276.

Ramos-Elorduy, J. 2003. Les jumiles, punaises sacrées au Mexique. pp. 325–353. In: E. Motte-Florac and J.M.C. Thomas [eds.]. Les "insectes" dans la tradition orale. Peeters, Leuven, Belgium.

Ramos-Elorduy, J. 2006. Threatened edible insects in Hidalgo, Mexico and some measures to preserve them. J. Ethnobiol. Ethnomed. 2: 51–60.

Ramos-Elorduy, J. 2009. Anthropo-entomophagy: cultures, evolution and sustainability. Entomological Research 39(5): 271–288.

Ramos-Erolduy, J. and M.J.M. Pino. 1989. Los insectos comestibles en el México antiguo. A.G.T. Editor, S.A., México D.F., Mexico.

Ramos-Elorduy, J. and M.J.M. Pino. 1990. Contenido calórico de algunos insecatos comestibles de México. Rev. Soc. Quim. Méx. 34(2): 56–68.

Ramos-Elorduy, J. and M.J.M. Pino. 2002. Edible insects of Chiapas, Mexico. Ecol. Food Nutr. 41: 271–299.

Ramos-Elorduy, J. and M.J.M. Pino. 2004. Los Colleoptera comestibles de México. Anales del Instituto de Biologia, Universidad Nacional Autómoma de México, Serie Zoologia 75(1): 149–183.

Ramos-Elorduy, J., M.J.M. Piono, R.C. Cerecedo and V.M. Duarte. 1981. Estudio de los insectos comestibles de Gerrero y valor nutritivo. Octavo Congreso Nacional de Zoologia, Memorias II, Escuela Normal Superior del Estado: 1107–1126.

Ramos-Elorduy, J., R. Flores and M.J.M. Pino. 1985. [cited from DeFoliart 2002].

Ramos-Elorduy, J., M.J.M. Pino, E.E. Prado, M.A. Perez, J.L. Otero and O.L. de Guevara. 1997. Nutritional value of edible insects from the State of Oaxaca, Mexico. J. Food Comp. Anal. 10: 142–157.

Ramos-Elorduy, J., M.J.M. Pino and S.C. Correta. 1998. Insectos comestibles del Estado de México y determinación de su valor nutritivo. Anales Inst. Biol. Univ. Na. Autn. México, Ser. Zool. 69: 65–104.

Ramos-Elorduy, J., E.M. Costa-Neto, J.F. dos Santos, M.J.M. Pino, I. Landero-Torres, S.S.A. Campos et al. 2006. Estudio comparativo del valor nutritivo de varios coleoptera comestibles de México y Pachymerus nucleorum (Fabricius, 1792) (Bruchidae) de Brasil. Interciencia 31(7): 512–516.

Ramos-Elorduy, J., M.J.M. Pino and V.H.M. Camacho. 2009. Edible aquatic Coloptera of the world with an emphasis on Mexico. J. Ethnobiol. Ethnomed. 5(11):

Ramos-Elorduy, J., M.P.M. Pino, A.I. Vázquez, I. Landero, H. Oliva-Rivera and V.H.M. Camacho. 2011. Edible Lepidoptera in Mexico: Geographic distribution, ethnicity, economic and nutritional importance for rural people. J. Ethnobiol. Ethnomed. 7(2): 1–22.

Ramos-Elorduy de Conconi, J. 1982. Los insectos como fuente de proteínas en el futuro. Editorial Limusa, Mexico D.F., Mexico.

Ramos-Elorduy de Conconi, J. 1987. Are insects edible? Man's attitudes towards the eating of insects. Sociala d Human Sciences in Asia and the Pacific, RUSHSAP Ser. on Occational Monographs and Papers, 20 (Food Deficiency: Studies and Perspectives) UNESCO. Bangkok, Thailand: 78–83.

Ramos-Elorduy de Conconi, J. 1991. Los insectos como fuente de proteinas en el futuro. 2nd ed. Editorial Limusa, S.A. de C.V., Mexico, D.F., Mexico.

Ramos-Elorduy de Conconi, J. and M.J.M. Pino. 1988. The utilization of insects in the empirical medicine of ancient Mexicans. J. Ethnobiol. 8(2): 195–202.

Ramos-Elorduy de Conconi, J., M.J.M. Pino, C. Marquez Mayaudon, F. Rincon Valdez, M. Alvarado Perez, E. Escamilla Prado et al. 1984. Protein content of some edible insects in Mexico. J. Ethnobiol. 4(1): 61–72.

Ratcliffe, B.C. 1990. The significance of scarab beetles in the ethnoentomology of non-industrial, indigenous peoples. pp. 159–185. In: D.A. Posey and W.L. Overal [eds.]. Ethnobiology: Implications and Applications. MPEG. Belém, Brazil [cited from Costa-Neto and Ramos-Elorduy 2006].

Rattanapan, A. 2000. Edible insect diversity and cytogenetic studies on short-tail crickets (Genus Brachytrupes) in Northeastern Thailand. Graduate School, Khon Kaen University, Khon Kaen, Thailand (in Thai) [cited from Hanboonsong et al. 2013].

Read, B.E. 1941. Chinese materia medica. Insect drugs. Peking Natural History Bulletin. [reprinted by Southern Materials Center, Inc., Taepei, Taiwan. Chinese Medicine Series 2. 1982].

Réaumur, R.A.F. de. 1734–42. Mémoires pour server à l'Histoire des Insectes. Vol. II and Vol. III [cited from Bodenheimer 1915].

Rehert, I. 1984. Cicada cuisine. Tymbal 1: 19.

Reim, H. 1962. Die Insektennahrung der australischen Ureinwohner. Veröffentlichungen des Museums für Völkerkunde zu Leipzig, Heft 13. Akademie-Verlag, Berlin, Germany.

Rengger, K. 1835. [cited from Yasumatsu 1948].

Revel, N. 2003. Présence et signification des insectes dans la culture Palawan (Philippines). pp. 123–135. In: E. Motte-Florac and J.M.C. Thomas [eds.]. Les "insectes" dans la tradition orale. Peeters, Leuven, Belgium.

Ribeiro, B.G. and T. Kenhiri. 1987. Calendário econômico dos indios Desâna. Ciência Hoje. 6(36): 26–35 [cited from Costa-Neto and Ramos-Elorduy 2006].

Rice, M.E. 2000. Fried green cicadas and a young African hunter. Am. Entomol. 46(1): 6–7.

Robson, J.R.K. and Yen, D.E. 1976. Some nutritional aspects of the Philippine Tasaday diet. Ecol. Food Nutr. 5: 83–89.

Rodrigues, A. DOS S. 2005. Etnoconhecimento sobre abelhas sem ferrao: sabers e práticas dos indios Guarani M'byá na Mata Atlântica. Dissertaçao (Ecologia de Agroecosistemas). Piracicaba: Universidade de Sãn Paulo. Brazil [cited from Costa-Neto and Ramos-Elorduy 2006].

Roepke, W. 1952. Insekten op Java als menselijk voedsel of als medicijn gebezigd. Entomologische berichten 14: 172–174 [cited from van Huis 2003].

Ronghang, R. and R. Ahmed. 2010. Edible insects and their conservation strategy in Karbi Anglong district of Assam, north east India. The Bioscan: Special Issue, Vol. 2: 515–521.

Roodt, V. 1993. The shell field guide to the common trees of the Okavango Delta and Moremi game reserve [cited from Malaisse 1997; van Huis 2005].

Roulon-Doko, P. 1998. Chasse, cueillette et cultures chez les Gbaya de Centrafrique. L'Harmattan, Paris, France [cited from Malaisse 2005].

Roulon-Doko, P. 2003. Les fourmis dans la conception des Gbaya de Centrafrique. pp. 73–86. *In*: E. Motte-Florac and J.M.C. Thomas [eds.]. Les "insectes" dans la tradition orale. Peeters, Leuven, Belgium.

Roust, N.L. 1967. Preliminary examination of prehistoric human coprolites from four western Nevada caves. Rpts. Univ. Calif. Archaeol. Surv. No. 70: 49–88 [cited from DeFoliart 2002].

Ruddle, K. 1973. The human use of insects: example from the Yukpa. Biotropica 5: 94–101.

Sachan, J.N., B.B. Das, S.K. Gangwar, K.A. Pathak and J.N. Katiyar. 1987. Insects as human food in north eastern hill region of India. Bull. Entomol. 28: 67–68.

Sangpradub, N. 1982. Edible invertebrates in the northeastern part of Thailand. M.Sc Thesis. Chulalongkorn Univ. Bangkok, Thailand.

Santos, E. 1957. Histórias, lendas e folklore de nossos bichos. Rio de Janeiro: Cruzeiro, Brazil [cited from Costa-Neto and Ramos-Elorduy 2006].

Schabel, H. 2010. Forest insects as food: a global review. pp. 37–64. *In*: P.B. Durst, D.V. Johnson, R.N. Leslie and K. Shono [eds.]. Forest Insects as Food: Humans Bite Back. FAO Regional Office for Asia and the Pacific, Bangkok, Thailand.

Schomburgk, M.R. 1847–1848. Reisen in British Guinana. 3 vol. Leipzig, Germany [cited from Bodenheimer 1951].

Schorr, M.H.A. and P.I. Schmitz. 1975. A utilização dos recursos naturais na alimentação dos indigenas da regiáo sudeste do Rio Grande do sul. Sáo Leopoldo: Universidade do vale do Rio dos Sinos. Publicações Avulsas n2 [cited from Costa-Neto and Ramos-Elorduy 2006].

Schwarz, H.F. 1948. The stingless bees (Meliponidae) of the western hemisphere. Bull. Amer. Mus. Nat. Hist. 90: 546 [cited from Bodenheimer 1951].

Seignobos, C., J.-P. Deguine and H.-P. Aberlenc. 1996. Les Mofu et leurs insectes. J. d'Agric. Trad. Et de Bota. Appl. 38(2): 125–187.

Sekhwela, M.B.M. 1989. A comparison of the nutritional value of phane caterpillar with of the food types. Botswana Notes and Records 1989, Gaborone, Botswana [cited from Moruakgomo 1996].

Sekiya, I. 1972a. Inago (*Oxya yezoensis* and *O. japonica japonica*). pp. 377–378. *In*: History of Plant Protection in Nagano Prefecture, Association of Nagano Plant Protection, Nagano, Japan (in Japanese).

Sekiya, I. 1972b. Characteristics of ground wasp (*Vespula flavipes lewisii*), and utilization of the wasp. pp. 940–941. *In*: History of Plant Protection in Nagano Prefecture, Association of Nagano Plant Protection, Nagano, Japan (in Japanese).

Setz, E.Z.F. 1991. Animals in the Nambiquara diet: methods of collection and processing. J. Ethnobiol. 11: 1–22 [cited from Costa-Neto and Ramos-Elorduy 2006].

Shaw, T. 1738. Travels, or observations relating to several parts of Barbary and the Levant. Oxford, England [cited from DeFoliart 2002].

Shaxon, A., P. Dixon and J. Walker. 1974. The Cook Book. Zomba (Malawi). Government Printer, p. 173 [cited from Malaisse 1997].

Shaxon, A., P. Dickson and J. Walker. 1985. The Malawi Cookbook. Zomba, Malawi. Blantyre Printing and Publishing Co. Ltd., Zomba, Malawi [cited from DeFoliart 2002].

Shennung Ben Ts'ao King. See Read 1941.

Shida, T. 1959. *Jibachi (Vespula flavipes lewisii)* in Musashino. pp. 77–145. *In*: K. Iwata, H. Hurukawa and K. Yasumatsu [eds.]. Insects in Japan 1. Life of waps. Kōdansha, Tokyo, Japan (in Japanese).

Shiraki, T. 1958. Hygienic Insects. Hokuryu-kan, Tokyo, Japan (in Japanese).

Shu, T. 1989. Food Culture of China. Sogen-Sha, Osaka, Japan (in Japanese).

Silow, C.A. 1976. Edible and other insects of mid-western Zambia. Studies in Ethno-Entomology II. Institutionen för Allmän och Jämförande Etnografi vid Uppsala Universitet Occasional Paper V, Uppsala Univ. Kungsängsgatan, Sweden.

Silow, C.A. 1983. Notes on Ngangala and Nkoya ethnozoology, ants and termites. Etnol. Stud. 36 [cited from DeFoliart 2002].

Simmonds, P.L. 1877. Les Richesses de la Nature. Lè Regne Animal. Transl. from the 2nd English edition. Grand [cited from Bodenheimer 1951].

Simmonds, P.L. 1885. The Animal Resources of different nations. E. & F.N. Spon, London, England.

Singh, O.T. and J. Chakravorty. 2008. Diversity and occurrence of edible orthopterans in Arunachal Pradesh, with a comparative note on edible orthopterans of Manipur and Nagaland. J. Natcon. 20(1): 113–119.

Singh, O.T., S. Nabom and J. Chakravorty. 2007a. Edible insects of Nishi tribes of Arunachal Pradesh. Hexapoda 14(1): 56–60.

Singh, O.T., J. Chakravorty, S. Nabom and D. Kato. 2007b. Species diversity and occurrence of edible insects with special reference to coleopterans of Arunachal Pradesh. J. Natcon 19(1): 159–166.

Sirinthip, W. and R. Black. 1987. *Xylotrupes gideon* eating bark of apple and pear trees in Northern Thailand. Trop. Pest Management 33: 236.

Sloane, H. 1725. A voyage to the Islands of Madeira, Barbados, Nieves, St. Christophers, and Jamaica; with the natural history. Vol. II., London, England [cited from DeFoliart 2002].

Smole, W.J. 1976. The Yanoama Indians: A Cultural Geography. The Texas Pan American Series. Univ. Texas Press, Austin TX, pp. 272. [Paoletti and Dufour 2005].

Smyth, R.B. 1878. The aborigines of Victoria. Vol. 1. John Ferres, Melbourne, Australia.

Sokolov, R. 1991. One man bites back. Nat. Hist. 7/1991: 70–73.

Solavan, A., R. Paulmurugan, V. Wilsanand and A.J.A. Ranjith Sin. 2004. Traditional therapeutic uses of animals among tribal population of Tamil Nadu. Indian J. Traditional Knowledge 3(2): 198–205.

Sparrmann, A. 1778. Voyage au Cap de Bonne-Espérance [cited from Bodenheimer 1951].

Spittel, R.L. 1924. Wild Ceylon. Describing in particular the lives of the present day Veddas. The Colombo Apothecaries Co. Ltd., Colombo, Sri Lanka.

Spruce, R. 1908. Notes of a botanist on the Amazon and Andes. 2 vols. Macmillan, London England [cited from DeFoliart 2002].

Srivastava, S.K., N. Babu and H. Pandey. 2009. Traditional insect bioprospecting. As human food and medicine. Indian J. Traditional Knowledge 8(4): 485–494.

Starr, C. 1991. Notes on entomophgy in the Philippines. The Food Insect Newsletter 4(3): 2 & 12.

Stedman, J.G. 1796. Narrative of a five year's expedition against the revolted negroes of Surinam, in Guiana, on the wild coast of South America; From the year 1772 to 1777. 2 volumes. London, England [cited from DeFoliart 1990].

Stefano Vanzin. 2003. [cited from Paoletti and Dufour 2005].

Stewart, J.H. and A. Métraux. 1963. The Peban tribes. Handbook of South American Indians Vol. 3, J.H. Stewart [ed.]. Cooper Square Publishers, New York, N.Y., USA.

Stone, J.L. 1992. Keeping and Breeding Butterflies and other Exotica: praying mantis, scorpions, stick insects, leaf insects, locusts, large spiders and leaf-cutter ants. Blandford, London, England.

Strickland, C. 1932. Edible and paralysific bugs, one of which a new species *Cyclopelta subhimalayensis* n. sp. (Hemipteron, Heteropteron, Pentatomida, Dinadorina). Indian J. Med. Res. 19: 873–876.

Styles, C.V. 1996. The ecology of *Imbrasia belina* (Saturniinae, Saturniidae) with reference to its behaviour, physiology, distribution, population dynamics, impact within mopane veld and utilization within South Africa. pp. 9–13. *In*: B.A. Gashe and S.F. Mpuchane [eds.]. "Phane", Proc. 1st Multidisciplinary Symp. on Phane, June 1996 Botswana, Dep. of Bio. Sci. & The Kalahari Conserv. Soc.

Sugiyama, Y. 1997. Insects eaten by Bemba people. pp. 234–270. *In*: J. Mitsuhashi [ed.]. People Who Eat Insects. Heibon-sha. Tokyo, Japan (in Japanese).

Sungpuag, P. and P. Puwastien. 1983. Nutritive value of unconventional protein source: insect. Prochanagan Sarn 1: 5–12 (in Thai) [cited from DeFoliart 2002].

Sutton, M.Q. 1985. The California salmon fly as a food source in northeastern California. J. California and Great Basin Anthropol. 7: 176–182.

Sutton, M.Q. 1988. Insects as Food: Aboriginal Entomophagy in The Great Basin. Ballena Press, Melno Park, California, USA.

Sweeney, G. 1947. Food supplies of a desert tribe. Oceania 17: 289–299.

Swezey, S.L. 1978. Barrett's armyworm: a curious ethnographic problem. J. California Anthropol. 5: 256–262.

Takagi, G. 1929a. Edible insects (2). J. Korea Mountains and Forests 47: 46–54 (in Japanese).

Takagi, G. 1929b. Edible insects (3). J. Korea Mountains and Forests 49: 33–39 (in Japanese).

Takagi, G. 1929c. Edible insects (4). J. Korea Mountains and Forests 51: 27–32 (in Japanese).

Takagi, G. 1929d. Edible insects (5). J. Korea Mountains and Forests 52: 8–11 (in Japanese).

Takagi, G. 1929e. Edible insects (5). J. Korea Mountains and Forests 57: 27–38 (in Japanese).

Takagi, G. 1929f. Edible insects (6). J. Korea Mountains and Forests 53: 24–27 (in Japanese).

Takagi, G. 1929g. Edible insects (7). J. Korea Mountains and Forests 54: 38–44 (in Japanese).

Takagi, G. 1929h. Edible insects (8). J. Korea Mountains and Forests 58: 42–54 (in Japanese).

Takeda, J. 1990. The dietary repertory of the Ngandu people of the tropical rainforest. African Study Monographs Suppl. 11: 1–75.

Tanaka, I. 1980. Foods in Shinano District in Japan. Shinano-Mainich Newspaper Company, Nagano, Japan (in Japanese).

Tarre, M.R. 2003. Harvesting *Hyles lineata* in the Sonoran Desert: A larval legacy. pp. 315–324. *In*: E. Motte-Florac and J.M.C. Thomas [eds.]. "Insects" in Oral Literature and Traditions. Peeters, Leuven, Belgium.

Taylor, R.L. and B.J. Carter. 1992. Entertaining with insects or The First Original Guide to Insect Cookery. Salutek Publishing Company, Yorba Linda, California, USA.

Tchibozo, S., A. van Huis and M.G. Paoletti. 2005. Notes on edible insects of south Benin: a source of protein. pp. 245–250. *In*: M.G. Paoletti [ed.]. Ecological Implications of Minilivestock, Potential of Insects, Rodents, Frogs and Snails. Science Publishers, Inc., Enfield, USA.

Terada, B. 1933. On the medicinal Coprinae. Japanese J. Phermacol. 16(3): 68–69 (in Japanese).

Tessmann, G. 1913/1914. Die Pangwe [cited from Bodenheimer 1951].

Thémis, J.-L. 1997. Recettes des Jean-Louis Thémis, Des Insectes à Croquer, Guide de découvertes. L'Insectarium de Montréal. Montréal, Canada.

Tiêu, N.C. 1928. Notes sur les insectes comestibles au Tonkin. Bull. Économique l'Indochine 31 (Nouvelle série): 735–744.

Tindale, N.B. 1932. Revision of the Australian ghost moths (Hepialidae), I. Records. South Australian Museum 4: 29–207 [cited from Bodenheimer 1951].

Tindale, N.B. 1962. Witchety grub. Australian Encyclopaedia 9. Grolier Soc. Australia, Sydney, Australia.

Tindale, N.B. 1966. Insects as food for the Australian aborigines. Australian Natural History 15: 179–183.

Toirambe Bamoninga, B. 2007. Analyse de l'etat des lieux du secteur des produits forestiers non ligneux et evaluation de leur contribution a la securite alimentaire en Republique Democratique du Congo. GCP/RAF/398/GER Renforcement de la sécurité alimentaire en Afrique Centrale à travers la gestion et l'utilisation durable des produits forestiers non ligneux.

Tommaseo-Ponzetta, M. and M.G. Paolett. 1997. Insects as food of the Irian Jaya populations. Ecol. Food Nutr. 36: 321–346.

Tommaseo-Ponzetta, M. and M.G. Paoletti. 2005. Lessons from traditional foraging patterns in West Papua (Indonesia). pp. 441–457. *In*: M.G. Paoletti [ed.]. Ecological Implications of Minilivestock, Potential of Insects, Rodents, Frogs and Snails. Science Publishers, Inc., Enfield, USA.

Torii, T. 1957. Note on the *zaza mushi*, a characteristic product of Ina-Tenryu, Nagano Prefecture, Japan. Shin Konchu 10(6): 26–29 (in Japanese).

Tsurufuji, S. 1985. Dictation on meal at Okayama Prefecture. Collection of Japanese eating habit 33. Association of Farmer's and Fisherman's Culture, Tokyo, Japan (in Japanese).

Ukhun, M.E. and M.A. Osasona. 1985. Aspect of the nutritional chemistry of *Macrotermis bellicosus*. Nutr. Rep. Intern. 32: 1121–1130.

Umemura, J. 1943. Pharmacology on insects. Description of medicinal and Edivble insects. Shobunkan-shoten, Nagoya, Japan (in Japanese).

Umeya, K. 1994. Insect dishes, I ate in China. Insectarium 31: 252–257 (in Japanese).

Umeya, K. 2004. Stories on Eating Insects. Soshin-sha, Tokyo, Japan (in Japanese).

Utsunomiya, Y. and K. Masumoto. 1999. Edible beetles (Coleoptera) from northern Thailand. Elytra 27: 191–198.

Utsunomiya, Y. and K. Masumoto. 2000. Additions to edible beetles (Coleoptera) from northern Thailand. Elytra 28: 12.

van der Burg, C.L. 1904. De Voeding in Nederlandsch-Indië [cited from Bodenheimer 1951].

van der Waal, B. 1999. Ethnobiology and uses of grasshoppers in Venda, norther Province, south Africa. S. Afr. J. Ethnol. 22(2): 103–109 [cited from Malaisse 2005].

van Huis, A. 1996. The traditional use of arthropods in Sub Saharan Africa. Proc. Exper. & Appl. Entomol. 7: 3–20.

van Huis, A. 2003. Insects as food in Sub-Saharan Africa. Insect Sci. Appl. 23(3): 163–185.

van Huis, A. 2005. Insects eaten in Africa (Coleoptera, Hymenoptera, Diptera, Heteroptera, Homoptera). pp. 231–244. *In*: M.G. Paoletti [ed.]. Ecological Implications of Minilivestock, Potential of Insects, Rodents, Frogs and Snails. Science Publishers, Inc., Enfield, USA.

van Huis, A. 2008. The future of edible insects in Africa. Proceeding of a workshop on Asia-Pacific resources and their potential for development, 19–21 February. 2008, Chiang Mai, Thailand. FAO Regional Office for Asia and the Pacific, Bangkok, Thailand.

Vara-asavapati, V., J. Visuttipart and C. Maneetorn. 1975. Edible insects in North-east Thailand. Res. Note No. 7, Mahasarakam: Univ. of Srinakarinvirot. (in Thai with English summary) [cited from DeFoliart 2002].

Verma, A.K. and S.B. Prasad. 2012. Bioactive component, cantharidin from *Mylabris cichorii* and its antitumor activity against Ehrlich ascites carcinoma. Cell Biol. Toxicol. 28(3): 133–147.

Villiers, A. 1947. Une Manne Africaine: Les termites. La Nature 1947: 239–240 [cited from DeFoliart 2002].

Wallace, A.R. 1852/1853. On the insects used for food by the Indians of the Amazon. Trans. Entomol. Soc. London (N.S.) 2: 241–244 [cited from DeFoliart 2002].

Wallace, A.R. 1869. The Malay Archipelago. Macmillan and Co., London, England [cited from Wallace, A.R.: The Malay Archipelago. Dover Publ., Inc., New York, N.Y., USA. 1962].

Watanabe, T. 1982. Story of Medicinal Insects. Tokyo-shoseki. Tokyo, Japan (in Japanese).

Waterhouse, D.F. 1991. Insects and humans in Australia. *In*: Div. Entomol. CSIRO [eds.]. The Insects of Australia Vol. I.: 221–235.

Wattanapongsiri, A. 1966. A revision of the genera *Rhynchophorus* and *Dynamis* (Coleoptera, Curculionidae). Dep. Agr. Sci. Bull. 1(1), Bankok, Thailand [cited from Dounias 2003].

Waugh, F.W. 1916. Iroquois foods and food preparation. Can. Dept. Mines, Ceol. Surv., Mem. 86, No. 12, Anthropol. Ser. Ottawa: Govt. Print. Bur. Canada [cited from DeForiart 2002].

Weaving, A. 1973. Insects: A Review of Insect Life in Rhodesia. Irwin Press Ltd., Salisbury, Wiltshire, England [cited from van Huis 2005].

Wellman, F.C. 1908. Notes on some Angolan insects of economic or pathologic importance. Entomol. News 19: 26–33.

Wijesekara, N. 1964. Veddas in transition. Colombo, M.D. Gunasena Publication [cited from, Nandasena et al. 2010].

Wilsanand, V. 2005. Utilization of termite, *Odontotermes formosanus* by tribes of south India in medicine and food. Explorer 4(2): 121–125.

Womeni, H.M., M. Linder, B. Tiencheu, J. Fanni and M. Parmentier. 2009. Oils of insects and larvae consumed in Africa: potential sources of polyunsaturated fatty acids. Oléagineux, Corps Gras, Lipides 16(4): 230–235.

Woodward, A. 1934. An early account of the Chumash. The Masterkey 8: 118–123.

Yagi, S. 1997. Entomophagy in Africa. Aera Mook 22: 111–115 (in Japanese).

Yamada, G. 1952. Study on the antipyretic action of "Extract of Cricket" (an effect on a rise of the body-temperature). Shikoku Medical Magagine 3(3): 51–54 (in Japanese).

Yasumatsu, K. 1948. Ants and Human Life. Yōyō-shobō, Tokyo, Japan (in Japanese).

Yasumatsu, K. 1965. Story of Insects: Insects and Human Life. Shinshicho-sha, Tokyo, Japan (in Japanese).

Yen, A.L. 2005. Insect and other invertebrate foods of the Australian aborigines. pp. 367–387. *In*: M.G. Paoletti [ed.]. Ecological Implications of Minilivestock, Potential of Insects, Rodents, Frogs and Snails. Science Publishers, Inc., Enfield, USA.

Yhoung-Aree, J. and K. Viwatpanich. 2005. Edible insects in the Laos PDR, Myanmar, Thailand, and Viet Nam. pp. 415–440. *In*: M.G. Paoletti [ed.]. Ecological Implications of Minilivestock, Potential of Insects, Rodents, Frogs and Snails. Science Publishers, Inc., Enfield, USA.

Zent, S. 1992. Historical and ethnographic ecology of the upper Cuao River Votiha: Clues for an interpretation of native Guianese social organization. Ph.D. diss. Columbia Univ. New York [cited from Paoletti and Dufour 2005].

Zhi-Yi, Luo. 2005. Insects as traditional food in China. pp. 475–489. *In*: M.G. Paoletti [ed.]. Ecological Implications of Minilivestock, Potential of Insects, Rodents, Frogs and Snails. Science Publishers, Inc., Enfield, USA.

Zhū, X. 2003. Resources of edible insects and their exploitation in human province. Development of Forest Science and Technology 17(2): 12–14 (in Chinese).

Zimian, D., Z. Yonghua and G. Xiwu. 1997. Medicinal insects in China. Ecol. Food Nutr. 36: 209–220.